Passibility

CLASSICS IN SCIENCE EDUCATION

Volume 3

Series Editor:

Karen C. Cohen

For further volumes:
http://www.springer.com/series/7365

All attempts to surmount – from within theoretical cognition – the dualism of cognition and life, the dualism of thought and once-occurrent concrete actuality, are utterly hopeless. . . . The detached content of the cognitional act comes to be governed by its own immanent laws, according to which it then develops as if it had a will of its own. Inasmuch as we have entered that content, i.e., performed an act of abstraction, we are now controlled by its autonomous laws or, to be exact, we are simply no longer present in it as individually and answerably active human beings.

(Bakhtin 1993, p. 7)

Phenomenology aims at ultimate clarification and justification of knowledge, both theoretical and scientific knowledge in the proper sense and that pretheoretical and prescientific knowledge by which we are guided in our life in the world of every day experience, and out of which theoretical and scientific knowledge grows.

(Gurwitsch 2010, p. 152)

Wolff-Michael Roth

Passibility

At the Limits of the Constructivist Metaphor

Springer

Prof. Dr. Wolff-Michael Roth
MacLaurin Bldg A548
V8W 3N4 VICTORIA
British Columbia
Canada
mroth@uvic.ca

ISBN 978-94-007-1907-1 e-ISBN 978-94-007-1908-8
DOI 10.1007/978-94-007-1908-8
Springer Dordrecht Heidelberg London New York

Library of Congress Control Number: 2011934504

Printed on acid-free paper

Springer is part of Springer Science+Business Media (www.springer.com)

Preface

> A thousand thoughts, a thousand ideas that I would want to reject, which I do not seek out, which I even pity, come through my mind. My reason is hardly ever asleep, it sees all of that, it moans, it blames and it approves, these are its only functions. If some good sentiment arises, do you believe that we should credit it [reason]? No, it does nothing but gives its approval, it uses all its power to maintain it. (Maine de Biran, in Naville 1874, p. 113)

The frontispiece and the present opening quote circumscribe the fundamental point of this book – theoretical reason is not the source of knowing and learning but a power that selects among the thoughts and ideas that emerge within the mind. This mind, as Bakhtin suggests in the frontispiece, needs to be understood from within life. When we approach it from within cognition, then the cognitional act comes to be controlled by its own immanent laws, as if it were only for itself. It is a kind of mind that no longer resides in the world, the kind of mind that Immanuel Kant and following him the present-day constructivists describe. This is also the kind of mind that those with an intellectual heritage in dialectical (historical) materialism – Mikhail Bakhtin, Lev Vygotsky, and Michel Henry – reject as suitable for understanding how human beings know and act in the world. This is a world of living labor where suffering, joy, and crisis are daily experiences. Yet in the constructivist metaphor of cognitive development and learning, these dimensions of daily experience do not appear.

In the introductory quote, Maine de Biran offers a different perspective on the mind, which primarily is but a(n) (un-) willing host. We are not only the *subjects of* ideas and thoughts but also, and primarily so, *subject(ed) to* ideas and thoughts, which suddenly come to and surprise us rather than are intended by us. We are but willing or unwilling hosts, approving, blaming, and bemoaning what arises from within ourselves. This position, which recognizes a fundamental dimension of human condition, has been worked out to some extent in phenomenological philosophy, but has yet to be acknowledged in the literature on knowing and learning generally and in the literature on the learning of science particularly. The existing literature is inhabited by an ideology of the intellect, an idealist ideology that makes reason and rationality the primary source of knowing and the agent of learning. Taking as its point of departure the world as we know it, the position is blind to the essentially passive constitution of everything that matters to and in our everyday lives. It not only is incapable of explaining experiences that arise from passibility but also, and importantly so, to the ways of learning and knowing that arise from passibility and associated experiences such as the passions, radical passivity, uncertainty, and otherness.

If ideas and thoughts come to us – i.e., are given to us so that we may select among them – then we cannot understand learning something new in terms of the *intentional* appropriation of something into our existing knowledge. The foreign/strange, precisely because it is foreign/strange, is invisible and therefore cannot be visualized, envisaged, and aimed at. That is, we cannot think learning in terms of a framework that already takes the new, unfamiliar, and foreign/strange as something available to be thought, considered, and intended. This is so because we cannot ever understand the learning of something absolutely unknown if we think it from the perspective of the known. To think the learning of something unknown we have to retain it as the unknown in the direction of which we think. If we were not to do so, we would do no better than Whig historians, who explain events in a teleological manner, from the a posteriori perspective of what we know today. But only hindsight has 20/20 vision, whereas learning means inherently engaging with the unknown, unfamiliar, and foreign/strange.

To get a better grip on what it means to learn something that we cannot even imagine what it looks and feels like, we have to think about learning from the perspective of the invisible, the foreign and strange, and the unknown. What does it mean to encounter something absolutely foreign/strange and therefore invisible? How does the invisible become visible, the unknown become part of the known, and how does something foreign and strange become familiar? At the same time, as this something becomes visible, the invisible continues to exist, cannot be absorbed into the visible, for otherwise there would be impossible – against everything we know today – to learn something new; as something becomes familiar, the foreign/strange only recedes. As a something becomes known as something, the unknown actually withdraws. That is, to understand the phenomenon of learning we need to think the unthought, to think the (currently) unthought from the position of the unthought rather than from the position of the newly thought that has arisen from the unthought. And we need to think what remains unthought in our newly thought thoughts. What makes our questions possible? What makes possible that which questions in our questions? As I was beginning to write this book, I set myself as the explicit goal to think the foreign/strange without subsuming it to the familiar, to think it in its radical form as that which forever withdraws. For if we were to know beforehand how our living body responds to the foreign/strange, answers would be nothing more than mechanical reactions. If life were such, we would not even require thinking.

These are some of the fundamental aporia and questions that I raise in this book. I do so because after more than two decades of reading and doing research on learning, I have become increasingly dissatisfied with the ways in which learning is thought, theorized, and researched. Increasingly I have realized that in our questioning, we adumbrate the real questions about how humans learn, remember, and know science. Why would a mind think a (scientific) thought? How/why does something like emotion mediate thought? And how/why would thought be able to bring about change of emotions? How does something as immaterial and without extension – whether it is called soul, psyche, or consciousness – have any effect on the extended, material body and put it into motion? Why does an unexpected turn of a story make us break ought in laughter, shaking our bodies to their foundations? Why does presenting a paper make young researchers sweat? Why does a person blush when caught telling a lie?

In pushing on, I have had to learn myself, that is, I had to overthrow what I have come to know and cherish. The more I knew about learning, the less I seemed to know about (the real issues of) learning. The more I learned about learning research, the more I have come to understand that our theories are no more adequate than those that we have had some decades ago. As more about learning has become visible, theorized, and discussed in the clearing marked by what is already known, all the more has come to be covered up and hidden from our questioning. In this book, I return to some of the more fundamental questions. Rather than asking questions such as "What do students learn in this curriculum?" or "How do students learn when provided opportunities to interact and reason with others?," I am asking questions about the *origins* of intentions, perceptions, discourses, and conceptions. All of these phenomena are currently taken for granted and the questions as to the origins of these phenomena are no longer asked or even available to questioning. And yet, as I have learned and increasingly come to know, learning researchers currently have no answers to these more fundamental questions. It is to this project – finding answers to the more fundamental questions about how human beings know and learn – that this volume contributes.

Throughout this book, I draw on books in their original French and German versions. All translations are mine, although I have checked, wherever possible and available, my translations with a translator's translation into English.

In this book, I use materials collected in a variety of research projects generally funded by grants from the Social Sciences and Humanities Research Council of Canada and the Natural Sciences and Engineering Council of Canada. The data fragments that appear in Chapters 3, 4, and 5 derive from an article that has appeared in *Science, Technology, and Human Values* (Roth 2009) and Chapter 6 includes a few paragraphs from an article that was initially published in *Educational Research Review* (Roth 2008). My thanks go to the publishers for allowing these materials to be used. I am grateful to Ken Tobin for access to the empirical data used in Chapter 10. My ultimate thanks go to my wife Sylvie, without whose patient support I would not be able to sustain the inquiries from which the present book evolved.

Victoria, BC
January 2011

Contents

Frontispiece	*v*
Preface	*vii*
Part A: Introduction/Deconstructing	1
1. De/Constructing the Blind Spots of the Constructivist Metaphor	3
Part B: Passivity, Uncertainty, Undecidability	23
2. Learning and the Erasure of Knowledge	29
3. Radical Passivity in Learning	43
4. Radical Uncertainty in Acting	71
5. Emergence of Duality	89
Part C: Otherness	105
6. Talking Conceptions Without Conceptions	111
7. Thought Follows Communication	127
8. Otherness of Self	141
9. The Nonsense of *Meaning*	159
Part D: Passion	179
10. Emotion, Motives, Motivations	185
11. Crises, Suffering, Joy	205
12. From Incarnation to Responsibility	225
Part E: Epilogue	243
13. Sublating the Constructivist Metaphor	245
References	267
Index	273

About the Author

Wolff-Michael Roth is Lansdowne Professor of Applied Cognitive Science at the University of Victoria. After graduate training in physics and statistics in the social sciences, he taught statistics and interpretive research methods for nearly 20 years. His research crosses disciplinary boundaries and is published in education and its subfields science, mathematics, and curriculum studies; qualitative research methods; sociology; applied linguistics; and epistemology. Wolff-Michael Roth has published over 330 peer-reviewed articles, some 150 book chapters, and nearly 40 books (as co/author or co/editor).

Part A

Introduction/Deconstructing

Chapter 1

De/Constructing the Blind Spots of the Constructivist Metaphor

> [B]eing does not resolve itself into empirical events and into thoughts that reflect these events or that aim at them "intentionally." (Levinas 1971, p. 153)
>
> One should say that in the body and as body, the sign demands the reality of the symbol: that is, the material union and co-presence of sense with the senses, the *body of sense* and the *sense of the body*. (Nancy 1993b, p. 194)

The two opening quotes articulate the tenor of this book. On the one hand, there is more to life than intentionality and the forms of (abstract) thought it gives rise to[1]; and there is a tight linkage between the senses and the sense we can make not only of our own bodies but also of the material world as a whole. This is so because the flesh of the world and our own flesh are but reflections of each other, objects reveal my flesh by means of their flesh (Sartre 1956). Together, the two quotes are the first traces of a "non-intellectualist" and "non-intentional" position that I articulate in this book. This is not a position against the intellect, but a position that opposes putting consciousness and mind before anything else – as is done in current theories of knowing and learning, including embodiment and enactivist theories. I develop such a position because

> [t]he strictly intellectualist thesis subordinates life to representation. One maintains that to will one first has to represent to oneself what one wills, to desire, one has to represent one's goal to oneself, to feel, one has to represent to oneself the object of the sentiment, and to act, one has to represent to oneself what one will do. (Levinas 1971, p. 143)

The strictly intellectualist position, focusing only on cognitional dimensions, misses not only the performative dimensions of human actions, but also its affective, valuative, and deontological dimensions (Bakhtin 1993). The performative, practice approach to cognition, which includes both the active and radically passive forms of life

---

[1] The emergence of this intentionality and the human forms of thought need to be explained. Constructivist approaches cannot explain this emergence but only the development of thought once intentionality exists.

W.-M. Roth, *Passibility*, Classics in Science Education 3,
DOI 10.1007/978-94-007-1908-8_1, © Springer Science+Business Media B.V. 2011

generally and of the living/lived body specifically, lies at the heart of my concerns in writing this text.

The intellectualist position underlies all current theories of learning (epistemologies). It is wanting because there is no place in these theories for other dimensions that characterize our human experience, most importantly, the experience of learning something that we do not know and forms of knowing that do not require representation such as the passions and the sense that comes with being/having a living/lived body (e.g., we do not have to cogitate walking, we just walk the walk). This intellectualist position has emerged during the Greek antiquity, and characterizes a form of thought generally known as metaphysics (idealism). Thus, thought is thought as something above and beyond the physical, that is, ultimately, above and beyond real, living/lived life. Most importantly, therefore, these intellectualist theories do not contain *internal* linkages between knowing and other aspects of life, such as emotions, desires, sentiments, passion, and so on. These intellectualist theories are concerned with constructions, externalities, which miss understanding and explaining the movement of the world and life: "All objective construction grows in *exteriority and it is submitted to its laws* . . . which assign to things the restrictive conditions of their existence, their spatiality, their temporality, their 'physical' properties, their causality, their finality, their utility" (Henry 2005, p. 184, emphasis added). Throughout this book I show how the externality associated with the construction metaphor misses a good deal of human experience, knowing, and learning.

Toward the very end of his life, just days before his death, Sigmund Freud – the master of the unconscious mind and artisan of a theory of the psyche as something immaterial – came to realize that there is more to the phenomenon that he has spent his entire professional career in researching and writing about. He writes: "Spatiality may be the projection of the extendedness of the psychic apparatus. No other derivation plausible. Instead of Kant's a priori conditions of our psychic apparatus. Psyche is spread out, does not know thereof" (Freud 1999, p. 152). In this posthumously published phrase, Freud notes that psyche is spread out, in other words, that it is body. But the psyche does not know thereof. In fact, it is this non-knowledge of its own constitution that makes psyche what it is. Freud tells us that psyche is precisely what escapes psyche, and it is this escaping that constitutes psyche. Recent phenomenological philosophers recognize the role of the living/lived body. Thus, "The 'unconscious' is the being-spread-out of Psyche, and which after Lacan nobody names *subject*, it is the singularity of a *local color* or an *incarnation*" (Nancy 2006, p. 22). The body and the incarnation of the psyche are absent from intellectualist theories – from Kant to Piaget and the modern day (radical, social) constructivists. Thus, "[a]ll these mundane explications of man that proliferate today confer to man properties that are those of things and forget his reality of a living being" (Henry 2000, p. 123). There is no schema of the psyche as an extended body, which moves, gesticulates, and displaces itself, and traverses all bodies (because all perception is immanent in and a perception of a living/lived body) and the material world in its entirety. We do not require the adjective "embodied," for its apparent necessity in some scholars' writings raises the ghost of the Cartesian dichotomy. The living/lived body is at the source, the origin of sense, because of its senses: *it* is the original structure of sense and the senses. It is, as I suggest referring to Sartre, the flesh of the objects that reveal my flesh to myself; and what and how we know (in, of science) is the result of this interlacement of the flesh.

In this introductory chapter, I outline the problematic and its solution. I begin by articulating some of the main tenets of intellectualism, here represented by (radical, individual, social) constructivism.[2] Against these tenets I set some of the phenomena that constructivism cannot explain; I articulate some of the reasons why constructivism cannot explain them, including the inherent, presupposed separation between body and mind. The problems include (a) the *learning paradox* (Bereiter 1986), which questions the possibility of a cognitive system to construct something of higher complexity than its own; (b) the impossibility to aim at (intend) the construction of knowledge that is inherently unknown from the perspective of the learner ("How can you aim at constructing knowledge when this knowledge itself is required for aiming at it?"); (c) the constitutive role of the living/lived body in knowing; (d) the role of passivity in learning from experience; and (e) the inherent otherness of knowledge (language) and self. I conclude by foreshadowing the inner coherence of the argument that unfolds in this book.

Intellectualism

Intellectualism – constructivism and associated conceptions of knowing and learning in mathematics and science (e.g., conceptual change) – has dominated educational research since the latter part of the 1980s. This epistemological theory is perhaps rooted in and at least consistent with a computer metaphor of thinking, whereby a given Self intentionally and rationally constructs itself such as to viably act in the material world. Any action, any word, any piece of art has been completed in imagination before it is expressed in the world (Kant 1964). In fact, "'epistemology' is but the title for the increasingly characteristic inability of modern metaphysics to know its own nature and its cause/reason" (Heidegger 2000, p. 73). Importantly, intellectualism is the central feature of constructivism and conceptions/conceptual change research that dominates science education. Over the years, the original framing of the theory has proven to be limited and limiting so that a number of external factors – emotions, motivations, sociality – have been added in the attempt to make something like "hot" and "shared" cognition out of an essential cold (unemotional) and individualistic

[2] I am actually concerned with all forms of intellectualism, including not only constructivist discourses but also those that come from traditional psychology – cognitive psychology, information processing approaches – and sociology, in which knowing and learning is reduced to the intentional actions of rational actors. Interestingly, although practice theory (e.g., Bourdieu 1997) does have an essential passive aspect – the body being open to the world and thereby subject to being fashioned by it – this passive aspect is not taken up in/by Anglo-Saxon scholarship drawing on this theory as a resource. Even embodiment and enactivist theories have this intentionalist orientation, which is the reason why philosophers recognize in these theoretical efforts the specters of Cartesianism (Sheets-Johnstone 2009).

process. In fact, as the embodiment of metaphysics, constructivism and the conceptual change theory that is built on it *cannot* explain that which fundamentally *constitutes* our lives, the sensual and the sensed, everything that makes us real human beings in flesh and blood. But these additions – hotness and sharedness – are only epicycles to a theory that has *essential* shortcomings. These do not allow the theory to take into account fundamental human experiences of knowing, learning, or experiencing (identity, emotions). Those brands of social constructivism in the literature that are said to have been derived from Lev S. Vygotsky do little to improve on the problems, because they situate the individual constructivist model in a social context that is said to precede the individual construction (i.e., construction first occurs inter-psychologically then intra-psychologically). In fact, those uptakes of the Russian psychologist's theories neither take into account their essentially materialist dialectical nor their concretely human dimensions, let alone the fundamentally Marxist basis of this line of work. Unless we develop a different theory in which these different constructive processes are thought as moments of a more encompassing phenomenon that gives rise to human experience of learning science, mathematics, technology, and the like, we will not make much inroad to improving education in these disciplines.

The founder of the ultimate intellectualism, the artisan of the intellect, is Immanuel Kant. He defines reason (*Vernunft*) as the capacity of humans to distinguish themselves not only from the world but also from themselves in so far as objects (*Gegenstände*) affect these (Kant 1956b, p. 88). It is even more powerful than understanding (*Verstand*), which itself is already higher than the sense that emerges when one is affected by things (and therefore when one suffers). Not only is sense derived from sense experiences but also understanding serves only to subordinate sense experiences to rules and thereby unify them in consciousness. Without the sense experiences, understanding would have nothing to think about, whereas reason goes so far beyond sense experiences that it can distinguish between sense experiences and understanding, and thereby shows the limits of understanding itself. That is, understanding is superior to the senses in that it not only submits them to order but also understands to integrate them into one consciousness. But it is inferior to reason, which exhibits spontaneity of such purity that it can transcend anything known and given to the senses. Reason is that power that distinguishes human beings from other beings. For Kant, this faculty is so powerful that it allows humans to think the causality of their own will – in terms of the concept of freedom. Freedom is the source of human autonomy. All actions of rational beings are founded on autonomy and the principle of morality in the same way that all phenomena are subject to natural laws. In this way, there is no realm of human experience that is not ruled by understanding and reason. But, to anticipate and summarize the fundamental message of this book, it is precisely the passivity that comes with the senses that leaves no play for the movement of such a freedom that would be able to assume it. Thus, "I am a threatened passivity . . . by a will in my will. In my action, in the for-itself of my will, I am exposed to a foreign will" (Levinas 1971, p. 212). Kant, being exclusively concerned with autonomy totally misses out on thinking the heteronomous aspects of human life and existence.

Modern day constructivists such as Jean Piaget take onboard the powerful, integrative, and synthesizing mind as core tenet of their epistemological theories. Thus, for example, to know

> is to assimilate reality into systems of transformations. . . . Knowing reality means constructing systems of transformations that correspond, more or less adequately, to reality. They are more or less isomorphic to transformations of reality. The transformational structures of which knowledge consists are not copies of the transformations in reality; they are simply possible isomorphic models among which experience can enable us to choose. Knowledge, then, is a system of transformations that become more progressively adequate. (Piaget 1970, p. 15)

In this definition of knowing, the intellect is the dominant feature of life. To know reality means to construct systems of transformation that correspond and are isomorphic to reality. How this metaphysical system of transformations is connected to anything physical or how the metaphysical system arises from and becomes independent of the physical is a problem that Piaget or the constructivists never resolve. Moreover, questions have been raised about the possibility of any cognitive system to develop viable, internally organized and connected structure that could be tested and improved by experiences in the world (Gurwitsch 2010). Thus, for Piaget and other constructivists, structure is the result of the organizing activity of a self-sufficient mind. This position, however, "prevents Piaget from recognizing organization as an autochthonous feature of sense-experience" (p. 49).

Intellectualism leaves a series of problems that tend to be recognized but that few if anyone dares to deal with. This is possibly so because it would mean abandoning the brands of constructivism that we currently have. One of the main problems of constructivism is the *learning paradox*, which refers to the question of how any cognitive organism can construct a mental organization more complex than its current one in a world that is of the same complexity as the mind (because observation is theory laden). Thus, the

> learning paradox results from the claim that very abstract and context-specific processual forms (i.e., schemes. conscious or unconscious mental codes) such as those underlying linguistic grammar, or Piaget's conservations, or formal reasoning may simply be acquired from everyday life, and the culture via learning without further ado. This raises a paradox, because to learn these very abstract and context-specific processual forms, the learner should either imitate and produce in his/her mind spontaneously the corresponding processual forms, or he/she must produce spontaneously the performances from which direct learning results. Consequently, if the theory lacks separate mechanisms for dynamically synthesizing the production of suitable performances in the absence of processual forms or schemes, prior availability in the subject's repertoire of the processual forms in question should be needed for learning to take place, since without these processual forms performance is not possible. This is a learning paradox, unless one assumes innateness, for it has as condition for learning the prior knowledge of the thing to be learned. (Pascual-Leone 1996, p. 85)

There have been responses to the problematic of the learning paradox (e.g., Steffe 1991). Referring to this work, two possible avenues of explanation have been offered (von Glasersfeld 2001). The "radical" approach would be to state that all data are "constructions" of the experiencing person and that there cannot be anything outside these constructions because from the beginning in the womb, the human being can only tests these constructions for viability in actual experience. The explanation is circular though, as there is no indication how the mental can ever make contact with the material world. The second avenue proposed invokes "schemes." But I already note above, schemes need to be explained rather than accepted. Schemes are the Kantian construct that bypasses the living/lived body. Thus, these responses to the learning paradox do not clarify how a child can be asked to learned something that the child itself cannot aim at, some form of knowledge that the adult teacher already has and that the child is supposed to construct. How do learners confront the unknown for which they do not yet have schemes simply because this unknown is beyond their horizon, foreign/strange, and invisible?

We must ask again: How can a cognitive organism intend learning something that is unknown and therefore *cannot be* the object of intention and therefore cannot be aimed at by the learner? How can I aim at learning a concept – a conceptual structure – when I do not know the concept so that I can take it as my aim? Learning really means appropriating the unintended, something *other*, something that lies outside of the horizon of the known, something that is *foreign/strange*. But how can I envisage learning something that I cannot see (understand)? Other problems relate to the nature of the conscious subject, which, in all forms of Kantianism (including Piaget's and von Glasersfeld's forms of constructivism), is supposed to exist prior to consciousness. Constructivism is built on the common Cartesian divide between the body and mind so that any attempt to base cognition on the experiences of a living/lived body are unintelligible (recall, Kant's project is *meta*physical, above and beyond the physical). Constructivism therefore has no means to explain the *integral* and *inner* (as opposed to external) relationship between cognition and emotion, which is such that knowing can change emotions and emotions change knowing; and these mediations occur not just as external factors, where emotions generally are taken to be limiting cognition. In my way of thinking about the relationship of the two, I take the difference between cognition and emotion to be *undecidable* – as anticipated by cultural-historical activity theorists (e.g., L. S. Vygotsky, A. N. Leont'ev).

Piaget (1981) does indeed talk about the relationship between cognition and emotion in an integrated manner. He views them as indissociably conjoined into one, as two sides of the same coin. But he does not view the two on the same level: "Affectivity would play the role of an energy source on which the functioning but not the structures of intelligence would depend. It would be like gasoline, which activates the motor of an automobile but does not modify its structure" (p. 5). Affect is what gives intelligence the drive, which makes the cognitive machinery run, whereas intelligence provides the structure of this cognitive machinery.

Philosophers of many brands (philosophers of difference, phenomenological, Marxist, and pragmatist philosophers) now agree that constructivism *cannot be correct* – in terms of its own discourse it is "non-viable" – as human consciousness could not have begun with a subject that constructs its cognition. *Being* always already is ahead of itself, producing order and orderly behavior prior to recognizing and conceptualizing this order. There is no order before schemes and the question is how schemes emerge from a situation without order – from the perspective of the learning organism. Moreover, we can recognize social (the other) and cognitive order only when it already exists, and this order has been produced – as Karl Marx says, praxis *always* precedes theory and societal being (collective consciousness) always precedes individual being (consciousness). The acting and knowing subject is as much the result of language and social interaction as self-consciousness and consciousness. All human sense making, therefore, ultimately bottoms out in experiences that are not of, and cannot be brought into, the realm of intellectual (linguistic) consciousness. The living/lived body, endowed with senses that allow it to make sense, constitutes the point where mind and nature, sense and energy, culture and nature change from one into the other and where the proper and the foreign are and come to be interlaced. *Otherness* (the *foreign/strange*), *passivity*, and *undecidability* therefore are phenomena that constructivism has no means of handling and theorizing. These are the themes around which I organize this book and that constitute my guiding threads through this work as a whole. These are also the essential aspects included in the more complete theory of cognition, knowing, and learning I propose in Chapter 13. Constructivism provides but a partial (one-sided) perspective on knowing and learning of science as we experience and observe it in the everyday world (including schools).

For this book project, I deliberately chose the *de/construction* of constructivism as my guiding idea because it allows me to make multiple intentions resonate simultaneously:

- First, I use the term deconstructing in the manner Heidegger and Derrida propose it: as de-constructing, *Abbau* (literally *un-build*, take down, take apart) such that what has been taken apart can be used to build something new (*Aufbau*) in which the old is as much a part as that which has been added. That is, I agree with the literary theorist Paul de Man (1983), who suggests that "deconstruction implies the possibility of rebuilding" (p. 140).
- Second, to deconstruct first and foremost means slow, careful, and deliberate reading. In this book, I make a case for resituating constructivism by slow and careful readings of classroom videotapes from mathematics and science lessons and of common, everyday experiences (e.g., seeing something for the first time, anxiety, anger, pain, foreign/strange), on the one hand, and equally slow and careful readings of a series of philosophers that loosely fall into the domain of phenomenology (phenomenological sociology, phenomenology of perception): François Maine de Biran, Edmund Husserl, Martin Heidegger, Maurice Merleau-Ponty, Jacques Derrida, Jean-Luc Nancy, Jean-Luc Marion, Pierre Bourdieu, Harold Garfinkel, Michel Henry, Mikhail Bakhtin, and Bernhard Waldenfels. In addition, I draw on Lev Vygotsky and Alexei Leont'ev in a dialectical rather than intellectualist reading that these Russian psychologists tend to receive in the West.
- Third, with this book I intend to provoke people on both sides of the discussion about construction as a metaphor of knowing and learning (i.e., friends and foes of constructivism). Those against constructivism, while initially delighting in the idea of a deconstruction of this theory (they might take deconstruction as destruction), may find that my deconstruction also deconstructs their own approach. (Above all, my aim is a sustained scholarly inquiry that has nothing to do with the vulgar critique of constructivism that science educators have become all too familiar with.)

The fervent defenders of constructivism will find a lot to grabble with, although the book as a whole constitutes an effort to give constructivism its proper place.
- Fourth, in using the slash ("/") in de/construction, the process of construction in the deconstruction is highlighted. Self-reflexively, it is a construction of the deconstruction, conducted in such a manner that it provides for a reconstruction.

Contradictions in and Problems with the Constructivist Metaphor

If it ain't broke, why fix it? (Popular saying)

There would not be a call for the next generation of science, technology, engineering, and mathematics (STEM) learning research if the kind of and frameworks for research that we currently have were not without problems; and the policymakers, who we "produce" in university education, design curriculum and evaluation/assessment practices that have little to do with what the science education community already knows about knowing and learning.[3] After having conducted STEM research along the entire lifespan – my youngest research participants were four years old, my oldest were retired scientists – and with 20/20 hindsight, I can now detect some of the shortcomings of the major frameworks that have guided STEM research over the past three decades even though I have contributed to some of these frameworks myself. Thus, in the course of my graduate studies and subsequent professional career, I have employed, contributed to developing, and ultimately critiqued (neo-) Piagetian theory, information processing, radical and social constructivism, and cultural-historical activity theory. In the course of this research, I have come to note contradictions and problems in these theories in part because of my continued interest in understanding what it feels to be a learner and how I, as a learner, make sense of my world and how I deal with the unknown. The purpose of this section is to articulate contradictions in and problems with existing theories.

[3] My wife works for the Ministry of Education of British Columbia. During the week that I am writing this footnote (September 2010), she returned home telling me about the great "innovation" at work: The policy makers have "discovered" individualized learning that is to occur at the rate most appropriate for the individual, focusing in his/her interests, and being situated in learning communities. She points out the irony of the fact that she had served as research assistant in the early 1990s on projects that showed precisely the opportunities for "meaningful learning" from individualized learning settings. She notes that it has taken the policymakers 20 years to catch up with research.

Theoretical Frameworks and the Official Curriculum

Constructivism and conceptual change theory have not worked in guiding policymaking and subsequent curriculum implementation. Today, official curriculum guidelines, such as the ones that the British Columbia government creates for science, are built on and perfused with behaviorist ideas. British Columbia is a good example to take because it has done quite well in international comparisons. In one of these, the 2006 Programme for International Student Assessment (PISA), the province – according to its public announcements (2007EDU0173-001572) – took fourth place in terms of the average scores for science, sixth place in reading, and 16th place in mathematics; in science and mathematics, there were only one and five jurisdictions, respectively, that scored in the range above students from this province.

But this province is by far not up to date concerning the orientation of its curriculum to the forefront of science education and learning sciences research on how students become STEM savvy. Already two decades ago, when I began my research program in British Columbia, the teachers I worked with declared that they could not teach according to constructivist principles as outlined in the curriculum guides. Perhaps not surprisingly, then, the prescribed learning outcomes in the most recent British Columbia guidelines for tenth-grade science are entirely built on Bloom's behaviorist taxonomy (MoE BC 2008). It articulates knowledge as "includ[ing] those behaviours that emphasize the recognition or recall of ideas, material, or phenomena" (p. 27); it distinguishes knowledge from understanding and application, which "represents a comprehension of the literal message contained in a communication, and the ability to apply an appropriate theory, principle, idea, or method to a new situation" (p. 27). The term "conceptual" appears only once in the document, and then it is in the context of traditional ecological knowledge and wisdom as part of "proven conceptual approaches which are becoming increasingly important to all [British Columbia] residents" (p. 14). This is so despite the fact that conceptions and conceptual change research has identified large numbers of "misconceptions" with respect to the curriculum content, for example, the physics of motion; acid–base relations and (balancing chemical equations); and ecosystems, populations, and food webs. Some of the forms of knowledge or competencies that the official curriculum document specifies include "Describe the elements of a valid experiment" or "Identify and use the most appropriate type of graphic, model, or formula to convey information, including Bohr model or diagram, convection model or diagram, Lewis diagrams, chemical formulae, line graphs of displacement, time interval, and velocity, diagrams (e.g., food webs/pyramids, nutrient cycles, plate boundaries)" (p. 39). But little if any of these forms of knowledge and competencies are useful in the everyday life of just plain folks and, for these matters, even in the everyday lives of most highly educated scientists.[4]

[4] I am a trained physicist (MSc), statistician, and social science researcher. I stay up to date on scientific research by reading *Science*. Yet none of my scientific knowledge helped when I found myself confronted with chronic illness and fatigue. I had to learn the required knowledge from scratch, for

Constructivism has been the main paradigm in science education for the past two decades, having relieved Piagetian stage theory from this position during the latter part of the 1980s. But the terms constructivism or constructivist do not appear at all in the document, and there are only two mentions to construction as an epistemological endeavor: "GOAL 3: **Knowledge** – Students will construct knowledge and understandings of concepts in life science, physical science, and Earth and space science, and apply these understandings to interpret, integrate, and extend their knowledge" (p. 12) and "with support and guidance from their teacher, students take responsibility for their own learning, constructing meaning for themselves" (pp. 33–34). It may not be surprising that constructivism and conceptual change have had so little impact on the official curriculum in British Columbia given that teachers wrote them; and it is precisely teachers who "usually are not well informed about the recent state of research on teaching and learning science and hold views of teaching and learning that are predominantly transmissive and not constructivist" (Treagust and Duit 2008, p. 322). But then we might ask, where have these science teachers been trained? Why have colleges of education generally and college/university-based science educators more specifically not been able to assist future science teachers in recognizing that these theories open up new possibilities for acting? The small impact constructivism has had in British Columbia is particularly surprising given that the science educators at the three universities that have prepared science teachers have been proponents of constructivist principles since the late 1980s.

Misconceptions, the Learning Paradox, and Culture

Existing theories cannot explain what Husserl (1939), as part of his agenda to describe the crisis of the European sciences, articulated as a major problem: how mundane, everyday language and culture constitute the very condition and resource of/for people in their ontogenetic development and of/for culture in its evolution. Thus, the constructivist approach generally and the conceptual change approach specifically fail to articulate some fundamental contradictions that come with their own presupposition including the roles of commonsense knowledge and culture and the problematic nature of intending (to construct) knowledge without knowing this knowledge that one is supposed to construct. Science educators often argue that students need to know this or that concept in their life. In fact, very little to nothing that can be "covered" in a school-like curriculum is actually useful in everyday life (Roth and van Eijck 2010). The very fact of the existence of "misconceptions" should tell us that these rather than the real, scientific conceptions are and historically have been the ones that are useful in everyday life. It is useful and beautiful to admire a sun*rise* or sun*set* in the desert rather than to think that the sun disappears because of the Earth's rotation. Experience-based notions, such as sunrise and sunset, often place agency

even the doctors involved failed to exhibit currently available knowledge and to enact a research process that could identify the causes of my illness.

with entities that differ from science. As a result, the sun is said to move rather than the earth.

Children learn their "misconceptions" by participating in societal and cultural life, that is, they appropriate them from their relations with adults. Societies and cultures reproduce their knowledge and practices precisely because they have been proven useful. Yet conceptual change researchers often write about their desire to "eradicate" "misconceptions." A simple Google search using the Boolean term "'Eradicating' AND 'misconceptions'" provides ample evidence of the prevalence to think that science teachers should work to get students abandon the discourses they have learned in and through their everyday participation in the life of society. Conceptual change researchers appear to be astonished to find out that "some misconceptions are difficult to eradicate" or that "teachers are taking very little to misconception perspective." Concerning the eradication of misconceptions, a very different understanding emerges when we think the knowing of science in terms of talking science. In this case, we think about students who talk in particular ways about certain phenomena; and these ways are enabled both by the language they use and by the way in which they have heard the language being used. Eradicating these ways would mean eradicating an aspect of language or a particular way of realizing the possibilities of this language that students are familiar with from their everyday lives. Eradication and replacement of ideas are undesirable goals that cannot but lead to misdirected actions. Whereas eradicating a misconception may sound like a reasonable goal, eradicating a part of language or the possibilities of linguistic expression truly is problematic.

Theoretical problems emerge within the very constitutive core of constructivism (Henry 2000; Marion 2004), which has Immanuel Kant as a major theoretician on the basis of whose work Piaget and von Glasersfeld established their theories of learning. One major problem with the idea of "constructing knowledge" is that those who construct something generally have a plan, a vision for the outcome of their constructive action. The Kantian subject is an intentional subject, in the process of constructing its transcendental knowledge. But construction workers have a plan: they know what the end product of their labor will look like. Learners, however, do not know the knowledge that they are supposed to construct: if this new knowledge is the end product of their learning process how can it serve as intentional object? That is, students cannot have a plan for engaging in the construction and they do not have an image of this product so that they cannot even know when they have achieved the construction (if we may call it such) as intended. This situation would have as equivalence construction workers who do not have a plan to guide them in their work: they would simply be putting together some stuff. Surely it does not make sense to build homes without having a plan. Why do we think students can construct something if it is inherently impossible for them to know the intended object? Would we call "painting" the action of a person who simply doodles – which has the dictionary senses of "scrawling *aimlessly*" and being idle – with a pen (e.g., during a boring faculty meeting)?

The verb "to construct" is transitive. This means it has a direct object; the verb would be incomplete without direct object. Students are said and asked to "construct knowledge," which means that "knowledge" is the direct object. This direct object is the "receiver" of the action, that is, it is the intended object. If knowledge is not intended, it makes little sense to talk about constructing it. It would make no sense to plan a curriculum expecting students will "construct" what it specifies. Because they do not know the knowledge to be learned, students cannot intend it – which is exactly the learning paradox to which constructivism has not offered a solution. The construction metaphor therefore misdirects us in trying to understand what happens in learning: something that we do not know comes into the clearing without having been intended.

Another issue often inappropriately thought and thought about is the role of tools in and of the so-called construction. Schools teach for yesteryear, for example, doing longhand division given that there is an abundance of calculators and computers. Thus, students are still required to calculate the speed of some entity when in fact there are many different devices that can be used to get this measure – including microwave motion detectors, laser radar, and speedometers. The cultural-historical approach tells us that former goal-directed actions are first reduced to invisible, conditioned operations and subsequently crystallized into tools (Leontjew 1982). Because of the tools, we do no longer have to know what our forefathers had to know but are now faced with much more complex actions still unimaginable when our ancestors were alive. A pilot of an Airbus A310 for a Canadian charter airline company recently told me that he had no college training, that most of the time his plane was flying on automatic pilot. More strikingly, he said that in his spare time he studied toward a Bachelors degree in psychology but that he was having a hard time with the required mathematics courses. Whereas in the early days of the unaided flight a pilot needed to know a lot of mathematics (geometry), today much of this knowledge has been crystallized into the tools available in the cockpit.

In sum, therefore, the constructivist metaphor has at least two major problems when it comes to intentionality. First, intention requires a direct object; it cannot aim at that which it does not know, whether this pertains to the conceptual realm or whether this pertains to the senses. Thus, vision cannot intend to see that which is unseen precisely because it is unseen; touch cannot aim at feeling what is what has not (never) been felt before. Second, intentionality cannot ever constitute its own condition of possibility – "in opening us to the world, rendering it manifest, it is incapable to assure the work of manifestation with respect to itself, to reveal itself to itself" (Henry 2000, p. 168).

Ptolemaic Epicycles: Affect and Motivation as Examples

In current theories of STEM knowing and learning, there is nothing about real, whole people with "personal needs and interests . . . inclinations and impulses" (Vygotsky 1986, p. 10). There are only sets of different dimensions – conceptions, affect, motivation, and interest – that apparently have very little to with real persons in living flesh and blood. The constructivist metaphor is not about people, who, when caught lying, may flush and experience a constriction in their chest; there is nothing about the experience of a teenager, who has fallen in love for the first time; and there is nothing about the changes that the death of a loved one bring about in the way we feel, think, believe, and act – or more generally, who we are in the world. Although Vygotsky had a stark warning about the separation of affect and cognition (thought), those who make STEM-education-related decisions at the level of the cultural authority (Ministry of Education) still conceptualize them separately. This is perhaps not surprising given that Western researchers, in the classical science tradition, tend to take systems apart, gaze at and theorize these parts, and then make conclusions about what these parts mean with respect to the whole. Thus, in science education the learning of science concepts during a teaching experiment or what someone says in response to interview questions about career motivations are theorized and used to make inferences about what is in the learners' minds without the consideration of the system (life) as a whole. Other aspects such as affect or motivation come to be tacked on to the main theory, much like epicycles came to be tacked onto the sphere-based theories of the universe (when really a scientific revolution was needed). In the following, I show how motivation does not have to be tacked on to a theory of cognition in the way that conceptual change theorists have done in an article that has become the rave of the field (Pintrich et al. 1993).

Psychologists and educators often wonder about how to "motivate" students to do science or mathematics. The question posed is how to get students to do in these classrooms what they do not want to do on their own. Teachers then might decide to do "fun" "activities" to get students to buy in, without really addressing the larger questions of why a student might or might not want to "learn" science. Motivation and motivational beliefs come to be tacked on as external factors to cognition, modifying it either from the inside – when students are "unmotivated" – or from the outside – when teachers are supposed to motivate their students. But in real, everyday life, we do not need to be motivated to do what we do. When I see so-called "unmotivated" students practice for hours until they successfully realize a particular move on the skateboard, I cannot but reject their characterizations on the part of researchers. Moreover, their motivation does not just come from the inside. Why would a human being be motivated to work at learning to do something as useless as a move on a skateboard? Yet their motivations do not come from their insides (alone); it is not "intrinsic" motivation that drives the skateboarders I observe from my living room window. In fact, the motives of skateboarding in which the people I observe participate orient their actions; and any attendant and anticipated emotions have control functions. These skateboarders learn something new when necessary, because it contributes to their increasing agency in the overall range of possibilities within the skateboarding culture.

That is, no external motivation – another way of saying this is "external control" – is necessary when people seek to expand their room to maneuver, their agency, in the pursuit of some goal by engaging in a "learning loop" that increases their possibilities with respect to the goal. Thus, an experiment conducted in the 1930s in the museum of Kharkov, where there was a workshop on building model planes, gives us a better appreciation of the relation between the motives of activity and motivation (Leontjew 1982). Initially few students attended and those who did were building very nice planes was all the while exhibiting very little interest in learning physics. Few students addressed themselves to the attendant resource persons and they hardly read the available physics content posters. But then the psychologists changed the structure of the activity, now formulating the *object/motive* to build planes that covered a certain distance. After building their first versions, many students found that their model planes crashed prior to making it to the finish line. As they intended to get their planes across the line, they began to seek ways of making them go farther; and, lo and behold, they became interested in the physics of flying. They began to ask instructors and read available materials. The psychologists reported a statistically difference of large effect size in the engagement with the content of physics (from less than 5 to over 20 minutes per student) and in the number of students taking the workshop (from an average of 6 students per day to over 40). It is not *intrinsic motivation* that allows us to understand what happened, but rather, the concept of *collective object/motive*, which the students have taken up; and in the course of their work, they felt their room to maneuver could be expanded by drawing on available physics resources.

The issue then is to develop a theory in which motivation derives from and is integral to core features rather than being tacked on, a feat already possible within cultural-historical activity theory. If we understand motivation as a derivative of affect (*e*motion) and motive, then there never is a problem in the way that psychologists formulate it. Whenever someone recognizes a way in which his or her room to maneuver may be expanded, they do so because it gives them greater control over their conditions, expands what they can do and how well they can do it. To understand the levels of engagement of students in our classrooms, we cannot understand it if we theorize motivation as a factor of individuals.

We can understand, however, what a student wants to learn, how a student learns, his "motivation," his orientation toward school tasks, his learning needs, and so forth when we look at science through his life as a whole. For example, one of my physics students (Tom) had the goal to enter one of the premier engineering schools in the country. Given that grades and test performances are used to determine entrance, it does not surprise that this mediated his learning, actions, and dispositions. He chose physics and chemistry not because he particularly liked the subjects but because successful completion of these courses guaranteed him access to the engineering school of his choice. In physics, Tom did not like experimentation, normally the "carrot" teachers use to "motivate" students. Tom did not like this part of physics because it led him sometimes to understandings that were inconsistent with the scientific canon, and this would lead him to incorrect answers on standardized tests. He told me that he had bought into the constructivist theories that we discussed in the classroom. But it was precisely the tenets of constructivism that he rejected experimentation: it too easily led him to construct knowledge that would not allow him to pass tests and entrance examinations. He knew that this would diminish his chances to enter the program of his choice. Thus, how he engaged in physics could not be understood within the context of physics, but his choice of physics and his engagement in the course became intelligible when we look at it through the lens of his life as a whole. Preferences, predilections, (structured structuring) dispositions and so on are the result of our experiences in the world, of life as a whole. To understand, we therefore need to study learning (and knowing) by beginning with life as a whole and by moving, if necessary, closer into the phenomena of interest. As shown in the example, to understand the future engineering student's dispositions for learning and learning environment, we need to understand Tom's life as a whole. Similarly, there are other examples that show how we need to begin with understanding the relation of people in and with life as a whole to understand their actions, behaviors, and dispositions within the school, which is only a constitutive part of life, and a one-sided representation of it.

Passibility, Affectedness, Passion

There is a whole range of phenomena that are central to our everyday lives and experiences but that lie outside of what the constructivist metaphor can explain including *passibility*, *affectedness*, and *passion*. These phenomena not only are central to our lives and experiences but they also enable learning to be possible in the first place. If the (human) organism did not exhibit the capacity to be affected, that is, were not subject to passibility and passion, no intentional construction of anything would be possible. It is not surprising that these dimensions are absent from conceptual change learning theory because the constructivist metaphor *excludes* these dimensions from the beginning. Thus, "constructivism – as an empirical epistemology – can provide a more or less viable model for *the construction of the experiential self*" (von Glasersfeld 1989b, p. 447). That is, from the outset, constructivism includes only *constructive* and therefore intentional aspects of knowing. It excludes all other aspects, including how the capacity to construct becomes possible in the first place. Thus, "the self as the operative agent of construction, *the self as the center of subjective awareness, seems to be a metaphysical assumption and lies, at least for this constructivist, outside the domain of empirical construction*" (p. 447). It is ironical that the concern with the nature and origin of a self is termed metaphysical by a theory that constitutes the pinnacle of metaphysics (the mind in and for itself). In the phenomenological literature, it is recognized that the objectivity of the object and the subjectivity of the subject are of one and the same *flesh*, that is, of the capacity to be affected. The flesh makes auto-affection possible, and auto-affection lies precisely at the origin of any intentionality, because only an auto-affected flesh immanently (without mediation by the [conscious] mind) knows that it can move and intend to be further affected in encounters with the world. And it is in these encounters with the world that the flesh, the living/lived body, comes to be further affected and changed – and thereby is led to development. That is, affectedness is associated with *radical passivity*, a passivity that has nothing to do with the intentional withdrawal from engagement, itself an active process and result of a decision.

Radical passivity is one with agency, which is a central but unrecognized aspect in theories of learning and development that make use of the constructivist metaphor. This is one of the central themes of the present book. It requires us to rethink epistemological issues in STEM learning specifically and in all learning more generally. To assist readers in understanding why radical passivity is central and has to be at the origin of a viable theory of knowing, consider this. Not only children but also adults decide to employ their senses to find out about some phenomenon. Think about yourself, for example, coming face-to-face with a new form of food, a wine or olive oil you have never tasted/smelled before. You decide to smell, pick up the glass and bring it to your nose. What made this decision possible? Even without reflecting upon it, your living/lived bodily self knew to reach out and bring the glass to your nose. But whence does this knowing to reach out come from? It is not from a construction, for children in the earliest parts of their lives, when there are no (linguistic) mental structures do reach out for objects. The source cannot be sensorimotor schema, as these are the *result* of reaching out experiences rather than their source – unless we want to begin with innatist explanations and all the theoretical problems that these come with. This is precisely the route that von Glasersfeld (2001) takes when he states that "the child has the innate tendency to search for 'Rhythms, Regulations, and Groupings' and to test its constructs for their viability in actual experience" (p. 144). That there is no innate tendency to search for anything was clearly and unequivocally shown in a study of congenitally deaf-blind children (Meshcheryakov 1979). Thus, innatist theories do not take us any further. Rather, we need to look into the emergence of any immanent knowing, whereby a living/lived body comes to know what it can do in an immediate rather than semiotically mediated way and whereby it is enabled during the self-affection of the flesh in originary non-intended movements. That is, this capacity is *given to* the subject rather than the result of an innate tendency to construct.

There is a second dimension of the smelling/tasting experience that involves radical passivity. We do intend the specific smell and taste of the wine or olive oil precisely because we do not know what it smells and tastes like. But if we do not know the taste and smell, we cannot *intentionally construct* the smell and taste. First and foremost we have to open up and allow ourselves to be affected. This not-knowing beforehand comes with uncertainty, and therefore, also with risk. Thus, the smell or taste of the unknown product may be completely negative, even painful, and can even be dangerous in the case of substances that a student might encounter in a chemistry laboratory. Not surprisingly, the standard recommendation is to wave the hand such that the smelling can begin with whiffs of odor rather than with the full, potentially dangerous experience of smell. In deciding to take a whiff, we expose ourselves to the new and unknown. We actively expose ourselves and we are also exposed to the unknown (passive): we are therefore vulnerable. The first whiff affects us prior to any construction or comparison. Vulnerability *precedes* knowing. Without vulnerability there would be no knowledge at all. It is only after having been affected that we can begin to think, classify, and relate the experience to something else.

In both instances, the intentional movement to bring the glass to our nose and in the opening up to be affected, agency is enabled by passivity, a passivity that is more radical than any non-action that we might choose as a form of action. The immanent capacity to move, the intention to move, and the experience of encountering the unknown that we cannot anticipate and therefore construct, all invoke passibility, our capacity to be affected. But because experiences thereby depend on affectedness, affect comes to precede cognition, even enable cognition. The new is experienced along a gradient from positive to negative tonality. Without such valuation, no orientation, no intentional movement would have emerged (Leont'ev 1981). That is, passibility and the passions precede anything that looks like the cognitive effort constructivists describe and theorize. In fact, passibility and the passions make cognitive development and cognition possible. Any viable theory of cognitive development, if only to be consistent with the emergence of humanity and human forms of knowing, has to take into account these heretofore unattended-to phenomena of human experience. As I show throughout the fourth part of this book, it is not just that we find ourselves exposed to the unknown that affects us. Rather, much of what we know emerges from these experiences rather than being constructed by the subject. We know pain because we feel it not because we construct it. We know what it is to be in love with another person because we have experienced it in flesh and body and not because we have constructed this experience in our mind. Emotions are not the objects of constructions but rather are passions that we *feel* and to which we are subject and subjected.

In sum, therefore, the constructivist metaphor focuses us only on the transitive aspects of learning and knowing, that is, to the role of the person as the *subject of* activity. However, the English language also allows us to make thematic the converse side of activity, passivity: the subject of (inherently collective) activity also *is subject to* and *is subjected to* collective activity. This framing further becomes plausible from the very etymology of the term subject, which derives from the Latin *subicēre*: *sub-*, under, below, and *jacēre*, to throw, cast. That is, the fundamental experience of the subject, as made thematic in the etymology, is one that phenomenological philosophers following Heidegger have come to term *thrownness* (Ger. *Geworfenheit* literally translates as "the state of being thrown"). Passivity is the originary experience, which not only enables agency but also accompanies it wherever we might care to look in the science classroom and whatever we care to study there.

Plan for the Remainder of This Book

Underlying the writing of this book is my intend of introducing a range of phenomena and concepts to the discourse about learning and knowing in science. The phenomena I have in mind all are part of our everyday life and experiences in the context of their performative dimensions. Yet they have not been attended to in past and current learning theories. In Part B of this book, I articulate different issues concerning radical passivity in learning and development by drawing on empirical materials from science classrooms and science laboratories in addition to conducting phenomenological investigations into the conditions of knowing and learning. Unlike conceived of in the constructivist literature, we are not only subjects of but also subject(ed) to actions and immanent knowing such that we consciously know them only after the fact. The strongest empirical evidence comes from my ethnographic studies in scientific laboratories, where scientists themselves suggest at the end of the day that they must have done something in the morning different than what they were thinking they had done (see Chapters 4 and 5). I exhibit the essentially passive aspects of learning, both in terms of what we learn and the learning process. It is precisely because we do not know something that we cannot intend it. In the second part of the book, and related to the issue of passivity, uncertainty, and undecidability, I show how the duality at the heart of the sciences – e.g., between knowing subject and the known, subject and object, process and product, theory and praxis – emerges in and from everyday actions.

Part C of this book is concerned with articulating how otherness (alterity) is the condition of knowing. That is, knowing is not something grounded in individual construction and subjectivity but is enabled by and *necessarily* requires otherness (e.g., *my* language always is the language of the *Other*). Invoking otherness constitutes a radical reframing of constructivist epistemology on many different levels. First, otherness means that what I can talk about and how I can talk about it always already is enabled by collective (cultural) possibilities and therefore never *merely* is my own. Second, because otherness is the condition for selfhood, this dimension is coextensive with the idea and phenomenon of radical passivity developed in Part B. That is, even my thoughts about who I am, the language in which I describe and understand my Self or my "motivational beliefs," invoke a radical alterity and therefore passivity with respect to myself. This has already been captured in an aphorism that Arthur Rimbaud (1951) articulated in the 19th century: "JE est un autre" (I is an other). Perhaps even more surprisingly, we come to understand our own thoughts as something we confront in our expressions produced for others, that is, we realize what we think only *after* having communicated it rather than before. It is precisely because of this alterity that we do not have to seek "meanings" behind words but rather take words as the very phenomenon of interest – the idea of meanings as separate from (somehow behind) words is itself an outcome of the metaphysical position of constructivism that we need to overcome. This is so because the very idea of "meaning" comes from a primitive understanding of language and there is no place for it in a pragmatic approach to language and cognition (Wittgenstein 1958). It is better to think of a word as something that accrues to significations rather than as an entity to which something else is attached by means of constructive processes.

In the Part D, I turn to the passions themselves – suffering, joy, and crises – as well as to passivity, emotions, and motivations. Central to a viable epistemology are not intentional constructive processes but the conditions for such processes. These conditions include the capacity to be affected, passibility, the unity of suffering and enjoyment. All of these phenomena cannot be understood through the concept of the body made thematic in the embodiment and enactivist approaches, because these do not distinguish between the ekstatic material body and the living/lived body. A table does not experience joy when it comes into contact with a glass of nice wine, nor does it suffer when it is loaded with the constituents of a feast. Living beings, however, do experience needs and fulfillment: they suffer (feel pain) and experience pleasure (being satiated after a meal). That is, only a living/lived body that is capable of being affected also is capable of knowing in the manner that makes possible not only constructivism but also theorists who evolve a constructivist discourse. Constructivism cannot provide such a language, as it is incapable of explaining the emergence of the subject of construction. The idea of the living/lived body has been captured in the phenomenological concept of the *flesh* (Ger. *Leib*; Fr. *chair*). The Latin word for the flesh is *caro* (genitive: *carnis*); it leads us to the word *incarnation*, from Latin *incarnāre*, to make flesh. It is in and through the flesh that life enables anything like humanity and human culture; and this process of enablement arises from the self-affection that occurs in and through the flesh. It is only when we turn to life and recognize how it affects *itself* in and through living/lived bodies that we come to understand the emergence of something that looks like human knowing and learning in the way we know it today.

In the final Part E, I provide a tentative synthesis of the ideas presented in the preceding parts to situate constructivism in a more comprehensive theory of life, its auto-affection that leads us to cognition via passibility. The latter enables the passions and emotion. From there we understand agency and the possibilities for the emergence of human forms of thought, including those that were developed by the ancient Greek and that underlie all forms of metaphysics, including those developed by Kant, Piaget, and the constructivists.

Throughout this book, my approach is descriptive and phenomenological in the attempt to understand knowing and learning from the perspective of the subject. We require this perspective, because the subject acts by taking into account what is salient to it. Any knowledge about some object that is not available to the subject "but derived from other sources, must not be permitted to intervene in the descriptive analysis of that perception and its noematic correlate" (Gurwitsch 2010, p. 226). In adopting this orientation, I aim at avoiding the "psychologist's fallacy," whereby acts of consciousness are confused with the object of this consciousness. A typical example of this fallacy is Piaget's assumption of the constant nature of a visual percept. Accordingly, constant sense data – constant within and between people – appear differently because they "are interpreted as organized by, and receiving structure from, extraneous nonperceptual factors or functions" (p. 49). It is precisely on this account that my own, phenomenological procedure differs from the way in which Piaget and the constructivists approach the relationship between sense data and knowing. Rather than basing accounts of knowing and learning on presuppositions, I work only from phenomena as they appear to the knowing and learning subject. The method will be evident in and through the analyses themselves.

Part B

Passivity, Uncertainty, Undecidability

> The ought-to-be of veridicality does not follow at all from the theoretical-cognitive determination of veridicality. . . . no theoretical determination and proposition can include within itself the moment of the ought-to-be, nor is this moment derivable from it. . . . the ought gains its validity within the unity of my once-occurrent answerable life. (Bakhtin 1993, p. 5)
>
> The discovery of an *a priori* element in our cognition did not open a way out from within cognition, i.e., from within its content/sense aspect, into the historically individual, actual cognitional act; it did not surmount their dissociation and mutual imperviousness, and hence one was compelled to think up a purely theoretical *subiectum* for this transcendent self-activity, a historical non-actual *subiectum* – a universal consciousness, an epistemological *subiectum*. (p. 6)

Constructivism is an epistemology, aiming to explain how human beings know and learn. In the opening quote to this, Bakhtin rejects the notion that theoretical cognition is relevant to the question of "validity within the unity of my once-occurrent answerable life," that is, the one and only life of my living/lived body. As a theory constructivism is silent about issues concerning the nature of the world, and therefore it also remains silent as to the nature of the living/lived body in learning. Bakhtin equally rejects the primacy of the theoretical cognition, which cannot account for the historical, individual act in a real world where no act can be undone. The result of the Kantian approach that Bakhtin critiques in these pages is a non-actual subject, an epistemological subject that has no hold in and on this actual world that we inhabit, a world in which we suffer as much as experience joy.

If the mind cannot know anything about the material world other than the viability of its constructions, it cannot know anything about itself, being part of the body that brings forth mind. But even in the cognitive domain, there are numerous phenomena that a constructivist epistemology cannot account for. Most importantly, the constructivist metaphor cannot explain how humans came to be able to think in the way they do. Thus, human forms of intentionality and directedness in thinking require the making present of another present, that is, require *re*presenting. In other words, whereas other living organisms – including the predecessors of the human species – exist in the presence, in the form of *Being*, humans exist in a state of dehiscence.

This dehiscence separates beings – i.e., the things of the world and *re*presentations – and Being. In a number of writings about the pre-Socratic philosophers – mainly Anaximander (c. 610–546 BCE), Parmenides (early 5th century BCE), and Heraclitus (c. 535–475 BCE) – the philosopher Martin Heidegger shows how what we now refer to as metaphysics emerged as the Greek began to think Being in terms of beings (*re*presentations). That is, the Greek began to think Being – that which is alive, in flux, and ever changing – in terms of representations, which are static and fixed. In this way, the Greek also evolved a way of thinking about thinking that was developed to its highest form in the work of Immanuel Kant. Not surprisingly, this philosopher has been referred to as *Logodaedalus* (Nancy 2008), the master artisan of the logos, mind.[1] This approach, consistent with metaphysics, still dominates present-day scholarly discourse in psychology, cognitive science, the learning sciences, and science education where researchers still wonder about the "grounding problem," how the mind is related to anything in the material world.

There are many phenomena, however, that do not require representations to be knowable, including those that I present in the fourth part of this book – e.g., the passions and emotions. Moreover, there is a form of learning that cannot be understood in terms of an intentional orientation. Thus, we cannot aim at, make our goal, or use to orient ourselves anything that is yonder our current horizon that circumscribes the visible and knowable: the unknown. It is impossible to take into account in my action that which I do not know and that which reveals itself only in the course of time *subsequent* to my actions. Beforehand, we do not even know if there is some *thing* that is anything. If it were possible to take into account and orient to something unknown, then we would not have expressions such as "being in the dark" if acting with respect to the unknown and the uncertainty it brings with it were as easy as the so-called "construction of meaning." If it were possible to take into account the unknown, then I would not have been able to observe scientists struggling to get their equipment to work even though it had worked the night before – an event that led me to ask "What do scientists do when they do not know what they are doing?" This is precisely the same situation in which students find themselves, having to orient to learn something they do not know and that they therefore are not in the position to aim at. Moreover, because students do not know what they are to learn, they cannot know whether anything they did learn corresponds to what they ought to have learned. They do not know whether they are any closer to the prescribed learning goal.

The verb "to learn" derives from the Proto-Indo-Germanic root *leis-*, to furrow, to pursue, to learn. The German translation of this verb, *lernen*, has precisely the same origin. Learners are on some track, on some trajectory, taking them to places that they do not yet know. The adjective "delirious" derives from the same root, spiritually being off the track (the German noun *Gleis*, track, derives from the same root *leis*). But when there is no track, then the person (learner) has to forge it, lay the garden path in walking without knowing beforehand where it leads, whether there will be a dead end

[1] The name Logodedalus is a reference to Daedalus, the master artisan and craftsman of Greek antiquity, who was responsible for the construction of the palace of Minos and who also constructed a set of wings for his son Icarus so that he could escape from Crete, where he King Minos confined him.

and whether the next step will engulf the foot in a sinkhole. Interestingly, however, the noun "lore," in the sense of the act of teaching, also derives from *leis-*, as do the German noun and verb for teaching: *Lehren* and *lehren*. That is, teachers know some tracks and can let students know whether they are off the track, heading off in the wrong direction, or have entered a dead end precisely when the students do not know where they are going. That is, learning means confronting the unknown, the foreign/strange; and it means experiencing passivity with respect to what reveals itself to the learner.

In this Part B, I provide examples of, discuss, and begin to theorize these phenomena as I observed them in everyday situations both in classrooms and in the world outside schools.

Chapter 2: Learning and the Erasure of Knowledge. One fundamental problem of the constructivist misunderstanding (should we say "constructivist misconception"?) of phenomena related to knowing and learning comes from the fact that those who know do not understand the non-knower's world and experience *sui generis*, as a phenomenon in its own right. Thus, it is not surprising that some phenomenological researchers have criticized Jean Piaget that he viewed the child as a diminutive adult (scientist), whose experiences, observations, and rationality are of a lesser kind. The same position is taken day after day when science or mathematics teachers ask their students to abandon their knowledge und to peruse carefully selected educational materials that are said to embody the scientific or mathematical principles to be learned by children (students). It turns out, however, that children's rationality is different because their access to and experience of the world is different; they live in different worlds (or rather, to be more precise, a different lifeworlds). How is it that Piaget and teachers do not appreciate the child's world and experiences *sui generis* but understand it in terms of their own adult rationality? The question is particularly interesting because Piaget and STEM teachers once were children, thinking and acting like them. If they no longer know how the world looks to a child, they must have forgotten what a child's world and experience is like. This is precisely the point that I make in Chapter 2. I draw on several experiences where I "discovered" entities that I had never seen before but in the same instance forgot that my immediately previous worlds had not been populated with these entities. I observe the same in videotapes from science classroom, where students, after perusing some materials for up to 10 lessons, all of a sudden discover one of its properties. I use these experiences and observations as materials for an extended reflection about *forgetting* that is involved in learning. There is, in fact, an active erasure of the worlds we have known even as we learn something new. I show how the same phenomenon shows up in the philosophy of Jacques Derrida – following some fundamental work by Heidegger – as erasure that comes with writing (learning).

There is a reason for treating this topic here in Chapter 2: I assist readers in becoming aware of the processes that do not normally allow them to understand the states of knowledge of learners and therefore lead them to misinterpret their classroom observations and recordings thereof. This gives us a first understanding how constructivism works only when one already assumes (has attained) the later states of knowledge. The constructivist metaphor works only in a teleological approach to cognition. In other words, a theory of the intentional construction of knowledge only makes sense when the builder already has a plan that includes the knowledge. But this is a poor model for understanding how a cognizing subject could have attained forms of knowing that are of higher complexity than the current form and the corresponding (i.e., homologous) complexity of its world of experience (the *learning paradox*). Construction is a limited metaphor because learners have no way of aiming at the knowledge to be learned, but to take aim they already need to know this knowledge before actually learning it. How could Christopher Columbus have *intended* to discover the Americas? Had he known their existence, he would not have needed to discover them; and if he did not know about the Americas, he could not have intended discovering them.

Chapter 3: Radical Passivity in Learning. In the philosophy of the sciences, it has become a truism to state that all observation is theory laden. That this also is the case among students has been the outcome of an experiment I conducted in an Australian high school physics class. After being asked to predict what they will observe in a particular experiment on rotational motion, all students precisely observed what they had predicted: five students predicted and observed no motion and 18 students predicted and observed motion. This raises a question for constructivism and conceptual change theory, where it is proposed that students be encouraged in the reconstruction of their conceptions by means of "discrepant events." How can events be discrepant if there are tendencies that subsume perceptual experiences in existing ways of observing and thinking about the world? That is, constructivism does not explain how *new* forms of viewing the world can emerge and how students thereby can escape from perpetually reproducing everyday ways of perceiving – and therefore of explaining – natural phenomena. Drawing on concrete examples, I articulate a phenomenology of perception that includes both agential and radically passive dimensions. Perception is more than constructing a version of reality: it is a process of making (provoking) visible and becoming visible. Accordingly, new entities *give themselves* as our perception comes to be restructured in the engagement with the initially unstructured textured surface, giving rise to new structures and new perceptual dispositions. It is not just agency that describes how we learn to look at a perceptual field to make a Dalmatian appear from a collection of ink splotches, and it is not just passivity of the perceptual apparatus onto which the Dalmatian imprints itself. The phenomenon of perception explodes the dichotomy of agency and passivity, inside and outside of the cognizing subject, or the intra-psychological and inter-psychological. This allows us to understand each of these phenomena as but one-sided manifestations of human perception in and of the world. I link this aspect of learning from experience to statements whereby the artists, creators of something new – students are artists in this sense because create *new* knowledge – do not know what they create until after they finished created it. In this respect, the painter/sculptor Alberto Giacometti says: "I do not know precisely what I see. It is too complex. So I have to try to copy the visible to make accountable what can be seen." Another artist, Quentin Blake, recently said this: "I didn't adopt this style. It adopted me." These descriptions of the creative process are much closer to what learning is about than the explanatory discourses that cognitive psychologists and constructivists have evolved for us.

Chapter 4: Radical Uncertainty in Acting. The constructivist metaphor is fundamentally based on the idea of intentional cognizing agents who build mental structures that allow them to cope in and with the world, that is, the mental structures are selected in terms of their viability rather than the quality of their correspondence with an outside world. (But see the shaping of perception in Chapter 3.) Because of the tight link between intentionality and agency, the constructivist metaphor is unable to explain a range of phenomena that have come to be known as the abysses between plans and situated action, theory and practice, or knowing-what and knowing-how. These phenomena can be observed not only among newcomers, for example, when novice cooks fail in their attempts to follow (situated action) a recipe (plan). They also can be observed among the most highly educated and practiced individuals, such as when engineers have trouble getting an "intelligent" photocopier to work (Suchman 1987), when engineers do not succeed in building some machine for a second time even though their first machine still works (Sørensen and Levold 1992), or when the biologists I studied realized after an entire day's work that they must have done their dissection in the morning differently than they had thought to have done it – and this although they had 30 years of experience doing this dissection! Why did these engineers and biologists *not* achieve a match between their intended and their situated actions? If scientists fail to bring into perfect alignment their intentions and actions, this is all the more to be expected to occur among science students asked to learn by doing this or that task (e.g., group discussion, laboratory investigation). In this Chapter 4, I use these two examples – the engineers and the biologists – as the data for a reflection about the source of the abyss between plans and situated action, leading me to theorize the gap. My answer is based on the work of two philosophers: Pierre F. G. Maine de Biran and Martin Heidegger. The former has developed a phenomenology of the flesh (living/lived body), which acquires patterned ways of doing things because of an internal self-relation of the flesh. Only after something has been learned are the changes available to a consciousness, but only partially because the self-affective relation of the living and physical body cannot be made the topic of talk (e.g., the "feel of the game" that comes with a practical sense cannot be articulated). Being is ahead of itself so that there is a dehiscence between the lived present and consciousness of the present, which requires the presence of the present or re-presentation. Being knows (in and through the living body). But it knows *immanently* before knowing (consciously) and *ecstatically*. In the philosophy of Heidegger, a gap is theorized between the present and the presence of the present in consciousness (technically, between *Sein*, Being, and *Seiendes*, beings). This gap, already theorized by Plato (as *khôra*), returns in the work of Jacques Derrida, where it is associated with a range of concepts, including *khôra*, writing (*écriture*), and *différance*. Here, I establish a link to Chapter 2, because writing, as any form of discourse, is linked to erasure. I also discuss erasure in terms of the work of Mikhail Bakhtin, who uses the trope of continual birth/death in explaining how language evolves as it renews itself in every speech act.

Chapter 5: Emergence of Duality. With the constructivist metaphor comes the dualisms of self-identical entities (e.g., mental structure, word, meaning), a presupposition that has detrimental effects on the ways in which we understand and theorize knowing and learning (Gurwitsch 2010). For example, the duality of the mind and world is inherent in the statements about mental models and the world. Throughout his work, on Glasersfeld states explicitly that constructivism is about knowledge not about Being and the world. A related duality pertains to the human body, which is only the seat of rationality but is not, other than providing experiences, an integral part of rationality, cognition, knowing, and learning. But any duality – including body/mind, mind/world, or practice/theory – is the result of theory (theorizing activity) not the origin of theory. Drawing on the examples provided in Chapters 2, 3, and 4, I show how and when the separation of the cognizing agent and the material body/world occurs. Thus, for example, whereas there is an originary unity between perceptual processes (observer) and the entities perceived (observed) (Chapters 2 and 3), this unity comes to be broken as things – Lao Tzu's *Tao Te Ching* speaks of "*the* 10,000 things" – become part of the world independent of the observer. It is the same process that leads to the independence of the world and the material things it contains that become independent Galilean objects. I also draw on a study that described the emergence of the first pulsar observed to become an independent Galilean object in the course of one night's work in an astronomical observatory. I show that at the level of the individual, Maine de Biran's philosophy – especially as Michel Henry articulated it in his work that establishes a material phenomenology – provides an explanation of how the world observed comes to be experienced as independent of the observer. Language, accountability, and most of all radical passivity are integral to this emergence of the dualisms.

Chapter 2

Learning and the Erasure of Knowledge

Over the years while I worked as a science teacher, I have come to note that one of the main problems of the constructivist metaphor and the associated conceptual change theories comes from the insensitivity to the world of children and those who know in ways that differ from the ways in which professional scientists, mathematicians, engineers, or technologists know. The latter establish certain norms for what it means to know and judge what others know in terms of the difference between the legitimized and the everyday ways of knowing. Thus, with respect to the sciences, children are said to have misconceptions; with respect to adult scientists' ability to think formally, Piaget elaborates a theory where children find themselves at one or more stages below, thinking, for example, *merely* in a concrete manner. As a high school science teacher I began to notice – without being able to immediately turn it into a plan of action – that students were not seeing in their experiments what I was able to perceive and wanted them to see. When I started teaching 30 years ago, I did not know about epistemology and the history of science. But I realized that if I truly took into account what students knew then I had to involve them and their knowledge in the curriculum design. I thought that I needed to take into account not only what they knew but also – and more specifically – what they did not know and could not know. I realized that I had to think learning from the perspective of the unfamiliar, strange, and unknown rather than from the perspective of what *I* knew. As a teacher, therefore, I increasingly involved students in designing their own curriculum, which always built on how they perceived and experienced the world. This began to give me access to forms of thinking that may have characterized my fellow teachers and myself when we were young, but which we must have had forgotten since. STEM curricula, on the other hand, are developed based on the inner structures of the respective field according to their core practitioners – e.g., organized around fundamental ideas – not around the lifeworld structures of children and older students.

Not only children and students experience their worlds to change, we adults do so, too. We may live for years in an area and then all of a sudden notice a particular aspect that we had never noticed before. For example, two years ago I bicycled

W.-M. Roth, *Passibility*, Classics in Science Education 3,
DOI 10.1007/978-94-007-1908-8_2, © Springer Science+Business Media B.V. 2011

Fig. 2.1 Despite its considerable size, I had not consciously seen this church in the course of my first 11 years of passing this spot on my way to and from the university. (© Wolff-Michael Roth, used with permission)

home from the university on the same road that I had taken for the preceding 11 years. All of a sudden I found myself startled: There was a church that I had never noticed before (Fig. 2.1). It is enormous, not even hidden. I asked myself, "How could it be that I have not noticed this church for 11 years?," "What is it that has brought this church into my consciousness now?," and "How is it that a church, which has just come into existence for me, all of a sudden obtains an existence that goes into the past?" (I did say, "I have not noticed before," and thereby accepted that it existed even though it did not exist in my conscious awareness.) At that instant I experienced a flashback to an experience that I had nine years before that while investigating what it means to notice something for a first time.

Noticing Something for the First Time

During a stay in the neuroscience unit of the *Hanse Institute of Advanced Studies* (Delmenhorst, Germany), I observed – while analyzing classroom videotapes of a static electricity curriculum for tenth graders – how some students spent hours of investigating something before taking notice of specific aspects of the tools and materials they had been using. For example, a group of students discovered after 10 lessons that the little glow lamp they were using since day one did not have a continuous filament but that there was a gap between what they later came to know as the two electrodes. They saw this gap between the electrodes for the first time while trying to figure out why they could not find the presence of charges where there ought to have been charges and where their teacher had shown charges to exist. I was intrigued by the question of how it is possible that students would not be aware for many hours of something like the gap between the electrodes even though they had been continually working with it.

To find out what it means to notice something for a first time and in an unfamiliar context, I designed an experiment. I would bicycle the same initially unknown route for 20 times, writing down in my field notes what I predicted I would see prior to leaving on the trip and recording what I remembered in the hours following the trip. Using a map, I selected a trip of about 20 kilometer length in a part of the countryside that I had not traveled before. Taking notes prior to the departure and upon returning was a way of establishing a record of what I learned.

During the experiment I began to realize that each time I took the trip, I became conscious of entities and phenomena that I had not seen before. On the third trip, for example, I noticed a sign with street name. The name struck me. I was trying to figure out why I felt struck. It took a few days riding past the sign until I remembered that this was the same name as that of a street in which I had lived some 20 years earlier. On the fifth trip, I became aware for the first time of white posts carrying numbers planted at regular intervals (about 100 meters) along the road. After having passed a few of these newly discovered posts and after having realized that they are about equally spaced, I concluded them to be markers that exhibited at 1-kilometer intervals the distance of the post from some origin; the intervening posts, I concluded, must be markers of 100-meter intervals. In each case of such noticing, the conscious awareness came only after a number of times taking the trip. The most important experience, however, and the one relevant to the ideas at the core of this chapter, happened to me on the seventh day. (Notice the passive voice – I did not plan this experience, it *happened to* me.)

On this day, as I am riding along on the country road K342, I notice all of a sudden, for the first time, two giant silos about 200 meters to the left across a field, the lower parts of which are hidden by trees. But the silos – easily visible in the satellite shot of Google maps – are enormous. Despite their giant size, it is only on this seventh day that their exteriority comes to stand out in my consciousness – a standing out, an *ek*stasis (ecstasy). Almost instantaneously, a number of thoughts and impressions emerge into my consciousness. How could it be that I have not seen these twin silos during the past six trips? How can it be that I have not noticed before these giant silos despite intending to notice as much as possible? I immediately think: if someone (a teacher, psychologist) had asked me a question that required knowledge of the twin silos, I would have failed miserably despite my intent to be aware and learn as much as possible during each trip. I would have failed in the past even though, prior to this trip, I have had the experience of riding along the same circuit for six times. I am thinking at the time about the experiments I have asked my students to conduct in the science laboratory, who must have been in the same situation: It is quite possible that they did not see what I thought they ought to have see in the same way that I did not see the twin solos prior to the seventh day.[1]

This experience, however, became even more interesting. On that day, as I am thinking about the twin silos that I had not noticed on the first six trips past this point of my circuit, a strange sense is creeping up within me. Something has been happening; something like a loss is occurring. "Just seconds ago," I am thinking, "I have had something and now it is lost." All of a sudden I realize that I have asked the question, "How could I not have observed the twin silos?" This question *presupposes* the existence of the twin silos in the world. That is, I am realizing that as soon as I have taken notice of the silos, they not only exist in my world, but they begin to have a permanent existence in the world *predating* my observation. Thus, I have come to realize that to be now, to exist in the now, "is to be from always and to be for ever" (Merleau-Ponty 1945, p. 483). The twin silos have not existed for me at first; their existence for me depended on my becoming conscious of them. Then, all of a sudden, they are things in a world that exists independent of my observations and independent of me. Even more serious, I notice at the time that I can no longer imagine a world without the twin silos. "What is it," I ask myself, "that allows something that has come into existence just now is taken to have existed way into the past?" When I think of this spot along country road K342, I can only envision it with the twin silos. (Which is what allows me to locate them again today, 11 years later, on Google maps.) Although I have not seen the twin silos previously, I can no longer think about this part of the world as it has been to me during the first six trips. I begin to realize that this has significance for teaching, even though at the time I do not have the means of understanding what has happened to me. I do realize that this process is part of a forgetting of the world as it has been is a form of amnesia that has significance for teaching. With the cognition of the twin silos comes a process of forgetting that there has been a time when the twin silos have not existed – *ek*sisted, literally, stood out (*ek-*, *ex-*) as a being (Lat. *esse*) in my consciousness. As teachers, we assume that students ought to see what is visible for us now, the twin silos, because we take them to exist even though they have not existed for us up to the first time when we became aware of them. As part of the experience with the twin silos, I come to realize that learning also means erasure of a world that has been and as it has been.

In theories drawing on the construction metaphor (Piaget, radical constructivism), perception is viewed as a form of organization in which sensorial data are either assimilated to an existing schema or, following an accommodation process, are assimilated to the new schema. This view, which is fundamentally based on a dualistic conception of cognition, has been thoroughly analyzed and critiqued as untenable in the face of empirical evidence (Gurwitsch 2010). Thus, even in the presence of a schema, "*the object thus presents itself in sense-perception itself,*

[1] Some readers might object suggesting that students are guided. However, the research I had conducted in an Australian classroom shows that even though students had seen a specific experiment previously and then discussed its implication for the theory of rotational motion, they continued to fail distinguishing instances of motion from instances of non-motion.

with no special organizing, assimilating, or interpreting activity intervening" (p. 99). We experience the world in perception, we do not have to interpret it – I could not have interpreted the sense data from the twin silos because I was not aware of them. Following that ominous seventh day, my stream of experience has changed and the twin silos are immediately given to me as soon as I pass this spot along country road K432. There is as little "interpretation" that has occurred after the event as there has been before. That is, my whole perceptual process has undergone restructuring so that what is given in my experience following this ominous seventh day is different from what was given in my experience preceding the first apperception of the twin silos.

This process whereby something takes/becomes a presence in and part of our worlds is not a necessary consequence of perceptual processes. We judge some of our perceptual (visual) experiences as having been figments of our mind. For example, there are instances when we come to think of shadows, figures, or people that we thought to have seen as deceptions, imaginations: spirits in the forest, thieves lurching around the house at night, and so on. Two of my neighbors, single women, often report having heard noises in their backyards at night; they keep their lights on all night even though it is unlikely that any human being would be in their backyards. In the instance of the twin silos, the process of seeing was adsorbed into the perceptual process whereas in my perception and reification of the twin silos the process has taken on an inverse trajectory right in and as part of seeing them for the first time. At the very instant of seeing them for the first time, my perception is adsorbed into the twin silos – they no longer depend on my perception. In becoming perceivable as such, they now can be seen by anybody. Today I know that this is so because all senses "project us outside in such a manner that everything that is sensed by us always is sensed outside of us, different from the power that senses them so that such a power is that of distancing, the Ek-stasis of the world" (Henry 2000, p. 161).[2] If – so I am soon thinking during the remainder of the trip – humans generally and teachers specifically forget the world they once inhabited as soon as they learn something new, if the new and previously unknown in fact not only erases what is known but also erases the memory of having known this, then we will no longer know what it means to live in a world of children. We no longer have the capacity to view the world with their eyes but will always see them and their acts in a world structured according to adult conceptions. We will always see children interact with some entities that may have one signification within our adult worlds, whereas these will have, if they exist at all, a very different signification in a child's world. Children and students will appear as behaving inappropriately in our world, whereas they are in fact not acting in the world as it appears to us. If children and students act intentionally, that is, if they aim at something in and with their actions, then we need to understand what *they* are aiming at rather than what they appear to be aiming at as seen with our adult

---

[2] That we take the world as independent of our perception is the result of the work of Galileo (e.g., Henry 2000), who established a theory in which objects are independent of the perceiver – thus the idea of the independent Galilean object.

eyes. We need to understand the children's and students' rationality, which means we have to understand the entities that they are reasoning with, their language and the things they make salient with it. I begin to think: For me, the researcher of learning, to know the significance of children's and students' discourse and actions, I need to understand how the world appears to them. To understand their intentions, the entities they are aiming at, I need to understand their horizons – their worlds and the possibilities these afford for expanding them. This is so because these worlds and the possible ways of expanding them will become mysterious when seen with my eyes when in fact it is rational within the world that they inhabit.

To summarize: on this day I have come to see the twin silos for the first time. But at the same time I notice them, the old world without the twin silos slips away, and with it, the memory of the old silo-lacking world. The existence of my new material world has extended itself backwards all the while I think of myself as having been blind to this newly discovered aspect of the world. In this instance, I have become aware of an aspect of the world, and this has happened to me even though I could not have intended it. I could not have wanted to learn about the twin silos because I was not aware of their existence. One of the serious questions that poses itself, then, is how a new entity – object, idea, form of thought – can emerge into consciousness if we have no way of aiming at the unknown at that which is radically foreign. How does something like the twin silo – the regular white posts, the street sign – become a reality to me if I cannot construct it because it is unseen? To some extent, it has not been through my (cognitive, intentional) agency that the twin silos have become a reality. But this reality, the twin silos, *has given itself*; it has shaped my perception such that from then on I could not but notice them. I could not but notice the twin silos in the same way that I can no longer pass the little church (Fig. 2.1) without noticing it and think about the experience of having lost a world in which it did not exist. In gaining a world, I loose a world. The birth of a world and the death of a world are but two manifestations of learning, of the dynamic of life.

Sense Impressions and Consciousness

> It is intentionality which characterizes consciousness in the pregnant sense of the term, and which at the same time allows for the characterization of the whole flow of experience as a conscious flow and as the unity of one consciousness. (Husserl 1913, p. 168)

In the episode described in the previous section, I had become aware of the twin silos. I had become aware of them only on the seventh trip. But optics, the physics of light propagation, suggests that the light coming from the twin silos would have fallen on my retina during the previous six trips, too. So why did I not notice the silos during those previous trips? Why did I not notice the little church as I passed the spot over a period of 11 years prior to the instant of becoming consciously aware of it? These questions point us to the need of separating impressions on the sense organs and being aware of an impression generally and these impressions specifically. One crucial point for our inquiry, therefore will consist in finding answers to the question of how perception comes to be configured such as to see the things that come to make our realities – the twin silos, churches, and anything else populating our worlds. The answer will have to be sought somehow between the knowing person, here me, and the world perceived; the perception is configured such as to make visible the world in a way that we share with others.

I had become aware of the twin silos because the light rays coming from them have left an impression on my retina. But the impression itself had not led to my perception. There has (likely) been impression without perception. But the impression was a precondition for the perception, and yet, as soon as the twin silos stood out and became entities (beings) in my consciousness, their presence no longer appeared to require the impression. Rather, the twin silos adsorbed any perceptual process and became entities in their own right. If I had asked farmers in the area, they would have told me of the existence of these silos predating my perceiving them. I could and now do point to these silos, and am able to show them to anyone in the aerial mode of a Google map. That is, these silos have an objective presence: they are present as objects to be acted upon and to be thought about. Although the being of these twin silos, as any other material thing in this world, depends on perceptual processes, they adsorb these processes and become independent: they become *independent Galilean objects*. However, this appearance of the world – the world outside of oneself in which the seeing deploys itself – never takes into account what is unveiled within it (Henry 2000). That is, what appear in our consciousness are beings; and what is unveiled in beings is Being. This twin fact is not taken into account or recounted when we talk about the world: the fact that there are processes that enable perception.

There is therefore a process at work by means of which sense impressions are exported outside of themselves, an effect that produces itself already within the cognitive subject. Although the twin silos came into existence in and by means of perceptual processes, they came to exist outside myself, they came to lead for me an independent existence transcending my perceptual processes. Standing out (*ek-stasis*), becoming aware, presupposes a process of dehiscence whereby Being (Henry refers to it as invisible Life) manifests itself in beings. With respect to the non-intentional strata of experience in the example of a sound experience, we know that it

> decomposes . . . into different sonorous phases in such a manner that once experienced, each phase slips to the "instant past," to the "just now past" (*soeben gewesen*). And this phase that just passed this instant slips in its turn into a past more and more removed. *Because this slipping to the past is given to an intentionality – the retention – it is the externalization in its primitive form*, the Ek-stase in its original surge, the Difference that one can also write Différance because it is nothing other than the pure fact of dif-fering, of spacing, of separating – the first space. This slipping of the impression outside of itself is the flow of temporality itself, its original temporization; it is the "flux" of consciousness. (Henry 2000, p. 75)

This transcendence, however, does not come on its own. It is not a result of the sense impression itself. It occurs with articulation in perception and language, with the phenomenon of *apophansis*, whereby language lets something be seen that shows itself from itself. In the course of subsequent chapters (e.g., Chapters 3 and 5) I return to this source of difference,[3] which also is the source of the apparent independence of world (beings) and consciousness. Becoming conscious of, the becoming present of the presence, presupposes a dehiscence, a diastasis (standing apart) of Being and beings, a form of difference with and from itself. This is one of the marked differences between the constructivist metaphor and my own way of thinking: the former is grounded in the self-identity of beings and the entire metaphysics that it entails[4] whereas I think cognition in terms of non-self-identity or difference-as-such.

Forgetting

In the experience with the twin silos, I was given something new, a new world. I have come to understand this world as a world containing entities that I had not noticed or known before. In the process, my old world, the one without the twin silos, has vanished. Generally this process and its outcomes go unnoticed. But my keen attention to the process was able to retain a trace, a cinder or ashes of what had been reality for me before. This type of forgetting is radical. But because this has been my world in which I had acted, it has left a trace: the forgetting is a sort of repression that neither annuls nor destroys but merely displaces a world.

The perception of something new, as I work out in greater detail in Chapter 3, involves a considerable amount of passivity and givenness. Because I had not known of the twin silos, I could not intend them to become part of my world. The twin silos, from my perspective, were given to me. Something new, heretofore absolutely foreign, strange, and unknown, has entered (into) my world and I cannot be anything other than a recipient, willing host, a donee. The existence of the twin silos is not the result of my intention – I could not have intended the twin silos because they have not existed for me and therefore could not be the object that

---

[3] In French, the difference between the two words *différence* (difference) and *différance* (a non-word) can be seen but not heard. In hearing, the difference between the two words is undecidable. And denoting this undecidability is the very purpose for the non-word *différance*, which thereby comes to be a word that denotes an inner difference.

[4] For example, classical logic is based on the opposition of some quantity *p* and its negation, *not-p*; a third is not given. At the same time, the following relations hold: $p = p$ and *not-not-p* = *p*. In the philosophy of difference, a thing is not self-identical, always differs from itself. This approach also underlies Bakhtin's perspective of language, which changes in the instant a person utters a word, so that it changes in the moment it is used, and, therefore, it is never identical with itself. Non-self-identical concepts and phenomena allow us to capture process, whereas self-identical concepts and phenomena capture stasis. But nothing in this world is stable: life is a process rather than a stasis.

orients an intention. Because they did not exist, the twin silos could not be the direct objects of transitive verbs in statements where I was the subject. This includes the verb *to construct*, which, as all transitive verbs, requires a direct object. This object has to be known or I cannot act in accordance with the transitive verb. Although I had not known (of) the twin silos, I found myself all of a sudden being conscious of them. I had not intended them to be or become objects within my conscious experience. There has been a donation: something has been given to me in my perception. But strangely, we normally do not notice or forget about the gift and the giving.

> For there to be a gift, not only must the donee or donor not perceive/receive the gift as such, have neither consciousness of it nor memory, no recognition; he must also forget at this instant and moreover this forgetting must be so radical that it exceeds even the categorical nature of forgetting in psychoanalysis. (Derrida 1991, p. 29)

Being unaware of the twin silos, I could not have intended their appearance in my consciousness. The gift of the twin silos came "out of the blue," or out of the forest. There was no *re*-cognition of *these* twin silos, because I never cognized them before. Most importantly to our present discussion, what was before actually has to be forgotten in a process of amnesia that is more radical than the process of forgetting (repression) as it is treated in the discipline of psychoanalysis. It is this forgetting that made me loose my previous, never-to-be-recoverable world.

> For there to be a gift event (we do not say act), something must come about/happen, in an instant, an instant that without a doubt does not belong to the economy of time, in a time without time, in such a way that the forgetting forgets, that it forgets *itself*, but in such a way that this forgetting – without being something present, presentable, determinable, sensible, or signifying – is not nothing. (Derrida 1991, p. 30)

Learning, therefore, means to receive the absolute gift. The absolute gift, the gift that could not have been foreseen and understood as such, the event in which the twin silos came to part of an individual consciousness that could not have intended it. This extraordinary gift always involves forgetting. For there to be forgetting in this sense, there has to be a gift. The gift then is the condition of the forgetting; and likewise, the forgetting is a condition of the gift. This forgetting is related to the movement of thought along its path, "the thought as path or as movement along a path is precisely what is related to that *forgetting*" (p. 32). Writing (*écriture*) – a process that I consider in its relationship to thinking at the same level as speaking – is a process that makes and allows us to radically forget. It is a process that erases previous knowing, which is an advantage if we think about how much we needed to remember if we tried maintaining a record of all the ways in which the world once appeared to us. No constructivist mind would be able to create archival records of our past that had such an enormous extent: what was, therefore, always is seen (teleologically) as it appears from how the world is today. The very nature of the sign as something that can stand for something else – i.e., of the sign in its *apophantic* function that allows something to show itself from itself – is from the beginning marked by effacement. Writing and talking imply learning and forgetting simultaneously.[5]

Science educators should find this way of thinking immediately appealing, because they know that as soon as we know something consistent with the current scientific canon, we forget what it means to understand the world in a non-scientific way. It is well-known in our research community that the scientific way of understanding the world is not immediately evident to the students or adults who refuse to abandon their "misconceptions," "naïve views," or "alternative frameworks."

In the essay "What is called Thinking?" (Heidegger 1954), a cultural-historical analysis of thinking as thought at the very beginning of philosophy, the question is led back to what the ancient Greek (Parmenides) might have thought when they talked about *ēon* (beings) *ēmmēnai* (be). It is precisely this expression, *ēon ēmmēnai* that calls us into thinking. The two parts of the expression are not identical, as already known to Plato, who suggested that between beings and Being is the *khorismos*, where *ē khôra* has the sense of "place" or "space." Being and beings, the things of the world that we are conscious of, take different locations, are spaced out. Thinking arises precisely in this spacing: "The incredible difference between an appearance and the appearing (between world and the experience) is the condition of all other differences, of all other traces" (Derrida 1967, p. 95). As we see in my account, the spacing is reduced to the instant. The spacing in which the cognition of the twin silos occurs is not a succession in time. Thinking about the twin silos arises from a spacing that itself is inaccessible because Being has been forgotten as soon as the phenomenon – here the twin silos – has surged into consciousness in a way that allows me to remember it as feature in the landscape and to find and recognize this feature the next time I come along country road K342. But in this movement from Being to beings, in the space denoted by the term *khôra*, we find the path that thinking both clears and takes. Being itself is inaccessible by consciousness, as all it can operate with are beings (*Seiendes*), *re-presentations* that make present something that is absent. Being thereby withdraws itself, turns away from human beings. This withdrawal, according to Heidegger (1954), is *Ereignis*, a difficult word to translate because it simultaneously denotes "event" and "propriation." Being affected by reality that consists of beings closes us off from what really matters: Being. It is the *Ereignis* of the withdrawal of Being that may be the most present of anything present and that therefore exceeds the actuality of the actual. It is precisely as gift that Being is signaled in the movement of *Ereignis*.

At the origin of my experience lies the experience of two giant silos being given to my awareness *as* "twin silos." The perception of the material things, their

[5] Both Vygotsky (1986) and Merleau-Ponty (1945) explain how in speaking, speakers discover what they are thinking: speaking means learning (about) one's thinking. Simultaneously, it is through thinking that we understand our speaking. Thinking and speaking are related, mutually constitutive processes that do not bear a one-to-one relationship. They are but manifestations of a higher order process: the ongoing cultural-historical activity.

becoming, figures against the ground (an indistinct forest and sky). Their articulation is associated with the ideational articulation of the figure as "silos." Here, the ground is something positive, having given rise to the very phenomenon that has become figure. The ground is the birthplace of the figure. Aristotle explained the function of speech (*logos*) to be *apophainesthai*, which, as the investigations in *Sein und Zeit* (Heidegger 1977) show, is used in the sense of *letting something be seen as it shows itself.* This something is precisely what is talked about. For the ancient Greek, *logos* and *mythos* were used in the same sense: making apparent that which appears in its Being. In the awareness of the twin silos *as* twin silos these became something that I could talk, write, and think about. But the experience shows us that it was not the concept of silo that brought the silos into my world – as some scholars might suggest following Kantian epistemology or the Sapir-Whorf hypothesis. Rather, in the *Ereignis*, in this event where the twin silos gave themselves to my conscious awareness, language also gave itself at the same time and with it the accountability of this experience. The twin silos are and have become objective precisely because they can be articulated, in perception and talk, and thereby we can let them be seen in the linguistic community more broadly. They are objective, accessible to any other human being, because I share with them the same possibilities of experiencing. The twin silos are not "just *my* constructions," they are objectively experienceable by all those subjects who have the same kind of subjectivity, arising from the same kind of living/lived body.

Some science educators might say that a teacher could have told me to see the silos and I would have seen them during my first trip. But these science educators forget that for the student both the phenomenon and the word (concept) are unknown. Even if a teacher were to utter some sound that students later come to know as a concept denoting something specific in their environment, they do not know what it refers to. This is the same situation most of my readers would find themselves in if, for example, I asked them to identify something and all they hear are the sounds [ɪ'mɑ:m baɪ'ɪldə].[6] It is only with some cooking experience and with the recognition of the sound as "Imam bayildi," Turkish for "The Priest fainted," that they might be able to pick out a dish with eggplant as the main ingredient. When we do not know a word, we cannot find what it is in our environment that the sound is designed to show itself from itself; and because we do not know the phenomenon, the sound cannot serve in a signifying function. I show in subsequent chapters of this book that even scientists find themselves in such situations. This is the case, for example, when they know whether a line is a significant signal (data) only when they also know what their object is, while they can know what the object is only when they also know that the signal is significant.

Already before *Sein und Zeit* was first published in 1928, it was noted that temporality and temporization are passively constituted (Husserl 2006). Phenomenological philosophers now agree that this constitution comes about in *khôra,* the

[6] This sound transcription follows the conventions of the International Phonetics Association (see http://en.wikipedia.org/wiki/Ipa) and is found not only in bilingual dictionaries around the world but also used by the Oxford English Dictionary.

dehiscence of Being and beings. It is in *khôra* that the sensed world comes to stand out in its exteriority, in terms of beings (*re*presentations). The very shift from Being (life) to beings is the origin of temporality and time, spatiality and space. We know that time and temporality are different during experiences of flow. We have experienced flow when we realize that we just passed three or four hours reading, gardening, thinking, writing, conversing, cycling, hiking, meditating, or whatever else has captured/captivated us without consciously noting the passage of chronological time and without making thematic what we have done. That is, when we are so absorbed into Being that the reflective consciousness that attempts to take the present as its object has come to a halt then temporality and time are experienced differently. In fact, there no longer is a passage of time: In absorbed activity, it is as if objective time has been put at bay: it no longer is a measure of what we are doing at the moment. (Even less is time exchangeable for money, as in the equation time = money.) There is no time in moments that are described by the concept of *flow*; and there is no (objective, clock) time during meditation. In both instances, there is Being: but it is not made presence in consciousness by means of beings (conceptions, mental structures). This does not mean that we are not relating to the world. We perfectly continue to stay on the road while being absorbed in cycling; we drink without spilling from a cocktail glass while absorbed in conversing; and we use the keyboard without a thought while absorbed in writing a book or an article. The point is that this activity itself is not an object of reflective consciousness. The knowing is *immanent* rather than transcendent; it is in the action rather than in the conscious mind. During such moments, we are absorbed into the doing and there exists nothing apart from this doing, no world that withdraws into itself apart from our doing. The world apart from us appears in consciousness during the event called *Ereignis*, in a process of propriation and appropriation.[7] But this simultaneously is a process of dis-propriation as well, as we lose and forget the world in which we have been dwelling before.

Children's World and Rationality

In this book, I am concerned with a phenomenological inquiry into the fundamental aspects of our experience to develop a theory of knowing that is more complete and more appropriate than intellectualism in all its variants. Phenomenology is suited to my task because it "aims at ultimate clarification and justification of knowledge, both theoretical and scientific knowledge in the proper sense *and that pretheoretical and prescientific knowledge by which we are guided in our life in the world of everyday experience*" (Gurwitsch 2010, p. 152, emphasis added). The inquiry into the latter form of knowing is important because it constitutes the material "out of which theoretical and scientific knowledge grows" (p. 152). As these quotes show, phenomenology supports us in appreciating pretheoretical and

[7] Heidegger relates the word to *Er-äugnis*, a neologism with the sense "capturing with the eyes."

prescientific knowledge for the purpose of understanding how these forms of knowing – which always already precede scientific forms of knowing – constitute the very ground upon which the latter emerge. In terms of a figure-ground analogy, everyday, prescientific and pretheoretical knowledge constitutes the ground from which scientific and theoretical knowledge emerges and against which it comes to figure as something special.

There is substantial empirical evidence that children and older students dwell in worlds very different from our adult worlds – as shown in the earlier mentioned experimental study in an Australian high school physics course. If my researcher colleagues and I had just looked at students' explanations without having ascertained what they actually had seen, we might not have understood them, especially not because from a scientific perspective, there was no motion to be observed of the kind that the majority described. That is, if we had not attended to the fact that the students were seeing motion, we would have assumed that they attempted to explain the absence rather than the presence of motion.

Research on balance beam problems also shows that children and adults unfamiliar with the physical device do not focus on the physical distances and weight but on other aspects that are salient in their perception and therefore in their consciousness (Roth 1998). For example, they might focus on the numbers marked on the balance beam and weight and try to add and subtract. Rather than reasoning about distances and weights to figure out when and under which conditions the beam will balance, these individuals focus on the relationships between the numbers they observe and attempt to figure out by playing around with the numbers to come up with a rule to predict when the balance beam will be balanced. If researchers think that the children and students act upon distance and weight, then the conclusions will be useless because the real objects of thought and action had been the numbers.[8]

The point is that teachers and researchers at some point in their own development have inhabited and looked at worlds that are very similar to the children, students, and other people to whom they have such a hard time teaching the opposite (scientific). Constructivist researchers state with a more than little complaint that misconceptions are resistant to change or that students with such conceptions resist the change efforts others mount to help them understand mathematical or scientific concepts. At no point have I seen in the STEM literature constructivist and conceptual change researchers ask themselves whether students actually have seen, acted upon (materially, discursively), and thought about those objects that are salient to the teacher/researcher. That is, teachers and researchers observe the students as if they were behaving in the adult world, when in fact the students act in (very) different worlds, populated by different entities, including language (*logos*), and with it, rationality (*logos*).

---

[8] As Gurwitsch (2010) shows, Piaget is fundamentally mistaken about perception because he assumed a constant world and perception. Piaget did not realize that the very perception changes when we come to see something new or see some entity in a new way. It is not that we *interpret* a perceptually fixed entity in a new way. Perception itself is changing.

The problem for science education teachers and researchers lies in the fact that they forget these other worlds in the very moment of cognizing the new. Thinking, as philosophers point out, is a process that leads not only to birth of new knowledge and concepts but also to the erasure, radical forgetting, and death of the old knowledge. This erasure occurs at the very instant of speaking, for example, when I say "I" in my solitary discourse at this instant of writing. In this case, I can only attribute a sense to the "I" when it implies that the phenomenon thereby indexically denoted – me, my body – is absent at the instant of the talk. The present can appear in the re-presentation, as presence in the present, as sign, only under the condition that it has already shifted with respect to that which it denotes, no longer present in the present but only in retention. This retention – where the recent past comes to be made present again by employing representations – always is preceded by non-conceptualized impressions (*Urimpression*) on a plane referred to as *hyle* (e.g., Husserl 1980), the matter or stuff that this world is made of. We can see in this instance that as soon as we think in terms of sign-mediated representation, the split between thought and the world is accomplished.

Chapter 3

Radical Passivity in Learning

> The body remains the organon, the instrument or the incarnation, the mechanism or the work of a *sense* that never stops rushing into it, presenting itself to itself, making itself knows as such and wanting to tell it there. (Nancy 1993b, p. 192)

In Chapter 1, I point to the need to think the learning of the unknown from the perspective of the unknown rather than from the perspective of the known. In the latter case, we transfer the known to the student who does not know or we allow her to construct the knowledge ("meaning") of what she does not yet know. But this does not get us to the enigma of learning. It does not get us the enigma of being asked to learn something that we do not see, that is not apparent to us, and that we therefore cannot intentionally construct. What might it mean to think learning from the perspective of the unknown? What would cooks make when asked to prepare Imam bayildi if they do not already know Imam bayildi?

Constructivist Learning from Experience Is Intentional

In constructivist epistemology, learning is the intentional production of mental structures that allow the individual organism to cope in the natural world.[1] But there is a question about how what happens in the mind is related to the material world that includes the body of the acting subject. If the acting subject is thought on the basis of the intentional mind, then the question of the role of the body in knowing and learning poses itself. How does a mind that knows nothing about the world but only its constructions direct the body that is of the same kind as the world? What does the constructivist metaphor tell us about passivity that we experience in

[1] The discourse of constructivist educators is actually ambivalent. Thus, the mathematics educator Pat Thompson told me that the construction does not have to be intentional (personal communication, September 9, 2010). But if students do not learn a concept intentionally, what *do* they learn? If they do not intent the concept but intend learning, what do they learn?

W.-M. Roth, *Passibility*, Classics in Science Education 3,
DOI 10.1007/978-94-007-1908-8_3, © Springer Science+Business Media B.V. 2011

and with our bodies, such as when afflicted with some illness? Moreover, the pain that we *feel* is not "constructed." Even though it is subjective, it is experientially real – something everyone who has felt illness can empathize with because we all share in the same possibilities of experience. Passivity comes with bodies because these are exposed to the world. But already for Kant, the artisan of the mind, the body had no role in cognition. Thus, although there are significant and sometimes precarious traces that are reminiscent of the importance of the body (and the hands in particular) for the knowing subject, "the overall picture of the disembodiment of the Kantian spatial subject and of the process of the constitution of spaces as such in the course of its transcendentalisation does not change" (Woelert 2007, p. 149). This disembodiment is taken forward to the modern forms of constructivism. For example, in the following definition of what it means to know reality, a disjunction comes to the fore between thought and reality:

> Knowing reality means constructing systems of transformations that correspond, more or less adequately, to reality. They are more or less isomorphic to transformations of reality. The transformational structures of which knowledge consists are not copies of the transformations in reality; they are simply possible isomorphic models among which experience can enable us to choose. Knowledge, then, is a system of transformations that become progressively adequate. (Piaget 1970, p. 15)

The systems of transformations constitute schemas. "Cognitive change and learning take place when a scheme, instead of producing the expected result, leads to perturbation, and perturbation, in turn, leads to accommodation that establishes a new equilibrium" (von Glasersfeld 1989a, p. 128). There is then a double instrumentalism at work. Initially, at the sensory-motor level, action schemes help "organisms *to achieve goals* in their interaction with the experiential world" (p. 128, emphasis added). At a second level, actions exist in an abstract form and help "organisms to achieve a coherent conceptual network that reflects the paths of acting as well as thinking which, at the organisms' present point of experience, have turned out to be viable" (p. 128). That is, the ultimate point of learning is the production of structures, and this production is goal directed.

Using a construction metaphor, we might consider a child to learn the word "cup" by associating the sound [kʌp] with the material object that we recognize as a CUP. There are three steps. In the first, the child focuses "attention on some specific sensory signals in the manifold of signals which, at every moment, are available within the child's sensory system" (p. 132). An adult aids this focusing, who provides general directions by, for example, pointing with the finger. "The second step consists in isolating and coordinating a group of these sensory signals to form a more or less discrete visual item or 'thing'" (p. 133). The parent's moving of the cup aids the process of isolation of the "thing" from the background of things. In the third step, the child has to isolate the auditory signal from the continuous background of noises; a parent's repetition of the sound [kʌp] aids in this process.[2] The child then has to associate the isolated visual pattern with the isolated auditory

---

[2] In the transcription of sounds, I use the conventions of the International Phonetics Association, which are independent of the ways in which particular languages transcribe sounds into words.

pattern. As a result, the child has begun to "acquire" "the meaning of the word 'cup'," where it ought to be "clear that the child's meaning of that word is made up exclusively of elements which the child abstracts from its own experience" (p. 133). I return to the question of meaning in Chapter 9 and pursue here only the question of the "sensory signals," their "association," and the action intention orientation in the constructivist account of learning.

In the constructivist approach, learning is understood from the perspective of the knower. Thus, the material object CUP and the sound envelope [kʌp] are assumed to already exist as distinguishable and distinguished entities for the child. Von Glasersfeld is a little finer in his description, recognizing that the figure that we denote by the word cup itself has to become apparent against a ground. In this approach, there are perceptual processes that select among a multiplicity of sensory signals those that have some regularity or that stand out. That is, there are signals that come from the material world and hit the sensory surface. The child isolates those that are significant, a process that is aided in making the corresponding signal stand out by moving an instance of material forming a cup and by repeating the sound [kʌp]. Then the child coordinates the two different sensory signals in its mind and thereby "constructs" the rudiments of the "meaning" of the word "cup."

Kant has prepared this constructivist approach to the question of learning by placing an enormous emphasis on the mind to bring about the mental structures. Thus, the integration of the sensory manifold cannot ever come about in and through the senses and therefore also cannot be contained in the pure form of sensual idea: "any connecting … of the manifold of perceptions or various concepts … is an act of the mind that we would denote with the general notion of the mind that we would denote with the general notion of *synthesis*" (Kant 1956a, p. 135). The unity of the mind and the manifold of sensual experiences are of different order. Between the sensual experiences, seeing a CUP and hearing the sound [kʌp] there does not exist anything but an association by chance so that there is "no necessity of a connection deriving from perception itself" (p. 216). The mind has to make this connection. And it is for this reason that Kant requires the faculty of *synthesis*: it is precisely because Kant has produced a difference between mental concept and sense-based intuition that he requires a powerful mind to hold it all together (Rorty 1979). The mind takes the entire charge of the integration. "The assumption that diversity is found and unity made turns out to have its *sole* justification in the claim that only such a 'Copernican' theory will explain our ability to have synthetic a priori knowledge" (p. 155). Thus, constructivism runs into the problem that it claims necessary truths about made ("constructed") objects as more intelligible than the one concerning found objects. This claim depends upon the (Cartesian) assumption that we have privileged access to the constructive activity. But such a privileged access does not exist. "For postulated theoretical entities in inner space are not, by being inner, any more useful than such entities in outer space for explaining how such knowledge can occur" (p. 155).

What the constructivist position leaves uninterrogated is the question how whatever arrives at the sensory periphery comes to be a signal in the first place.

This Kantian presupposition that also characterizes constructivism was subject to critique already during Piaget's time (von Weizsäcker 1973). Thus, we might "try and depict the measured and calculated orders and proportions of objective bodies in an imaginative representation and compare this with the perceptions" (p. 166). But how far will this take us? How much will it tell us about perception? This process of depicting and comparing "tells us nothing about the genesis of the perceptions, and it does not reach beyond the representational capacity, but is, so to say, locked up within it" (p. 167). This question is of particular importance, as that which the organism perceives is a function of its current perceptual organization. It is not that the things (beings) are constant and our interpretation therefore changes; rather, the perceptual process changes as a whole (Gurwitsch 2010). Even if it were possible that there is some unexpected result – where the question is about the expectation, an intentional orientation to the world – how would the organism change its perception given that it can only work with the current mental organization? How is it possible to escape the apparent gridlock set up between the homology of mental structures (schemes) and perceptions?

A very different approach to these problems arises from phenomenological philosophy generally and the phenomenology of perception specifically. Accordingly, the order and integration of the sense experiences derives from a self-organizing process that structures itself through a self-reference, that is, through auto-affection. "The sensory properties of my gaze, my touch, and of all my other senses together are the powers of one body integrated into one action" (Merleau-Ponty 1945, p. 267). Therefore, we do not require the syntheses of a powerful mind to pull diverse experiences (sound, vision) together but these diverse experiences are those of a unique and integrated body. Thus, for example, in an analysis of reading online science materials I show how reading is a self-organizing process that changes as it engages with some text, which becomes a particular kind of text only as a result of a self-configuring reading process (Roth 2010a). When we pick up a text, reading initially does not know beforehand what is being read. Is it to take the text literally, as in an experience report, or as a parody, as in a comedy or spoof (think of what the media publish on April 1 of every year)? As it unfolds, however, reading is confronted with particulars that may change it so that rather than reading an experience report it turns into the reading of a spook. That is, reading is configured and configures itself as it engages with the thing to be read (e.g., images and text). In a sense, it is the (multimodal) text that configures reading as much as reading configures itself. But if this is so, then the configuration cannot be reduced to the mental cogitation. Moreover, because reading does not know beforehand what kind of text it is in the process of encountering, it cannot intend its own organization beforehand. The organization itself has an essentially passive component: it is the result of a transaction between reading and the text in the world. This organization cannot be intended beforehand because to do so would require knowledge of the text that only is given after reading is complete.

A Phenomenological Exploration

> How, on the basis of what does one construct an image of something that we know nothing about? (Henry 2000, p. 120)

Learning is not all intentional and we cannot intend that which is new precisely because, as something new, it lies beyond our current horizon. In other words, we cannot intend to learn what we do not know precisely because to intend the concept would require knowing it. We cannot have a plan to learn something unknown precisely because it is unknown and therefore cannot constitute the object counterpart of a subject's intentional action. A student who does not know the physics concept of momentum cannot *intend* learning or constructing it, because the very theoretical concept of construction implies an intentional orientation. Thus, a carpenter constructs a house because she has a plan and she knows from experience how this plan relates to the house that will emerge from her work. A recipe in the kitchen leads to the dish as intended only when the cook already knows what the outcome is or can be. Otherwise, and many will have made the experience, the result of a first try at a recipe is often some failed dish rather than something to be presented and proud of during a formal dinner.

In the following subsections, I present some investigations that readers can easily reproduce. By engaging in these investigations, readers actually live the immanent knowing and learning rather than "constructing" it. These investigations reveal that there is more to learning than "constructing" "meaning" from given structures, which, in the present instance, are of the visual-spatial and tactile type. The investigations show that precisely because we do not know, we cannot intend perceiving and therefore learning something. More importantly, we cannot intend constructing whatever it is because we do not know what there is to be seen in the first place, if there is anything to be seen at all.

Vision, Take 1

To begin our inquiry into the passive dimensions of learning, take a look at a type of puzzle that most readers likely are familiar with (Fig. 3.1).[3] Upon first sight, the

[3] There are many other examples I could have used, some of which are presented and discussed in *Even Odder Perceptions* (Gregory 1994). Of philosophical relevance is the duck-rabbit (see figure) made famous by Wittgenstein (1958) as part of his discussion of the difference between seeing as and interpreting.

Fig. 3.n3 The duck-rabbit became famous in Wittgenstein's discussion of perception.

Fig. 3.1 An example of the givenness of perceptual objects.

image consists of black splotches on white ground. In most situations of everyday life, we might pass over unnoticed such configurations – we do not notice configurations even if they are of familiar type (Chapter 2). If, however, for one reason or another we suspect or are being told that there is something hidden in the configuration, then we might look at it with the intent to find something, but we do not know what to look for and therefore, we do not know *how* to look at it. We may turn the figure around or we may move it farther away and up close. We may move our eyes about in the attempt to see something emerge to become figure against everything else as ground. Then, all of a sudden, an image appears. In this case, we may see a little dog, or, in fact, two little dogs, one upright and the other one upside down. We could not provoke, aim at, interpret, and learn about these images of a little dog, precisely because we did not know there were these images. We were not in a position to construct a mental structure about something that is not even present in our minds; and when it appears it is not because we intended it. It is, in a sense, given to us in our perception. Now if I had told my readers about finding a little dog, then this would not have fundamentally changed the process. This is so because, as made thematic in the concept of *apophansis*, the words/language would have been an index for the little dogs to give themselves from themselves. Again, there is both an agential "search" and a self-giving, an auto-donation, of the dogs that originates in the lines that guide my eyes and allow them to disclose the dogs. However our gaze may wander about the image, it is the image that awakens a certain motor intention of my head and eyes and my gaze thereby comes to be structured in its search for structure. What I see is as much given as it is found. As soon as I have located the dogs for a first time, a glance is all that is necessary to perceive them. Perception now is structured such that we can see what we initially could not see.

In many such figures, we may not find anything unless someone else tells us what there is to see. Thus, knowing that there are little dogs hidden (or Holstein cows) configures our image-reading process to find something that could be seen as a (Dalmatian) dog or a cow. Or, a situation we frequently find between parents and children, one person may follow the trajectory of a finger that eventually turns to outline the little dog. Once we have seen it, we can always see it. Just directing our eyes toward the image, the little dog will stare at us and we may, if we care to reflect, ask ourselves how come that before noticing the dog for a first time, we could not see it. This is so because our image-reading process has been configured so that at this time we know something that is completely external to us. Now the eye immanently knows how to move to make the little dog appear time and again; it does so without semiotic mediation, without involving conscious thought processes so that its memory constitutes something like immemorial memory. But we could not have intended configuring the process, as we did not know that it is a little dog that we could/might find. We could not intend seeing a little dog because we did not know that it was a little dog that we might be able to see. Thus, the perceptual process resulting from our experience with this part of the world is not entirely intentional. It cannot be intentional because the resulting image-reading process emerges unpredictably and undecidably from a transaction between earlier image-reading processes and the world. The world makes the eye follow it and thereby configures the eyes' movements that allow us to see a dog rather than a cow. In one sense, therefore, the world gives itself to be seen in a particular way because it contributes to the configuration of perceptual processes.

Prior to seeing the little dog for the first time, it does not exist for the observer. It comes into existence in and through the perceptual process. I move my eyes to look for something. But this intentional movement presupposes my capacity to move. Before I can intend moving my eyes, "I" have to know that "I" can move my eyes. But such knowledge does not come from thin air. If I know that I can move my eyes, I must have learned this from movement. It is precisely from some primordial movement that I develop a capacity that "I can" "move the eye," it is from some primordial movements that the eye develops the immanent memory that it can move. That is, the eye movements that lead to the first emergence are associated with the resistance within the muscles and from what they move; the effort associated with working against the resistance leave traces in the body, traces that it remembers as the possibilities of an "I can" – or perhaps more appropriately and tongue in cheek, "eye can." The movements constitute an enactment of the "I can," and it is precisely the "I can" that figures in remembering the movement. This is a form of "immemorial memory of a flesh that keeps within it all its powers" (Henry 2000, p. 208). It is the memory of and for this movement that allows us to *re*cognize, see again, the little Dalmatian dog after we have seen it once. But despite this dependence of the existence on the perceptual process, we immediately come to experience the separation of perceptual process and entity such that the dog is out there in the world once we have seen it. It no longer is a function of my perceptual process, because now I am in a position of teaching others to see it so that the dog exists objectively, as object among the world of objects. That is, the perceptual process has been adsorbed into the object itself. Because the structures in the world direct the eyes of others in the same way, eyes that are biologically structured in the image of my own, they can see the same. It is precisely this sameness of the underlying eye movements that all human beings share – and this is what makes the world intersubjective and therefore also objective for all. The world is interobjective.

How does the little dog (Fig. 3.1) come to be seen? How are perceptual processes self-organizing such that something that we recognize as a dog comes to be a reality? The double bind is that there is an unseen that the observer looks for, which remains unforeseen up until the instant of a first seeing – unseen and thus unforeseen and unforeseeable (Marion 2004). The *unseen* – which is not to be thought as an invisible that inherently could not be seen – remains unapparent until the decisive moment that it disappears with the seeing of the seen. The unseen does not assist us in foreseeing what we come to see. The unseen bursts into the visible by surprise, and we see the little dog rather than nothing or a Holstein cow. But this unseen, or rather, what comes to be seen, the visible itself has to provoke both visibility and the aim/intention that renders it accessible. "The visible precedes the aim: this is what must be rendered *visable* [aimed-at] by us, since we did not expect it" (p. 33). Because it was unseen and unforeseeable, that which we come to see has to have given itself so that it could be aimed at and become visible. The event can be glossed as *the unforeseen appearance of the unseen in the visible.*

We then experience the seen as objectively present. The perceptual structure denoted by the term "little dog" comes to take its own existence in the world that we inhabit with other human beings. But the shape is there *for us*, human beings rather than in the object (here the image) in and by itself. We know this to be much more easily in the case of colors. For physicists, there is just a continuum of electromagnetic waves, a small portion of which is experienced by a majority of human beings in terms of the rainbow colors (technically, we, as the primates, are trichromats). Those human beings who do not perceive in this same manner generally are thought to be deficient, for example, individuals who are (red-green) color blind or "afflicted with" "blue cone monochromatism." Most animals are dichromats and would experience objects, if they were to experience the same objects at all, differently in terms of the color distribution. Color perception and discrimination is increased for those human beings (mostly women) hypothesized to be tetrachromats. That is, these tetrachromats – mostly women – have phenomenologically different color experiences and spectral delineations (color bands) in the part of the electromagnet spectrum accessible to humans.

The appearance of perceptual objects therefore is not just a function of the mind, there are other aspects involved in what we see and how we come to see it. Being outside of the mind, they are not subject to the constructive processes; rather, we relate to them in radical subjectivity. There is nothing in the mind that allows us to predict seeing a little dog and there is nothing that could predict the appearance of a configuration that is being seen for a very first time, such as this happens in the (fine) arts or sciences (think of cubism or abstract painting).

Touch

In many ways and for many reasons, vision and visual experiences do not constitute a good starting point for thinking about how we know and learn, for the interactional aspects between the subject of knowing and the object (world) it knows about is all too easily obfuscated. Especially among scientists and like-minded philosophers, mind (knowledge) came to be seen as a mirror of the world – an idea that goes back to the ancient Greek when Aristotle (1986) suggested that "the sense is the recipient of the perceived forms without their matter, as the wax takes the sign from the ring without the iron and gold" (424a [p. 187]). Although the visual sense has come to be the dominant one in our culture – thus, the mind as mirror of nature rather than an impression – touch was for Aristotle the sense that distinguishes humans from animals:

> For in the other senses man is outstripped by many of the animals, but in point of touch his accuracy exceeds that of the others by a long way. And it is also for this reason that he is the most intelligent of the animals, an indication of which is the fact that even within the race of men it is in accordance with this sense-organ that individuals are well or badly endowed by nature, and in accordance with no other. (422a [p. 180])

Two millennia later, the hand, the primary touch organ, also became the distinguishing feature in the philosophy of Karl Marx because of its capacity to make tools and to transform the world. Here, then, both the pathic aspect of the hand as the place where nature leaves its imprint and the agential aspects where human intentions get transformed into nature-transforming actions, are both highlighted. In this section, I show how the two moments, the pathic and the agential, are irremediably intertwined. Learning from and about nature always already has pathic moments that remain un(der)theorized when we think about learning using a constructivist metaphor.

To explore the sense of touch, we actually ought to do it in the dark where our sense of vision is eliminated so not to contaminate our ongoing investigation. School children in their early years frequently begin their science and even mathematics lessons by exploring substances that are hidden in boxes or bags (e.g., Roth 2011). To sense the nature of a mouse pad surface, it does not suffice to place one's fingers on it, but I actually have to move my hands while sliding the fingers across the surface (Fig. 3.2). We immediately can make a number of observations. First, there is an essential agential moment to learning about the surfaces by means of touch. I instantly and without a second thought move my fingers along the surface of the objects I encounter. As a result, I have a sensation at the surface of my skin. It is smooth for the mouse pad that I use right now. This sensing already exhibits the affective being-affected, something has come close and thereby relevant to me. But this smoothness is not the sum total of my sense impressions. Rather, "it is the manner in which the surface uses the time of our tactile exploration or modulate the movement of our hand" (Merleau-Ponty 1945, p. 364). If the hand moves across a different mouse pad surface, the sensation changes. If I push harder onto the surface, a change also occurs. I begin to sense

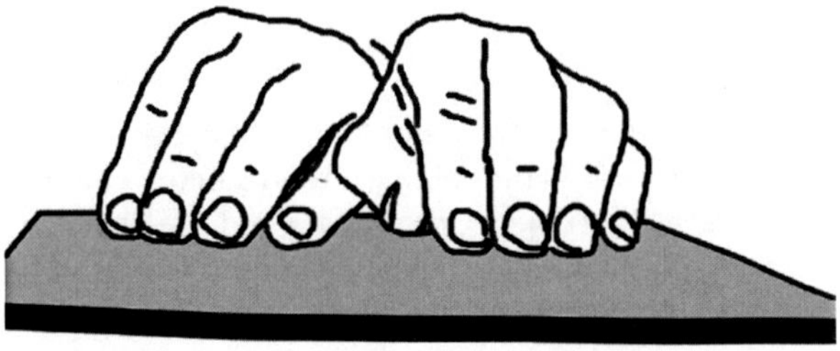

Fig. 3.2 Finding out about the texture of a surface through touch.

the effect of moving along the surface as affecting me on the inside. If I stop, the sensation stops.

For Kant, this shift in sensation had meant a shift of consciousness, now emphasizing the (inner) sense organ rather than the external object so that external representations (*Vorstellungen*) are changed into internal ones. "The style of these modulations defines an equal number of appearances of the tactile phenomena, which cannot be reduced one to the other and which cannot be deduced from an elementary tactile phenomenon" (Merleau-Ponty 1945, p. 364). If you find creating this sensation is difficult with the hand and fingers then think about what you do when you feel an itch somewhere. You scratch or rub yourself hard on a corner of a wall or doorframe. The itch is on the inside of your skin and moving along an object removes the itch. That is, depending on your intention, to feel or to scratch, the interaction of moving along an object that you already know lies outside or inside of your skin. That is, the intention changes what happens and what you learn (about surfaces, about how to deal with itches). But in any event, the sensation created first and foremost is an auto-affection, created at the interface between an intention to sense or scratch and the touch, which requires a movement of my hand (body) along an object. I cannot anticipate the contents of my touching unless I have seen it before. This result therefore is a pathic experience. That is, to understanding the real living act, we require the concept of auto-affection, which is the "self-reflection of life in motion, of life in its actual aliveness" (Bakhtin 1993, p. 15).

Auto-affection, in fact, is the universal structure of experience as such. "Only a being that is capable of symbolizing, that is to say, to auto-affect itself, can let itself be affected by the other in general" (Derrida 1967, p. 236). Auto-affection is the condition of experience in general, a possibility that we may also call *Life*. Auto-affection emerges from the encounter of the touching and the touched, the touching-touched, in the "minute difference that separates acting from passioning" (p. 235). This separation, actually, is a *différance*, because the difference between acting and passioning is undecidable, perception and self-movement pertain to the same (dialectical) unit and they constitute one another.

Thus, when I conduct this experiment in the dark and with unfamiliar objects (e.g., after having asked someone to put items into a bag), then what I am embarking on is not all about intentions. To feel by means of touch, I have to open up to

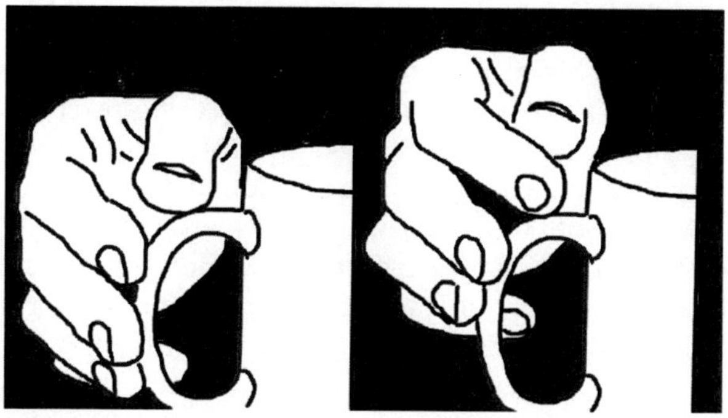

Fig. 3.3 Exploring shapes by means of the sense of touch requires the movement of the hand and opening up to be affected.

allow myself to be affected – even though the movement itself is already part of my "I can." As I do not know what to expect, I cannot but allow the world to act upon me all the while I intentionally move my hand across the mouse pad surface to sample its texture. That is, although I intend to sample the surface, I actually have to allow the surface to affect me, as it is only through this affection that my touch sense functions. Sensing is essentially pathic, where I open up to the world to affect me. We do in fact say "I am touched" in the passive voice when something emotionally affects us.

When I intend feeling the hardness of the surface, I actually stop and now push down on the surface using my finger like a stylus and push with arm and hand. In the case of a mouse pad, I can sense how the surface folds around the finger, which moves a bit into the surface and then I can feel resistance within my muscles to the effort they expend pushing downward. That is, what I initially sense is a resistance not of the substance to my finger but within the muscles that enact the downward push. When I do the same with the desktop, my finger does not penetrate and there is an immediate resistance I feel in the muscles of my arm and hand.

Yet I can have another form of experience once I change my intention to feel the shapes of things. For example, I may run my fingers along a coffee mug with the result that the fingers begin to move with respect to each other and the hand moves with respect to the remainder of my body (functioning as a frame). For example, as I move my hand along what I know to be the handle of the coffee mug, the thumb and index finger begin to move with respect to the middle finger, which itself approximately stays in the same position for a while (Fig. 3.3). From these changes, I know by experience how to name the underlying shape (topology) as a curve – or, after feeling the other parts of the object, as the handle of a mug. Again, if I do this experiment in the dark, my intention to feel shapes requires my hand to move and to open up to receive the impressions that arise from the interactions between my hand and the object. Because I cannot anticipate (foresee) what the shape is, I cannot but be a welcoming recipient of what happens to me as I move my fingers and hand *following* the shapes. The shapes of the objects determine the direction of my hand movements. "I make myself passive with respect to [the objects] and that they are revealed to me from the point of view of this passivity, in and through it" while at the same time, "in my desiring perception I discover something like a *flesh* of objects" (Sartre 1956, p. 392). Importantly, what I feel is not the shape. What I feel are changes in the position of my hand and fingers with respect to each other and with respect to my overall body as a frame of reference.

In all of these explorations by means of touch, we notice the interplay between the agential and pathic moments of learning. The perceptions *are given to me*, in person, in the flesh, as Merleau-Ponty says, not to an abstract mind. The smooth and rough surfaces of the different mouse pads or the handle of my coffee mug resonate within me. They are first and foremost experiences in and of a living body, which integrates the auto-affection of its own movement and the unanticipated sensations that are aroused in the process. Although I can intend to learn (find out more about the world I inhabit), I cannot intend and directly aim at the contents of this learning. I cannot construct the surface because the sensation is an entirely pathic experience. Because I do not know what a surface texture will be, I cannot construct its nature. Whereas I can intend to learn by touching, requiring the ability to move my hands, I cannot in a strong sense construct the knowledge about the surface as I can only open up and let it (the surface) affect me. Learning arises from this concurrence of affection and auto-affection in movement. "The perceptual . . . is always given with the feeling, with the phenomenal, with the silent transcendence. Yet *a Piaget absolutely ignores this*, has completely converted his perception into a cultural-Euclidean perception" (Merleau-Ponty 1964, p. 262, emphasis added). The philosopher therefore poses us the task to accurately describe how perception masks itself, how it makes itself Euclidean – which is precisely part of the task I accomplish in this book. In contrast to Gestalt psychologists, Piaget views perception as probabilistic in nature, so that a sensorial datum comes to be the product of an equilibration that depends on innate factors that are external one with respect to the other and that interfere with one other.

From very different theoretical backgrounds, theoretical biologists, physiologists, and phenomenological philosophers have come to the same conclusion: perception is self-movement. Perception and self-movement emerge from each other: they are made from the same cloth. This is so because the flesh (living body) is not the sum total of tactile sensations and kinesthetic experiences but of a unitary "I can." The flesh is the source of an auto-affection that is at the source of the "I can," which subsequently allows intention to emerge. This auto-affection also precedes any sensorimotor schema. The living body, the flesh, is the condition of this self-apperception; and this self-relation allows the self of movement and perception to coincide and to weave with and from it the world that we know. The flesh is the living/lived body, which is not the body thought by mind (soul) as its own, but the sensible in a double sense: it is what I sense, the sensible world including the body, and what senses, a body given to itself in auto-affection that is further affected through its senses.

Vision, Take 2

In a previous subsection, I point out that vision is not a good starting point for developing an understanding of learning by means of sense experience. This is so because it appears as if the retina functioned like a surface that receives light from the outside and which therefore functioned like a two-dimensional projection of the world into (onto) the organism. In computer models of graph interpretation, for example, the software, which also embodies a form of theory, extracts in this way features from the retina (Tabachneck-Schijf et al. 1997). In the present subsection, I show that vision is not a passive process but, like the hand and its sense of touch, is fundamentally built on movement. To begin this investigation, take a look at Fig. 3.4 before continuing to read. What do you see?

Gazing at this figure, most people tend to see (a) a hallway that appears to move into the page and is closed off by a square wall or (b) the bottom part of a wedge section that ends with a square on top and appears to protrude from the page. Others may see (c) an assembly of four two-dimensional shapes that includes a square, two parallelograms, and an isosceles trapezoid, (d) the bottom of a rocket or ballistic missile (the trapezoids being the fins), or (e) the bottom of a robot. I am concerned here with the two spatial ways of seeing. Why are there spatial shapes that I may see even though the drawings are but lines on a flat piece of paper? Why would I see something spatial even though the retina is flat? Something is happening at the level of my eyes that I am generally not aware of and that makes depth appear where there is none. In response, it has been said that "depth is born beneath my gaze because the latter tries to see *some thing.* But who is this perceptual genius at work in our perceptual field, who always tends to the most determinate?" (Merleau-Ponty 1945, p. 304). Whatever we first see in the figure, it likely appears as if it were just there, on the paper. Yet, and this is the central point of this demonstration, the image itself is the result of a dialectic of agency and passivity.

Let us stop for a second and pursue some questions that arise at this point. How is it possible for us to see one or the other of these figures? How is it possible for

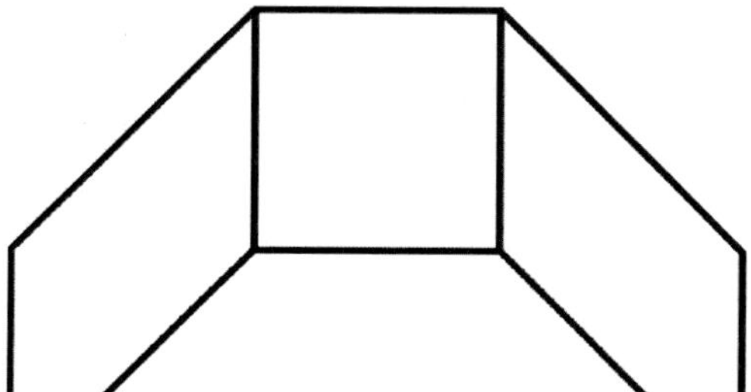

Fig. 3.4 Depending on the focus and eye movement, different organizing figures can be seen in this drawing.

us to recognize, or rather, re-cognize a hallway? If we see the figure as such, prior to any intention – which we cannot have prior to seeing the figure – then there is something in the movement of the eye, guided by the lines, allowing us to remember and *re*cognize. There is flesh to the objects of the world in the sense of Sartre precisely because without the movements of the flesh, there are no objects at all. The memory that affords *re*cognition then lies in the auto-affection of the movement of the eye muscles, and the way in which the visual sense is affected. The subjective movements of the eyes, as the sense organs in other situations, that is, the kinesthesis, are necessarily linked to the images we see and cognize again. Associated with the kinesthesis is a memory of the organs themselves. This is a memory prior to and deeper than any memory mediated by representations, the entities that allow us to make present again something currently absent. The latter is the case when you attempt to remember the figure after having turned the page or after having put the book aside, perhaps in the evening, when thinking about what you have learned during the day. But remembering the image itself is built upon the immanent memory of the living body itself. Scientific research now provides evidence that the (mirror) neurons firing when we move some body part also fire when we recognize such movements in others. This firing and memory are co-extensive.

In the following, second step of the demonstration, you first have to come to see the two spatial shapes: move from seeing the hallway to seeing the truncated wedge and back. Attempt to flip the two perceptions back and forth as quickly as you can within the flicker of the eyes. Once this flipping is continual and as fast as possible, attempt to identify what your eyes are doing. You may find that such flipping can be achieved very quickly when the focal point is close to the left bottom diagonal line. If the focal point is on the trapezoid and runs toward the square, you see the hallway; moving the focal point across the diagonal and letting the focal point run to the left along the diagonal, you see the truncated wedge. That is, the locus of the focal point and the movement of the eye that follows allows you to see a hallway or a truncated wedge, the former reaching into the page, the latter strutting out of the page. The agential moment of the perception lies in the choice of focal point and the movement of the eyes that follows; the pathic moment lies in the corresponding impression of the lines upon the retina and the experience this gives rise to. This is the same process that occurs in the duck-rabbit figure (Fig. 3.n3), which gives you a different perceptual experience as a function of running the focal point from the eye to the right (nose-mouth of the rabbit) or the left (beak of duck). In the case of the hallway/truncated wedge, the perception of shapes is the result of experiences in a three-dimensional world that we inhabit with parallel lines (e.g., rail road tracks, road sides) that seem to approach each other in the distance and with spatial shapes.

The demonstration that it is in fact the movement of the eye that allows us to see the spatial shapes is more difficult and requires a little training. One way of attempting to do this is by training oneself to see one of the three flat appearances, for example, the bottom of a rocket. You notice that this is possible, for example, by choosing the focal point in the square and above all by preventing the eye to run along one of the diagonals. But even just viewing a flat shape already involves the movement of the eye, a movement of which we are generally unaware. The eyes are in fact moving – the technical term for the movement is *saccade*. With some training, you may be able to look at the figure without the eye moving at all. This is the difficult part because our eyes are constantly moving. Psychologists therefore have devised an apparatus to fix an image to the retina so that a line always falls on the same photoreceptors. But once this happens – you may have achieved this by staring, for example, constantly at the same intersection of lines – you notice that the figure dissolves into a constant grey. That is, when the eye stops moving (or more generally, when the figure stops moving on the retina), any image you may have perceived will dissolve.[4] This then leads us to the important conclusion that (a) without movement and (b) without an impression upon the senses, we only experience diffuse ground or do not see anything at all (as seen in blindness). This is precisely the case we observed above in the phenomenon of touch – without movement there is no sensing and, after a while of dull sensation, even this disappears. Without the movement of our eyes – the one that allows us to see the little dog (which itself required us to move our eyes until we find the right focal point) – there is no little dog. That is, even just to see the dog the eye has to move between it and the diffuse ground against which the dog constitutes the figure.

What happens is this: Whatever figure we see emerges with respect to everything else that remains diffuse. The figure does not stand before the ground just like that and on its own but, as our exploration shows, the figure | ground relation has to be produced continually by the eye that is moving between the two. Moreover, when I see the figure of the hallway, when I feel the surfaces of the mouse pads, my perception is not saturated by the current sensation. For my actual perception of a thing under this or that angle relates the image or touch properly perceived to all the other possible perceptions that are expected or anticipated. Thus, "every actuality implies its potentialities" (Franck 1981, p. 47). That is, we learn from the world only in and through the dialectical relation of agency and passivity. We do not generally control immanent processes (eye movement), to which we are as much subject as to the impression that the world leaves upon our senses. This is especially evident in the case of the little dog (Fig. 3.1), where the dog, to appear at all, appears against all the other black and white areas that make the figure. The saccade tends to move the focus from what is figure to ground and back to figure. This is also the case in the preceding exploration of the sense of touch, where we can shift intention to surface texture, topology, temperature, or other aspects with these other aspects always falling into (remaining in) the more-or-less-diffuse perceptual ground. And just as with the pain we experience when we touch

[4] Here I am only interested in the phenomenological aspects of the phenomenon, not the natural scientific. The latter has to do with the fact that rods and cones only respond to differences in luminance. If there is no movement, then the photoreceptors come to be saturated and no longer perceive anything.

or rub along a surface too hard, we experience pain when the light that falls on our retina is too intense.

On Learning the Unknown

I begin this exploration in the agential | pathic nature of learning from sensual experience with a puzzle. For the little dog (Fig. 3.1) to become figure, the eye has to find the right focal point and then stabilize the figure by means of saccadic movements, that is, by moving away from it to the ground and returning to it. But not knowing that there is a little dog, the learner cannot intentionally direct her eyes to the focal point. She cannot intend the little dog to appear. So for a while, which may be in fact very long if unaided, there may be little order in the visual field. The drawing itself provides the resources for seeing, it guides the eyes until the little dog appears, all of a sudden, completely unanticipated, unforeseen, and unforeseeable. We cannot but open up and ready ourselves to cognize something for a first time something that we had not known before and that for this very reason we could not aim at learning. From this perspective, learning has essentially pathic dimensions both in the sense that we cannot aim at (a) what is to be learned or what we will have learned when we are done and (b) how to go about learning what there possibly is to be learned. Moreover, because we do not have an image of what there is to be learned, we cannot have a plan of what it is that we will know once we have learned it and we also have no way of controlling our learning process. The general term for referring to the control of learning processes is metacognition. That is, these reflections now suggest that there are limits to metacognition when an individual is in a true process of learning, that is, comes to know something that she has not known before and therefore could not have intended to know and acquire.

Important in the present context is the fact that perception is not interpretation – seeing-as, feeling-as, hearing-as, smelling-as, and tasting-as. Following Wittgenstein (1958), we ought to say "I see a little dog" rather than "I interpret the perceptual field and as a result see a dog." This is so because immediately prior to the dawning of this aspect, I had no way of knowing *what* it is that is to be expounded, translated, rendered clear and explicit, to note but some of the senses in which the term "interpreting" is employed. Interpretation requires an object, and the absence of the object in the state of not knowing is precisely what we are concerned with here. This is why might say with Wittgenstein that "'seeing as . . .' is not part of perception" (p. 197e) and when you exclaim after looking at Fig. 3.1 for a while, "I see little dog," you both report and express perception and visual experience. "And for that reason it is like seeing and again not like" (p. 197e). Most importantly, "since it is the description of a perception, it can also be called the expression of thought" (p. 197e). The thought, however, did not precede the perception, because we did not know that there was a dog. Thought then, rather than preceding the expression follows it: "If you are looking at the object, you need not think of it; but if you are having the visual experience expressed by the exclamation, you are also thinking of what you see" (p. 197e).

Seeing the little dog for the first time means thinking it for the first time in connection with the drawing. You have learned something; your eyes have learned to see a little dog in this part of the world (even if it is only in the pages of this book). This learning, as process or product, could not be anticipated. The process could not be anticipated because the movement of the eyes and focal point is in part the result of engaging with the world at hand, and when the little dog emerges, it gives itself as much as it has been the result of the voluntary and involuntary eye movements when you investigated the image. In the same way, you were a recipient of the first visual experience you have when looking at Fig. 3.4. You could not foresee and intent the results but you are subject to the interactions between perceptual processes and the world. (The question of how the world comes to appear independent of our perceptual processes itself requires an investigation, which is the topic of Chapter 5.)

Once we know it, we see the little dog (Fig. 3.1) without problem. We switch back and forth between duck and rabbit (Fig. 3.n3) or between hallway and truncated wedge (Fig. 3.4). When someone sees no or only one aspect where there are two (duck-rabbit) or more that are possible seen (Fig. 3.4), then lack of familiarity is one of the main criteria: "What tells us that someone is seeing the drawing three-dimensionally is a certain kind of 'knowing one's way about'" (Wittgenstein 1958, p. 203e).

Uncertainty of Objects

Many science educators might think that observation is unproblematical in the natural sciences. Even many scientists might think that way. Yet when ethnographers spend some time in a laboratory, then the nature of perception may be recognized as something much more problematical. Over a period of five years, I conducted an ethnographic research project in one biology laboratory that conducted – coincidentally – advanced research on fish vision. In this laboratory, what could be seen – under the microscope, on a computer monitor – was continuously in question especially during the early parts of an experiment or introduction to the laboratory of some new equipment. This was so because the scientists could not intentionally look at what they did not know how it would look like.

Ethnography of an Advanced Biology Laboratory

The scientists in the study I conducted were working with retina that they had excised from salmonid fishes. They wanted to find out how much light would be absorbed in the rods and cones – i.e., the photoreceptors – that make up the retina.

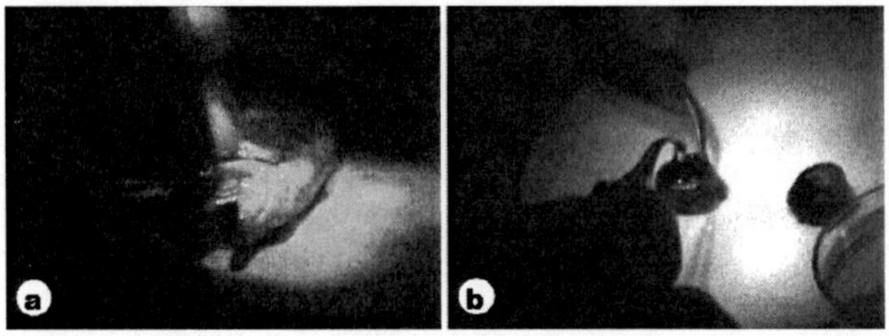

Fig. 3.5 a. After spending a night in a bucket sealed from light, a fish is sacrificed and placed on the laboratory table, where, under near-infrared light, the eyes are removed from the body. **b**. The eyes are hemisected. The eyeball is removed to allow access to the retina, which, requiring some "fishing in the blind" is pulled in pieces from the back of the eye socked and placed in the MEM-containing dish to the right. Two pairs of tweezers are visible, one holding the eye socket, the other pulling a piece of retina.

They wanted to know the amount of light absorbed not just in general terms but also separately for different wavelengths (colors) of light along the visible spectrum. The photoreceptors in the retinal tissue, once removed from the living animal, no longer regenerate once light has fallen onto and transformed them. They no longer absorb light as they normally do. All parts of the experiment therefore have to be conducted in near darkness, using only a red light source so faint that it takes the scientists between 45 minutes and one hour to adapt to see anything at all (the infra-red mode of a digital camera was used for the recordings exhibited here). Once dark adapted, the scientists first sacrifice the fish (dark adapted between two hours and half a day), "excise" the eye (take it out of the socket and head) (Fig. 3.5a), "hemisect it" (cut it in half), and remove the retina (Fig. 3.5b). A piece of retina is placed on a microscope slide fixed on a microscope stage, "macerated" (cut into tiny pieces with a scalpel) (Fig. 3.6a), and covered by a drop of a tissue culture medium that contains the basic nutrients and minerals required to keep cells alive ("minimal essential medium" or "MEM"). A cover slide is placed on top of this preparation and the whole assembly is sealed with paraffin wax. This piece of retinal preparation constitutes the transformed natural object – or actually little tiny pieces of it that are visually represented on the computer monitor – about which the scientists represent some aspect in their graphs, such as, in the present case, absorption. The slide is mounted in the microscope (Fig. 3.6b), where it can be manipulated to move in x- (left-right) and y- (back-forth) directions. In this position, it also sits in the path of a beam of light from a xenon lamp, which radiates with nearly the same intensity across the visible spectrum.

Each data run (several can be done with each slide) unfolds as follows. To obtain information about each photoreceptor in the retina, two measurements have to be made. In the first, a light pulse is made to traverse the slide at a spot where there are no cells; the person operating the microscope asks to take a "reference." In the second, the pulse is made to go through a cell. In this situation, the interior of the cell is aligned with the crosshair and the person operating the microscope asks to take a "scan." Because more light (normally) is absorbed in the cell than in

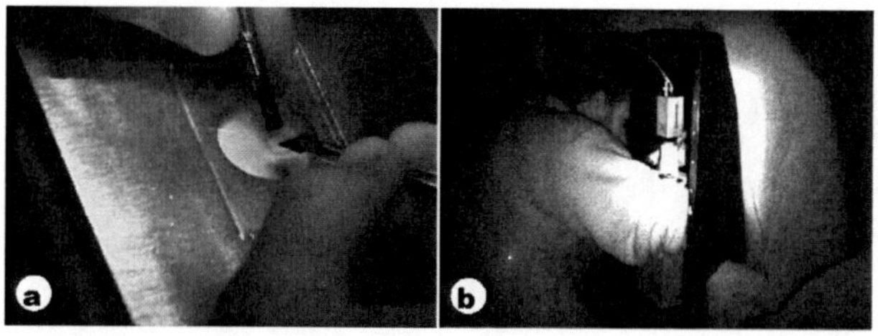

Fig. 3.6 **a.** Pieces are cut from the blob of retina, placed on the stage holding the microscope slide; the piece of retina (here visible as a *grey circle*) is macerated "to spread the cells across the slide." **b.** The stage holding the slide with the sample is mounted into the microscope, which has micrometer screws to move the stage in minute amounts in left-right- and backward-forward directions.

the surrounding saline MEM solution, the difference in the intensity of the two light pulses is attributed to absorption in the cell.

During the early part of my research, the scientists searched for photoreceptor targets by looking through the ocular of a microscope. Later, they purchased a camera that provided them with a visual image on the computer monitor (e.g., Fig. 3.7a). The absorption spectrum (Fig. 3.7b) is generated in the following way. After the beam of light has traversed the microscope slide, it is made to fall onto a grating. The grating diffracts the light so that different wavelengths travel at different angles until they fall onto the charge-coupled device (CCD), the same kind of device that is responsible for the recording in digital cameras. Each "tube" in the honey-comb-like device receives photons corresponding to a different wavelength so that the intensity of each wavelength can be measured. The difference between the two arrays of tubes generated by the reference and the scan is a measure of the amount of light absorbed in the photoreceptor (Fig. 3.7b).

Perception in an Advanced Biology Laboratory

In the series of events described below, scientists *produce* a chain of outcomes, which has the reference linking a sign, on the one hand, and an object as one of its qualities, on the other hand. The chain begins with the living fish in the hand of the scientist (Fig. 3.5a) and ends with a final graph on the computer monitor (Fig. 3.7b), always involving material action, matter, and light. Despite the scientists' presupposition that they directly deal with reality, access to the "things themselves," the unstructured materiality underlying the research object is never given. Scientists always interact with and act towards the object in the way it is given to them, mediated by the images from the various microscopes and graphical displays: consecutive inscriptions therefore are separated by ontological gaps. Nevertheless, *they* presuppose the continuum, because their material actions transform material, and even the tools and instruments that mediate the actions are characterized by their materiality. For scientists, therefore, there is no room for interpretive

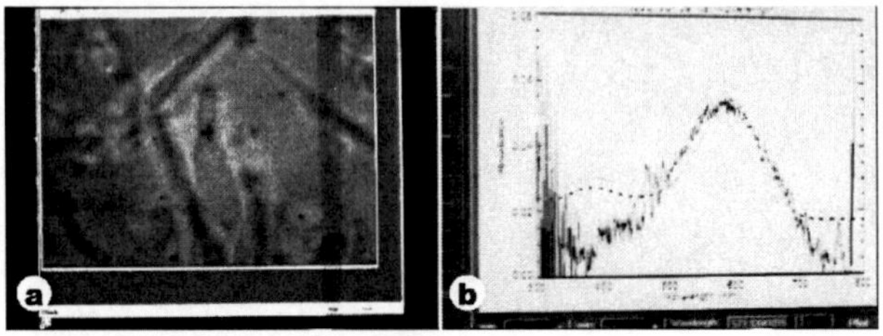

Fig. 3.7 **a.** Rod and cone-shaped retinal cells under the microscope. One "cone" is aligned with the crosshair, where the sampling light beam will traverse and be partially absorbed. **b.** In the ideal case, the absorption curve (here "a beauty") is consistent with the green member of a double cone (peak of curve lies in the green region). The assessment further supported by the comparison with a reference graph, which matches in its shape and "lambda-max half bandwidth."

(re-) construction: there are only materiality and effect-producing material actions. This continuum, however, may be disrupted by interference and disturbance without it being apparent and known to the researchers. That is, although some practical action has the structure "doing [*X*]," where *X* is a gloss of what they have done, and although all participants agree that *X* has been done, the scientists may find out later that they have not done *X*. A scientist may say "I macerated the retina" when in fact he did not do it at all.

Problems that interfere with the goal of producing a usable graph for a specific photoreceptor may arise at various stages in the experiment along a chain of actions. The understanding of the entire process of scientific production – which takes the scientists from seeing some initial state *A* to seeing its representation *G* – and the possible causes of interference are crucial to the interpretation of G; scientists would not be able to interpret a graph (G), for example, without understanding all aspects of the process of getting to the graph beginning with the fish. Here, a fuzzy image may be the result of the preparation of the slide, problems with the CCD, problems in the optics, or trouble with the computer. The classification of the visual image of the photoreceptors may be inconsistent with the scientists' expectation or with information that they get from the graph. The uncertainties these scientists experience with respect to the graphs they produce may be due to the preparation (absence of signal, shift in location), trouble with the generation of the data (noisy data, temperature variation in the CCD), or inappropriate amplification and interpretation of the graph itself. Without recourse to the way in which the animal (i.e., *A*) is continuously transformed through their actions and means of production employed, scientists would not have the resources upon which they actually draw in their interpretations of the many graphs (*G*) that they face during any one day in the laboratory.

Visually available things may turn out not to be what they appear to be. That is, an observational action at one stage in the experiment may turn out to be something entirely different. For example, although the scientists may think they are looking at a particular type of photoreceptor (a cone), subsequent measurements of an absorption spectrum may be inconsistent with the previous observation, and in fact, may question this observation as having revealed something very different

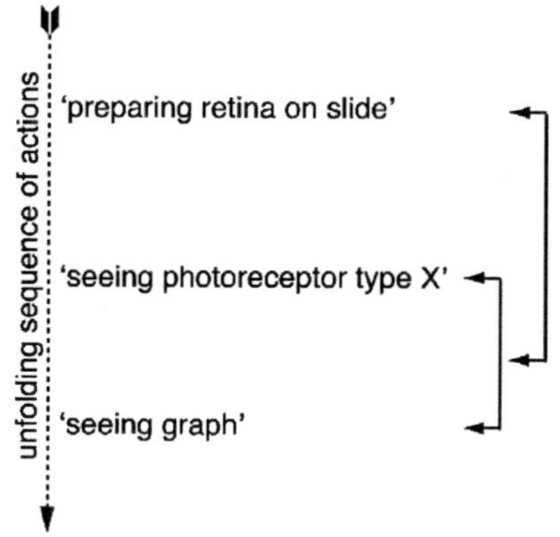

Fig. 3.8 Scientific discovery work consists of sequences of actions. The discovery of a scientific fact requires each action to be consistent with all other actions. Uncertainty in any action raises doubt about the whole process. The vertical bidirectional relations denote the ontological reflexivity between material and observational actions.

(e.g., a broken rod). That is, the observation (action) and the entity it yields have been destabilized; it is (ontologically) uncertain just what observation has identified (Fig. 3.8, lower action pair). In other words, a wedge has been driven between the *work* of seeing and the descriptive adequacy of the *gloss* denoting what is seen.

In one situation recorded on videotape, Simon interprets a circular image as denoting a cone seen "from above" rather than from the side as it usually would be seen (Fig. 3.9a). Seen from the top, each cone should give an image like looking from above at a bottle with a big, bulgy bottom. However, although the absorption spectrum initially looked like a "textbook case" (Fig. 3.9b) clean of any artifacts, subsequent processing leads Ted to hypothesize that the resulting absorption curve was "too narrow." The outcome of the work glossed as "(not) putting up a reference curve" revealed that the visible graph indeed was much narrower than what would have been expected had the visually identified entity been a cone (Fig. 3.9c). That is, in this case, although the initial inspection of the graph confirmed the real and objective nature of the object, closer inspection using one of the available productive means (a reference curve) destabilizes their perception of the curve and object as definitive, rendering them uncertain and thereby downgrading them from facts into artifacts.

In this situation, Simon and Ted do not know what the entity is; they simply discard the recording without further questioning themselves what it might have been. At other moments in the video record, they evolve hypotheses about the possible sources of an unexpected signal. For example, after making one recording, the absorption spectrum does not show one of the desired peaks (Fig. 3.10a). Ted then points to an area next to one of the cells (Fig. 3.10b) suggesting that the strong signal in the infrared is coming from "dark matter," a term that the lab members use to denote material not always evident in the visual image but showing up as "absorption" in the graphical display.

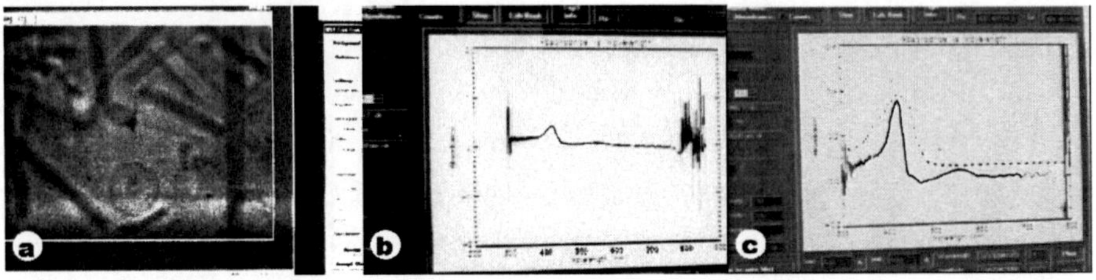

Fig. 3.9 **a**. Simon perceived the circular entity (vertically below crosshair) as a cone seen from the top (rather than from the side). He decided to take a recording. **b**. The graph seemed to be a "textbook" case of a retinal cell normally absorbing in the UV region of the light spectrum. **c**. The test with the reference spectrum (*dotted*) confirms Ted's suspicion that the bandwidth of the sample curve is too small to correspond to a cell. Thus, what they had seen previously could not have been a cone.

At other times, an absorption spectrum seemed to exhibit a peak – especially if further processing ("doing a baseline") would take away the background curve on which the "real" peak sits like a rider on a horseback. In the following situation, a rod appears to be clearly visible under the microscope. After having measured the reference and sample spectrum and displayed the difference, doubt is raised about the graph and, inherently, about the nature of the object under the microscope. The microscopic image of the cell is consistent with a rod (Fig. 3.11a). However, the absorption spectrum (Fig. 3.11b) turns out to be problematic as the following episode shows.[5]

Fragment 3.1

```
01 T:  yeah.
02 M:  a bit broad, ha?
       (0.75)
03 T:  its very broad, yea (0.45) its very hard to define a-
04     (1.32)
05 S:  strong=
06 T:  =straight line through it.
       (0.43)
07 S:  travel a bit further down.
```

In this episode, the utterance in turn 02 glosses the spectrum as a bit too broad to constitute the absorption line of a rod. Ted agrees (turn 03) and then elaborates that it would be difficult in this case to define a straight line that would constitute the baseline for the absorption. By indicating that he is continuing to search for

[5] Throughout this book, I employ the transcription conventions established for conversation analysis and frequently employed by linguistically oriented ethnographers: ((points)) : double parentheses enclose transcriber comments, often actions visually available but not recorded on the sound track; (1.20) : pause in tenth of a second; define– : n-dash indicates sudden stop in talk; = : equal signs are used when the normal conversational pause between two speakers is absent but no overlap occurs; [] : beginning and ending square brackets in consecutive lines are used to show overlapping talk; .,? : punctuation is used not in a grammatical sense but to indicate pitch changes speakers use to structure their utterances.

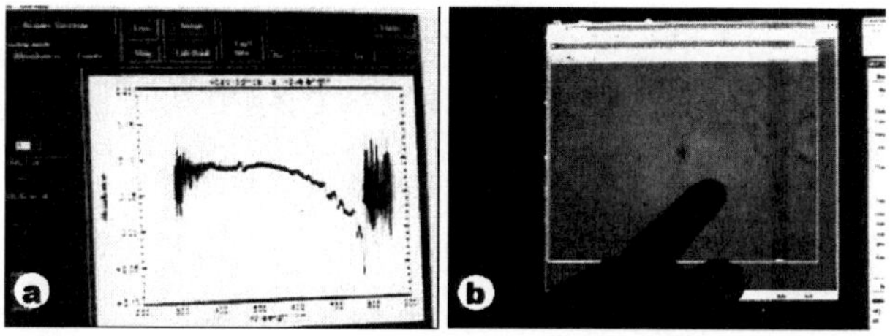

Fig. 3.10 After seeing the graph in (**a**), Ted points to unexplained "dark matter" (it is dark because it absorbs the near infrared light used to search for cells on the microscopic slide (**b**). This dark matter corresponds to high peaks and elevations on the right end in the absorption spectrum. "The problem here is the right environment".

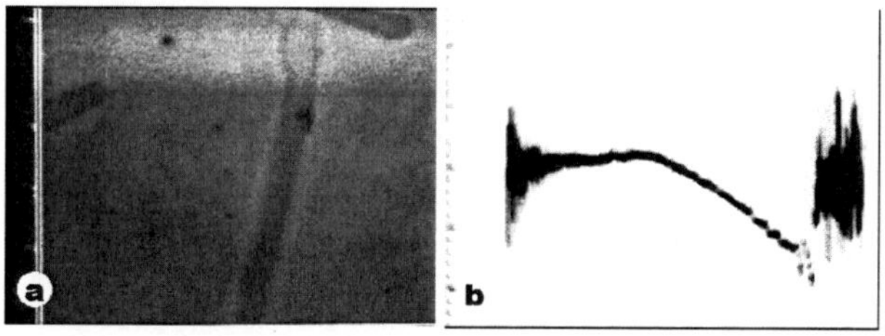

Fig. 3.11 **a**. A rod-shaped photoreceptor has been aligned with the crosshair, indicating the place where the sampling beam would hit to measure the amount of light transmitted. **b**. Although there is a peak, it cannot be attributed to a cell because "it's very broad" for being a signal. Although it is unclear what it is a signal of, the scientists continue presupposing that the entity is not of interest to their study.

another cell, Simon exhibits and articulates understanding that the spectrum presently displayed is not of interest for further investigation. In addition, moving on in search for another photoreceptor, he also indicates that what he has seen as a rod may not have been one, or it may have been one that for some unknown reason does not show the expected absorption and therefore is not of interest. The observed absorption spectrum and the entity under the microscope, as rendered in the microscopic image, therefore do not stabilize each other. They are inconsistent, though at this moment, the two regular members of the biology laboratory are not interested in finding out why the two forms of observational action are incongruent. At other times in the video records, the graph is not desired, but it stabilizes the observation in the sense that there is a possible reason for the graph to look in the way it does. This can be seen in the following example, which emerges when Simon finds what he takes to be a cell suitable for recording an absorption spectrum.

Ted points to the location on the cell that Simon is to align with the sampling beam (Fig. 3.12a). Having already recorded the reference spectrum, they now measure the spectrum of the beam traversing the cell. When the difference between the two spectra is displayed, the following interaction ensues.

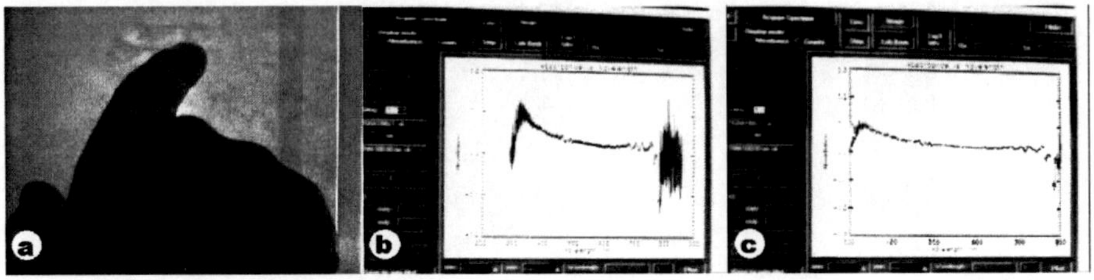

Fig. 3.12 **a**. Ted points to a rod from which a measurement was to be taken. **b**. The difference between the reference and sampling beam in un-enhanced version. **c**. Same graph as in (b), but after removal of the near infrared absorption.

Fragment 3.2

```
08 T:  yea that is what I was thinking about. ((Orienting toward
       Fig. 3.12b))
09     (4.42) ((Transforms the graph, to become that in Fig.
       3.12c))
10     [c] pretty flat (0.59) pretty bleached.
       (0.33)
11 S:  alri[ght. ]
12 T:      [Okay.]
```

The episode begins when Ted first articulates the flatness of the curve and then suggests that it is due to a "bleached" cell (turn 10), that is, a cell that should no longer absorb because the light absorbing molecules normally present have already been transformed and therefore no longer absorb. Simon utters a sign of agreement (turn 11), followed by Ted's positive assessment (turn 12). Such repetitious statements of the same words or ideas tend to stabilize discourse and thereby confirm the participants that they "are on the same page." As a result of this interaction, the perceptual identification of the cell has stayed intact, but its state has been redefined: it *is* a particular cell but it is a bleached one. The graph and perceptual identification therefore are consistent but identify an entity that is not of interest, as the graphs of interest can no longer be recorded from this cell. The observations have stabilized, but the graph is problematic only in as far as it cannot be used for any publication.

Radical Passivity in Scientific Perception

In this ethnographic study of one science laboratory, we can see that (visual) perception amounts to more than what meets the eye. Perception is not a matter of gazing at something on a microscope slide and recognizing what it is. When scientists work on the cutting edge, that is, when they do not yet know something that they later report as their phenomenon, they find themselves in a double bind. To know what they are gazing at they need to know of its existence; but to know of something as existing, they need to know what they are gazing at. They cannot intend to see what they do not yet know. It is for precisely this reason that they tend to struggle as much as high school science students without having the recourse often taken by the latter, which is to ask, "am I on the right track?" It is only when some initial phenomenon, a particular graphical feature, a particular perceptual feature in the probe, consistently appears and disappears with experimental manipulation that scientists come to know that they are seeing something and what this something might be. Here, experimental manipulation has the same effect that the saccadic eye movements have on ordinary perception. A figure has to be constant and has to be found again after having looked away. The very early stages of the research are difficult, because the scientists do not know what to look for, and when they seem to see something that is salient, they do not know whether it is real or an artifact. Whatever emerges is not intended but is received by the investigator who is a willing host of what gives itself to be seen and known.

Theorizing Learning from the Perspective of the Unknown

Constructivism ever since Kant has failed to show how something that is given to the senses has a content fit to be described by consciousness (or Descartes' *cogitationes*). How could cognition see anything if it did not know about the image in the first place? Before my hand can move to reach out and learn by touch, before my eyes can gaze at some segment of the world and learn by vision, my hand and eyes need to know that they can move. The agent cannot construct this knowledge because it precedes all constructive abilities, all intentionality, and all intellectual consciousness. I know how to move my eyes because my eyes know how to move on their own; and this immanent knowing, knowing without representation, is the result of originary movements in which the living/lived body (the flesh) auto-affects itself. The auto-affection of the flesh leads to an immanence of its powers that makes the living body an exceptional location of original memory. This is not memory in the classical sense, such as the power to remember something by rummaging through the contents of mind, but a memory that is entirely in the body itself. Thus,

> all the movements executed by the hand, all the positions that it has taken as it traverses the solid, can be repeated at will in the absence of it. These movements are the signs of diverse elementary perceptions relative to the primary qualities, inseparable from the resistance; they can therefore serve to recall ideas, and this recall is executed by means of the available signs that constitute the memory properly said. (Maine de Biran 2001, p. 311)

The kinesthetic experiences constitute a form of memory prior to and more basic than the memory that is mediated by representations. These movements constitute sensory-kinetic archetypes or kinetic melodies (Sheets-Johnstone 2009).

Immanent memory, precisely because it is unmediated by linguistic consciousness, is given to the subject. Intentional movement requires knowing that I can move, for without knowing that I can move, I cannot have an intention to move. Because we do not have to ask ourselves, "Can I make this movement" before intending the movement, the "I can move" constitutes a form of immanent knowing on which the movement intention is built. Thus, this intention itself is something given rather than constructed by the agent. Immanent knowing, auto-affection, and givenness, which are the results of a primordial experience of movement, lie at the heart of agency. That is, givenness and passivity *enable* and make possible agency. It is for this reason that this form of passivity is radical. It cannot be intended although it is foundational to agency.

We all have had experiences of memories that cannot be reduced to mental representation. I personally became alerted to these powers when I attempted to call my doctoral advisor after having not seen her for years. But I had not recorded her phone number anywhere and could not remember the digits. As I was sitting at my desk and taking the phone into my hands to call directory assistance in Mississippi, my right hand dialed the area code. But rather than continuing with the directory number (i.e., 555-1212), my fingers continue on their own and I am struck by the familiarity of the dial tone. I let it ring. After a brief moment, someone picks up: "Hello?" I recognize her voice. In this case, my hand "remembered" the number that my sign-mediated memory had forgotten. My hand dialed a number that had gone from my mind. Once launched into dialing, my hand and fingers were driven by a kinetic melody. Coincidentally, my ears recognized a dial tone; they recognized something as immediately familiar, even though I had forgotten the telephone number. The ears, too, recognized a melody that my conscious mind had forgotten. Other individuals I know have similar experiences with sounds, odors, or tastes that make present experiences, phenomena, and people that they had long forgotten about. In the classical sense, memory, however, means representation, an intentionality that gives us events as something that has been present at some time in the past and that can become present only in and because of *re*presentation. Remembering the digits of a phone number constitutes memory of that kind, which, however, hides another form of memory much deeper and more profoundly anchored in my body. It is the memory of my fingers that remember each time the sequence of the keys depressed and the memory of my ear that recognizes the sounds of a melody. In both of these memories, there is no representation involved, no thought of any kind. Rather, it is the auto-movement of a power of touch dialing "revealed to itself in the pathic auto-donation of my originary corporeity" (Henry 2000, p. 207). This kind of power is carried and kept within the living/lived body as the principled possibility from which it is never separated and which it never loses because it is nothing other than originary memory.

This displacement of memory, from representation to the immanently experienced living/lived body, is the result of the extraordinary intuition that Maine de Biran has had. It allows the body to know before any intentionality of knowing intervenes. Anything that we are conscious of requires an immanent capacity with respect to which conscious experience is absolutely passive. It is precisely the condition for recognizing an object or situation without any other requirement than the movement (kinetic melody) itself. This movement requires no sign other than itself, no representation. When I see a hallway or a truncated wedge, it is not because I desire to see them. I look and all of a sudden a truncated wedge protrudes from the page or my eyes are following a corridor into the depth of the printed page. The little dogs, unintended as they were, all of a sudden become salient to me, willing host to whatever offers itself to my perception. As soon as we see the dogs our eyes can find them on their own by following the movement that first revealed the outline that made the image appear. They eyes remember the movement, the associated kinetic melody, so that each time we return to the figure, two little dogs emerge from the page.

In this manner, we have begun to turn our theorizing of learning from that which is known to the unknown. For the unknown we cannot aim at seeing, learning from, or constructing something ("meaning," "knowledge," "conceptions") thereof. The foreign/strange, the unknown, cannot be constructed but has to reveal itself precisely because it is unknown. At the very instant that something reveals itself, is given to our perception, the unknown, the foreign/strange has already receded, is as inaccessible as before. The newly seen – the learned, that which has appeared – has not brought us any closer to the unknown and unseen. The newly seen has not even revealed more and thereby decreased the unseen and unknown. Any true object of learning therefore will always remain unknown and thus incapable of directing the learner. Learners might intend to learn something, but precisely because they do not know the something, they cannot aim at it to bring it into the light of the known. Moreover, anything revealed as such in the clearing that allows us to know does not clarify the darkness beyond, which, precisely because unseen (unknown) constitutes the ground of knowing in the same way that the perceptual ground constitutes the condition of the perceptual figure. Without the unknown we would be unable to distinguish the known in the same way that without the unseen (ground) we would be unable to distinguish the figure. In a strong sense, therefore, both seeing and knowing presuppose the unseen and unknown. The seen and the unseen, as the known and unknown, co-implicate each other. At the same time, the unseen and unknown – precisely because unseen and unknown – imply a radical passivity with respect to perception and learning. We cannot intend what is given to us in the radical passivity that underlies our Being.

Chapter 4

Radical Uncertainty in Acting

> All attempts to force one's way from inside the theoretical world and into actual Being-as-event are quite hopeless. The theoretically cognized world cannot be unclosed from within cognition itself to the point of becoming open to the actual once-occurrent world. But from the performed act (and not from the theoretical transcription of it) there is a way out into its content/sense, which is received and included from within that actually performed act; for the act is actually performed in Being. (Bakhtin 1993, p. 12)

In this opening quote, Bakhtin suggests that from within thought, we cannot understand how human beings act in the world. On the other hand, from within the performed act rather than from its theoretical description, we can understand the content and sense of the act. This content/sense *is received* and *included within* that actually performed act – it is not from a construction within the mind, the theoretically constructed and cognized world that we understand how people act and why the act in the way they do. In particular, we do not understand why there is a radical gap between thought, on the one hand, and acting in the world, on the other hand.

In the preceding Chapter 3, we see how perception and learning are associated with radical uncertainty because to intentionally look for or learn presupposes seeing and knowing the unseen and unknown. The investigations in that chapter show that seeing and learning are understood differently when considered from the perspective of the unseen and unknown. The investigations reveal in addition that any seeing and knowing also presupposes a power to act, which is itself the result of some originary movement that enables us with an "I can." But this "I can" is of a different kind than the one we generally associate with intentional action. It is due to our original body, which we immanently know "and as such is not the product of constitution in the Husserlian sense of intentional acts but is directly known in internal transcendental experience" (Sheets-Johnstone 2006, p. 368). It is well known that there is a gap between the plan of an action and the situated action that it describes. Even highly trained and experienced golfers shoot bogies, even highly trained soccer players miss the goal during a penalty shot, and even highly trained and experienced tennis players make double faults. In all of these

W.-M. Roth, *Passibility*, Classics in Science Education 3,
DOI 10.1007/978-94-007-1908-8_4, © Springer Science+Business Media B.V. 2011

cases, what the athletes might intend to do and what they actually do falls apart. This is also and especially the case when novices use instructions to do something for a first time: putting together a piece of IKEA furniture, cooking a recipe for a first time, or operating a new television or computer system. Why would there be such a gap between my intent to act in a way and the way I actually act?

There is very little if anything that we can find in the constructivist literature to make thematic let alone explain this gap. A little over a decade ago, there existed less than a handful of studies concerned with the question of how students in school science bring about the investigations from instructions that they follow (e.g., Roth et al. 1997). All existing studies have been conducted within phenomenological and ethnomethodological frameworks. It is as if the problematic did not even exist within the constructivist literature on learning from laboratory investigations. Thus, for example, the website of the *National Association for Research in Science Teaching* features an article on inquiry and laboratory strategies, which takes following instructing as unproblematic: "The commercial approach primarily required following instructions exactly as stated" (Leonard 1989). In the cognitive approach of constructivism, the subject comes to a situation with a particular knowledge and then acts by implementing the intentions or goals that are derived from the framework. It is very difficult to understand, within this approach to knowing, where there should be a gap between episteme, theoretical knowledge, and practical action (pragmatic knowledge). Why is this so if the cognitive agent has abstracted forms of operations and derived abstract operations? Why is the same cognitive subject unable to go from the abstract operations back to the concrete ones? Why may someone who has theoretical knowledge be unable to bring about the corresponding practical knowledge in concrete situations?

A reader with constructivist inclinations might hastily suggest, in the case of the recipe or assembly instructions, that I "interpreted" the instructions differently then they were intended, that the "meaning" I constructed differed from the intended "meaning." But this suggestion presupposes the gap between plans and actions, between descriptions of the world and the world, between words and their significations ("meanings"), and between *Being* and *beings*. In this chapter, I continue to articulate a different approach to this question. My answer is grounded in the difference between immanent knowing in movement – i.e., *kinetic melodies* – and the explicit knowing that constitutes plans and instructions. Here, movement is not a transcendent phenomenon, known by means of a representation, but is immediately known to itself (Henry 1965). I suggest that we act because of our immanent knowing; there is no intermediary between some mind and the world. The gap is between this immanent knowing, which is the result of originary movements (i.e., *Being*), and explicit description (i.e., *beings*). I suggest that how these descriptions relate to the immanent knowing is an empirical matter to be established after the action has occurred.

This chapter is about the gap between plans for actions and the situated actions themselves. I have not seen any (social, radical) constructivist or, for these matters, cognitive psychological explanation other than one based on deficit. Thus, one handbook chapter on the constructs of mind states that

> [w]e have no satisfactory account of actions disconnected from mind except perhaps to refer to developmental immaturities, as when we speak of *disorders of impulse*, in which the person acts before thinking or, like a child, fails to control impulse; or, like the sociopathic adult who has failed to internalize proscriptions, the person does not experience guilt and is therefore not restrained by it. (Lazarus 1989, p. 111)

In this classical approach, the gap between intended and situated actions is the result of some disorder. In the approach I take, the gap is constitutional. That is, the body knows to act on its own. Conscious thought is but an impulse that sets in motion the kinetic melodies of the body. The degree of overlap between situated action generated by the knowing body and the cogitated plan can be established only a posteriori.

The issue is real and is not a question of deficit. Even though we normally walk perfectly well, we may stumble; and this does not make us or our walking deficient. Again, we need to find an explanation that does not begin with an ideal conscious mind–practical action relation. Rather, we need an explanation that begins with the presence of the gap as the constitutional feature of actions to be explained. There are frequent instances in everyday life where we are confronted with some plan that we are asked or want to convert into a course of action. Examples include (a) receiving directions for going somewhere in a city especially when we do not know it; (b) following a recipe from a cookbook; (c) doing a laboratory experiment in school or university; (d) assembling furniture (e.g., from IKEA) from a sheet of instruction that are either in verbal or, frequently nowadays, in pictorial form; (e) learning to operate a new piece of software from a manual; or (f) using a new, fully electronic kitchen stove when we previously have used one with manual dials only. It appears to lie in these situations themselves that we need to begin asking questions about the instructions (plans). Just within this past week, I was confronted with two such instances (a) when my wife talked about the trouble she had experienced getting the self-cleaning option to work on our new kitchen stove and (b) when my mother tried to bake a bread from a recipe I had sent her. In the first instance, my wife talked about how she spent "hours" trying to find out why the oven told her to "remove the trays" after she had already removed all of them; in the second instance, my mother wrote to me about how a bread that she had baked according to my instructions had not risen enough so that it was far too dense when it came out of the oven.

Actions and Accounts

The relationship between actions and descriptions thereof, that is, accounts of action, can be articulated as the relation between the work of doing something and the way that that which is done becomes accountable. Thus, the expression "doing [following instructions]" consists of a pair of irremediably bound moments of practical action (Garfinkel and Sacks 1986). The first term, "doing," denotes the work involved in acting, and the second term, "[following instructions]" denotes the situational particulars that allow observers to recognize what is being done. The work itself tends to be invisible but usually comes to the foreground when there is trouble. For example, getting in line to purchase a movie ticket, for a bus, or to access a bank teller might be an unproblematical action in our everyday life. Yet it requires work to line up. That there is real lived work required comes to the fore precisely when there is trouble in the line up. For example, consider the situation that there are two or more ticket windows for different movies and it is unclear where the line-ups are for each. Already, the recognition that the "line-ups are unclear" exhibits our knowing about line-ups and how they ought to appear. We may then ask, "Is this the line for *Breakfast at Tiffany's*" or whatever the movie is that we intend to see. In the bank, we might notice trouble when the first person in line does not move even though there is a teller calling for the next person. Our monitoring of the line and our noticing that a teller is calling and that the first person is not moving both are part of the work of queuing, the "doing" of the "doing [queuing for movie tickets]" for which "[queuing for movie tickets]" is a proper gloss (notational particular). Finding the "end of the line," "the first person in line," the "open till," and so on are aspects of this work. Importantly, the text "queuing for movie tickets," always "is meaning differently than the speaker can say in so many words" (p. 172), that is, there is an inherent gap between what is said – i.e., the gloss – and what is done – i.e., the work designated by that gloss. At the end of this chapter, I suggest that this work derives from immanent knowing of the living/lived body, which necessarily is invisible precisely because it derives from the unmediated "I can." It is invisible in the same manner that the canvas of a painting is invisible even though without it there would be no painting: without our living/lived body there would be no knowledge of the world. That this body and its work remain invisible is not a reason to leave it out of our theories – which is precisely what the constructivist metaphor does. In the following subsections I articulate studies concerned with (a) the difference between retrospective accounts of actions, for example, clinical records, and the work that has produced these records and (b) the work of following instructions, which function as prospective accounts.

On the Gap Between Actions and Accounts Thereof

I am certain that every reader has at numerous occasions heard complaints from those facing trouble while following or after having attempted to follow instructions. Perhaps the reader has produced such complaints him/herself. The prevailing view appears to be that there exist good instructions, which, when "followed exactly as stated," will lead to the intended results. That is, these instructions would be extensive and explicit allowing the user to bring about the desired action without problems and to minimize situations in which questions about what is meant by them. Of equal interest is the observation that someone experienced in a domain will find it nearly unproblematic to follow the instructions for a heretofore-unknown course of action. For example, an experienced (hobby) cook generally will find it relatively mundane to follow a new recipe; as I (an avid hobby cook) know from experience, there are frequent occasion where I "know" that something cannot be right or where I "adapt" the new recipe to make it work because it does not make sense in the way it is written. That is, based on my extensive experience with cooking, I read and enact a recipe such that it will work, even if the instructions appear to be otherwise. In fact, the recipe appears to presuppose this knowledge and familiarity – as I have first-hand experience when the instructions for pruning pear trees in a textbook on the topic made sense only when I already knew how to prune (Roth 2004). It was only after (a) taking a course which allowed me to follow an experienced pruner around and (b) pruning in the way he did that I found the written instructions "meaningful." Before that I was frustrated and could not convert the written instructions into a set of situated actions despite the many photos and drawings that the book presented. Once I had pruned a few trees, I found that the book explained pruning beautifully. Again, if I had not been a keen observer, the process of amnesia might have made me forget about the time when the instructions were impenetrable for me.

The dehiscence between situated actions and their accounts, that is, their abstract-sense aspect derives from the way in which we understand unrepeatable living/lived actions only in terms of the implementation of prior plans. But such a plan, the "abstract-sense aspect, when it is not correlated with inescapable actual uniqueness, has the character of a project" (Bakhtin 1993, p. 44). As project, "it is something like a rough draft of a possible actualization or an unsigned document that does not obligate anyone to do anything" (p. 44). That is, to understand actions, we must not come from the outside, the theorized and described action (plan, instruction) but we need to come from the inside of a situation as a whole. Thus, for high-performance (professional) athletes, there is no dehiscence between plans and situated actions: They do what has to be done within the specific, once-occurrent context of the present, living/lived situation as a whole. Their thinking is a "*participative (unindifferent) thinking* [which] is, in fact, the emotional-volitional understanding of Being in its concrete uniqueness on the basis of a non-alibi in Being" (pp. 44–45). Because practitioners know with their bodies, which operate according to very different principles than the thematizing mind, their descriptive accounts of what they have done is no better than the accounts provided by some observer (Bourdieu 1980). What generally deceives scholars about the relation between theory and action is the fact that "the agents voluntarily draw on ambiguous language of the *rule*, of grammar, of moral, and of law to explicate a social practice" (p. 174).

An early description of the relationship between a set of instruction and the courses of actions that followed was provided as the result of a sociological study of the process of coding the records produced in a psychiatric outpatient clinic (Garfinkel 1967). The purpose of the study was to determine the workings of the clinic on the basis of the records it produced. Two sociology graduate students examined nearly 1,600 records for items to complete a coding sheet. The study shows that to accomplish their task the coders not only assumed but also required knowledge of the very organized ways of the clinic's operation. That is, the practices of the outpatient clinic could not be derived from the records, because reading and classifying the records required an understanding of the practices in the clinic. The coders therefore consulted their knowledge/understanding "*regardless of whether or not they had encountered 'ambiguous' folder contents*" (p. 20, original emphasis). To better understand the phenomenon of coding, the researcher assumed that the two coders, rather than being in error, might be considered as following proper procedures in some coding game. In the process of getting at what they got, the coders enacted considerations such as "'et cetera,' 'unless,' 'let it pass,' and 'factum valet' (i.e., an action that is otherwise prohibited by a rule is counted correct once it is done)" (pp. 20–21). These are "ad hoc" (Lat. *ad hoc*, lit. "to this") considerations and their enactment might be referred to as *adhocing* – literally acting *to this* end, for the purpose at hand. Not only did *adhocing* occur when the coders coded, but also when they read the coding instructions in respect to their relevance to actual events in the clinic. *Adhocing* was an invariant part of following instructions in the coding of the clinical records no matter how elaborate and definitely the instructions were written. In fact, *adhocing* is required to fill in what the decontextualized descriptions leave out; *adhocing* therefore is essentially determined by contingencies and contextual particulars.

Now some readers might still be tempted to think about the possibility to write instructions such that no ambiguity remains and, therefore, that there is no place left for ad hoc features. However, we can find some stark advice in the literature:

> To treat instructions as though *ad hoc* features in their use were a nuisance, or to treat their presence as grounds for complaint about the incompleteness of instructions, is very much like complaining that if the walls of a building were only gotten out of the way one could see better what was keeping the roof up. (p. 22)

That is, *adhocing* is required if the user is to grasp the relevance of an instruction to the particulars in any actual situation. Any instruction has to be made to work given the particular contingencies of the setting; and these contingencies cannot be calculated in advance. Even an infinite amount of information cannot describe a situation in its entirety so that there is an inherent gap between situated action and its description. Because of this gap, no law can be considered closed but always has to be treated as subject to revision; no law has direct and unmediated relevance to a particular situation but is subject to the (incalculable) decision of the court. Expertise and practical wisdom refer us to the competencies required for accomplishing the minimum distance between abstract descriptions – plans, recipes, instructions, laws – and actions in setting. This, then, is a first explanation for the pervasive feature of failure and breakdown in the following of instructions, recipes, plans, and laws as intended. Positively viewed we may state that what people do – when producing actions intended to be consistent with some description – is always correct in some "game." To understand what they do especially when they fail to produce the desired or anticipated course of action requires us to understand this "game."

On the Work of Following Instructions

There are a number of studies that can be read in terms of the relation between descriptions of and recipes for action, on the one hand, and what is actually being done (i.e., the living/lived work), on the other hand. It turns out that problems that occur in attempts to follow instructions do not only occur when relatively inexperienced laypersons are involved but also emerge in the case of scientists. Underlying the success of science, there is an the assumption that experiments can be, and in fact are, repeated on a regular basis and thereby replicated anywhere in the world. However, learning a new practice such as getting a commercial polymerase chain reaction (PCR) analysis to work, building a TEA laser according to published building instructions, or copying an engineering practice is not easy even for highly trained scientists and engineers. Thus, for example, one study reported the case of a competent researcher (elsewhere identified as a postdoctoral fellow), who, despite training in another lab, had difficulties getting PCR to work in her own laboratory (Jordan and Lynch 1993). Much of her struggle concerned developing singular applications of PCR for her own laboratory. In her effort to get PCR to work, she relied on formal descriptions, word of mouth, customized primers produced at a university facility outside the lab, standard equipment, and formal recipes. With all of these information sources available, she was largely left to wade through a morass of possibilities for adapting these resources to the task at hand.

The researchers provide further descriptions of how PCR is made to work (Jordan and Lynch 1998). They show the ways in which a particular protocol that instructs novices on how to do a particular PCR analysis is a literal example of a cookbook recipe requiring the same kind of knowledge. The recipe makes reference to equipment, ingredients, temperature, and other features but leaves out what happens at a molecular level parallel to the macroscopically described actions and equipment. The researchers suggest that the instructions are systematically vague, leaving open, for example, what an "appropriate volume of appropriate DNA" refers to. This vagueness is not to be seen as a deficiency but rather as a property of the PCR recipe that makes it applicable to a variety of situations. The instructions remain inherently open and underspecified even in the case of the successful patent application designed to stake out the ownership over the process not only at the present day but also, prospectively, to cover future lines of inventions. This description provides a "formal account of a procedure enacted by a competent practitioner under delimited conditions" (p. 783). But even in the case of such detailed descriptions, according to the patent law written for the ideal type practitioner of ordinary skill in the art, the instructions are not sufficient to determine what happens when someone does what the instructions appear to say. This "problem," as another study suggests, does not disappear even when the instructions are modified or adapted to the user, such as in the case of an intelligent photocopier that provides and modifies assistance as users go along (Suchman 1987). Users still need to act such that they can find the relevance of an instruction in their own situated practical action. That is, rather than *determining* action, (written) instructions tend to be resources for determining after the fact whether what has actually been done fits the descriptive account provided beforehand.

These opening paragraphs already imply that the essential problem lies in finding a course of action such that it minimizes the differences between canonical descriptions of objects and actions and the concrete, situated actions and actual objects at hand. This suggests that there has to be an inherent openness and underdetermination, because recipes are intended to cover many or all situations and people that they instruct and the specifics of the situation in which the instructions are to be enacted. In fact, in order for scientific knowledge to be ubiquitous, laboratory instructions have to be valid for any laboratory in the world. It is precisely because the instructions have to be general and generic, that is, abstracted from the specific situation, that they require work of bringing the generic to bear on the specific. It is precisely because they are generic that any instruction has to leave out the identifying specifics that characterize *any* actual situation.

We can also understand these problems in terms of the difference between *Being* (i.e., life) and *beings*, the manifestations of *Being* (life). These manifestations themselves are not the same as Being but only what literally shows itself clearly – the Latin root *festāre* derives from the Proto-Indo-Germanic root *dhers-*, "to dare" – at and is palpable in hand (Lat. *manus*). The manifestation makes Being stand out, *ek*sist, but always only one-sidedly, each manifestation exhibiting a different dimension in the manner that particle and wave are two manifestations of light. The question of what light *is*, its Being or essence (Lat. *esse*, to be), is a moot one. How light manifests itself depends on the situational particulars (experiment); but these manifestations are only *ek*static exteriorizations of the phenomenon not the phenomenon itself. In the same way, how work manifests itself depends on the situational particulars and these manifestations do not constitute the essence of the work. Therefore, based on their knowing of the living/lived work, phenomenological studies can reveal to us what possible experience there will be (i.e., manifestations); but from the manifestations (i.e., thematized experiences, instructions), we do not know the work.

An occasion for studying the relation between instruction and situated action arose when a sociology graduate student assisted an undergraduate chemistry student paralyzed from the neck down in completing a quantitative analysis laboratory course. The undergraduate student instructed the sociology student – who knew nothing of the chemistry at all – what to do. The study describes

> numerous problems which were encountered in attempts to translate the step-by-step instructions in the lab manual to a course of embodied procedures. Many of these concerned sequential issues. Where the manual specified a linear sequence of discrete steps, the embodied enactment of those steps enabled, indeed necessitated, that several steps be done simultaneously. "Doing more than one thing at a time" entailed students' working-out of arrangements where they could interrupt a coherent sequence of steps while moving to actions in another sequence. Students had to find out for themselves where in the course of any sequence such interruptions were practicable, and where the latency periods provided "time" for alternative activities. While the formal instructions were indispensable for the students, their adequacy was not accomplished by reading a

> page of writing, and idly wondering about its meaning. Instead, the sense of what the instructions *instructed* was found by turning to the lab bench and bodily engaging a complex of equipment to *perform* chemistry's events. (Lynch et al. 1983, p. 212)

That is, this study shows that the written instructions are inherently and unavoidably inadequate because they do not describe the *work* of doing an experiment, because this would have required describing the living/lived engagement of students with the equipment. The sense of the instructions cannot be fully founded by further text, elaborated in wondering about the "meaning" of the text. Rather, the sense of the instructions, that is, the sense of what the instructions *actually* instructed the students, could only be found in the performance of what the instructions described. The important point is that extant rules and procedures are not self-contained or foundational, they do not inherently determine actions even in the case of the most experienced person. Rather, they depend (are contingent) on and are derived from the situation and situated action to which the rules stand in a mutually constitutive relation. It is always only after the fact that the relationship between plans (instructions, recipes) and situated action can be established, which points us to an inherent gap between the two orders. This is not a function of linguistic competence, but that to understand an instruction, the order of action itself has to be understood and brought to bear on the situation. The contingency of knowledge and action are essential resources that make knowledge possible and give actions their sense.

Uncertainties in/of Practical (Material) Action

In the previous section I note that just what an instruction instructs can be found only in the actual living/lived performance itself. Some critics of the kind of research reported in the preceding section may claim that the problem really lies in reading, comprehending, and understanding written instructions. Such critics might be tempted to claim that if the subject knew or interpreted precisely what the instruction meant, there would not be a gap between the agents plans what they wanted to do and what they were actually doing. To find the relation between the prospective plans subjects have and their situated actions that follow from the former, we have to eliminate the point of possible critique. That is, we have to investigate the relation between situated actions and the plans that the agents themselves have had prior to acting. An indication is provided in the following account of how experienced canoeists get ready for running a series of rapids (it might just as well be a world cup level skier preparing for a slalom course, which they can only look at but not run prior to the race).

> The plan might go something like "I'll get as far over to the left as possible, try to make it between those two large rocks, then backferry hard to the right to make it around that next bunch." A great deal of deliberation, discussion, simulation, and reconstruction may go into such a plan. But, however detailed, the plan stops short of the actual business of getting your canoe through the falls. When it really comes down to the details of

> responding to currents and handling a canoe, you effectively abandon the plan and fall back on whatever embodied skills are available to you. The purpose of the plan in this case is not to get your canoe through the rapids, but rather to orient you in such a way that you can obtain the best possible position from which to use those embodied skills on which, in the final analysis, your success depends. (Suchman 1987, p. 52)

In this situation, the plan that the individual subject has is but a form of orienting toward action and to deal with the details as a matter of course by drawing on the "I cans" of my living/lived body. But, the critic might say, in this situation the subject is not in control of the setting – raging rivers are notable for their ever-changing nature – and that this would be different, for example, for expert cooks working in their kitchen, and, perhaps especially so for scientists working in their laboratory doing what they have done for a couple of decades. The study of biologists at work reported on in Chapter 3 gives me an occasion to investigate precisely this phenomenon. As this section shows, even scientists with many years of experience of enacting a particular practice do not know what they are doing until they have the possibility to see and grasp the outcomes of their actions.

Material things are often considered to be objective because of their materially factual permanence. Material actions, though realized in and through the bodies of human subjects, have, among others, an ephemeral quality in addition to leaving more (e.g., modified material objects) or less durable traces (e.g., sound waves) in the material world. The materiality and durability provides actions with their objective nature. But actions are not certain. Even experienced scientists may think they are doing and have done something which eventually turns out not to be that way at all. At this point, experienced practitioners may question their observational actions, doubting what they see, but they normally take their material actions for granted in the sense that they take them as aligned with the goals (intentions) that had brought them forth. If an action has not realized its goal, it is reproduced often with some slight modification (researchers "try again," thereby implying that it will work this time).

There were many instances where the biologists I studied, after several hours of work, began to question what they had really done at some earlier point, or simply noted that something they had done previously must have been inappropriate. They questioned the nature of their actions only after confronting some contradiction later on; otherwise they assumed to have done what they had intended to do. Such contradictions existed in, for example, the absence of photoreceptors on their monitors when they were looking at the prepared microscopic slide or in the non-alignment of visual image and graphical representations of the absorbed light. The following episode exemplifies how a sequence of actions previously taken as having done what it was intended to was rendered problematic while searching for photoreceptors.

After scanning the microscope slide for a while without finding a conical photoreceptor for recording an absorption spectrum, the professor and laboratory owner Carl raises questions an action that has occurred much earlier during the day. His present observation raises doubts about a dissection he had done; that is it questions "dissecting" as the appropriate gloss of a preceding action. At this point he raises a question about the correspondence between the actual work, the "doing," and its gloss "[dissecting a fish eye to extract retinal pieces]."

Fragment 4.1

```
01 C:  now struck out on that one. this dissection really
       bothered me. that dissection really bothered me and i dont
       know why.
02 T:  you want to go back to the old way you did it? where you
       just cut and took the lens out.
03 C:  yea; i am gonna do that.
```

Initially, Carl has conducted the dissection and has extracted eye and retina from it (see Figs. 3.5 and 3.6). He has continued to take retinal pieces, "macerated" them, and mounted them on microscope slides. He then has sought cells to do absorption measurements on them. That is, the dissection initially and throughout much of the day remained unquestioned. Carl has continued the preparation and measurement as if the dissection had been appropriate. He has done such dissections for the past 30 years and there has not been an occasion to doubt that he actually has done what he had intended to do. It has been a dissection that evidently had achieved what it was supposed to achieve. Now, however, after many hours of working with the retina extracted, Carl puts the action into relief – it apparently had not done what it was supposed to do: "This dissection really bothered me" (turn 01). In saying this, he changes the nature of what he thought he had done ("dissecting") to the object. This change occurs at some later point in the day when one of the graphs he produces turns out not to be what he expected it to be. Carl suggests that he does not know why the dissection had bothered him (turn 01). However, about one minute later, after I ask what it is that bothers him, he does provide an after-the-fact explanation, "Well, I couldn't visualize the fold in the eye. And therefore I couldn't open up the slip and get into the retinal . . . have really a limited view of what I am supposed to be going after, and that dictates the success" and "I didn't use a razor blade in the macerating process. And I notice that I don't get those fringes of photoreceptors that . . . that I have grown to like somewhat."

In this instance, Carl has worked all day without that what bothers him now had bothered him. The bothering of something comes to be thematic and becomes relevant when he "struck out on that one" (turn 01). That is, he has worked all day as if the dissection had been as it should – why would he spent (i.e., waste) an entire day working with a sample that is troublesome and therefore useless as data source? Would he, we might ask, have designated the dissection as bothersome if the data collected corresponded to expectations? It is with hindsight that something is made and becomes salient. This hindsight views an earlier episode through all the experiences – i.e., the "retentional adumbrations" – between it and the present instance, bringing these together "in all-inclusive coincidence" (Husserl 1991, p. 98). Initially, therefore, he treated the relationship embodied in the expression "doing [dissecting a fish eye to extract retinal pieces]" as unproblematic. But at the end of the day, Carl recognizes that something about the work ("dissection") has bothered him, which is to say that "dissecting a fish eye to extract retinal pieces" no longer is a proper gloss of what he has actually done. That is, although his plan was to engage in "dissecting a fish eye to extract retinal pieces," what he has actually done differs when considered with hindsight after a day of work looking for appropriate retinal pieces.

Another instance of the gap between the work and its notational particulars occurred on the day following the preceding events. The instant concerned the problematic status of the minimal essential solution ("MEM"). After the research associate Ted had made a new bottle of it, the actions denoted by the (emic, native) gloss "preparing the MEM solution" came to be questioned. Throughout the episode, Carl uses the microscope to look at the slide, whereas Ted and I are focusing on the computer monitor. Carl begins the fragment to question the pH value of the solution that they had had in the previously prepared "stock," which he thought to have been 7.2 (turn 01, Fragment 4.2). Ted responds that the pH had been 7.42, which constituted only a two-hundredths difference with what he currently had produced (turn 02). He goes on to explain that it had been the stock before the previous one that has had the pH of 7.2 (turn 06). Possibly pursuing a line of questioning to find out whether an error might have occurred during the preparation, Carl asks about the total quantity that Ted had prepared (turn 08), which Ted confirms as having been one liter (turn 09). Responding to my question, Carl explains that he is not finding any cone-shaped photoreceptors and seeing rods shrivel up, which means that there is possibly something wrong with the MEM solution. This renders the work glossed by the notational particulars of "drawing a conclusion" uncertain (turn 18).

Fragment 4.2

```
01 C:  was it seven point two last time?
02 T:  no; seven point four two (0.33) and ze same time (1.31)
       well ts ze same only about two hundredths differences of
03     (0.58)
04 C:  yea.
05     (1.38)
06 T:  it was only the time before i had the seven point two.
07     (2.12)
08 C:  and you made up a full liter; right?
09 T:  yea. (0.92) its always one (0.24) of zese (0.24) jars of
       ze (1.81) em=e=em.
10     (0.58)
11 C:  um.
12     (3.45)
13 M:  do you see trouble?
14     (1.20)
15 C:  well; (2.81) ts s what i=m not seeing (0.54) cones and i=m
       seeing rods (0.16) shrivel up pretty quickly; which means
       that the osmolarity of the medium is: (2.77) possibly
       sUSpect.
16 M:  um.
17     (10.95)
18 C:  its hard to draw conclusions.
```

```
19      (8.40)
20 C:   just not seeing the (0.39) number of photoreceptors i
        normally see.
```

In this situation, the scientists collaborate to find a possible cause for the problem of locating the object of interest, cone-shaped photoreceptors. Not finding cones, or taking a long time to interpret, is not problematic in itself, though annoying. Any number of actions could have not done what they were intended to do and thereby have led to the absence of the cone receptors. This is the case when Carl suspects that his dissection is problematic or that the way in which he has prepared the eye prior to taking the retina from it. In the present case, the search for the culprit action takes on a different direction. There is an additional perceptual clue provided by the preparation itself – other entities currently not of interest, the rod-shaped photoreceptors, are identified as "shriveled up" (turn 15), which points him to the "MEM solution" and its preparation rather than some other source. That is, a problem is identified in the work for which "preparing the MEM solution" is the notational particular. At question is whether what Ted has done, the work, corresponds to the particulars – Carl articulates the particulars to be a "[pH of] 7.2," whereas Ted states them to be "7.42," which is only two-one-hundredth of a pH different to what he has previously made. That is, although they had been certain having prepared the MEM solution as intended, it now turns out that there is a problem with it. If redoing a suspect part of the preparation leads to a finding, then this is a strong indication to the scientists that the action identified was the culprit. Redoing here functions to establish causality between "making new MEM" and its intended and anticipated outcome "seeing cones"; this causality clearly establishes material continuity and contravenes uncertainty.

These fragments from my research in the scientific laboratory show that even scientists note at some later stage that what they had not done what they thought to have done earlier. That is, at the very instant of doing a dissection, at the moment of cutting this rather than that way, the scientist does not know what he *actually* is doing. That Carl does not do in the way he intends to do becomes aware to him only subsequently. In the same way, although Ted mixes the MEM solution in some way, it is only on the next day that it becomes apparent that he has not done what he has intended to do. In both instances, the scientists become aware of the discrepancy between what they intended to do and what they actually have done when the outcomes of their action lead to unexpected observations. In the first instance, not finding cones on the microscopic slide is what triggers Carl that something is wrong, which he attributes to his dissection. In the second instance, Carl sees cells that are shriveled up, which points him to a problem with the MEM solution. There is a temporal diastasis, a shift, between the doing and the knowing what one has done. This, in fact, turns out to be a fundamental feature that characterizes the very possibility of knowing.

To summarize: this section shows that there is a gap between plans and situated actions not only when the plans come in the form of instructions from others that the subject turns into actions. The implications are far-reaching, for if the subjects do not know what they do at the very instant that they are acting, then this has to have effects on the way we understand how the subjects make sense of the world (in constructivist speak, how the subjects construct viable knowledge). At the same time, when they think about their actions, elite athletes perform less well than if they rely on their immanent knowing. That is, as the present case studies show, the mismatch between the actual work and its account does not become apparent immediately. In both instances, the problematic nature of the relation between the work ("doing") and its notational particulars ("dissecting a fish eye to extract a retinal piece," "preparing an MEM solution)" arises only at some later point. This also means that any account can and may have to be revised; and such a revision may occur at any point after the work has been completed. Any account therefore is inherently open to questioning and reframing, a fact illustrated in two cases of police conduct – the "beating" of Rodney King, which the court decided was nothing but an appropriate control that the police used in the handling of a recalcitrant citizen and the "unfortunate death" of Robert Dziekanski at the hand of Vancouver police officers using a Taser gun. In both instances, the accounts of what had been done were repeatedly revised even though and precisely because there existed videotaped records of the respective episodes.

On the Origin of the Gap Between Plans and Situated Action

In the two preceding sections, I describe empirical findings all of which point to an essential gap between ("mental") plans and situated actions, theory and praxis, or instructions and corresponding actions. I suggest that the relationship between living/lived work and its prospective and retrospective accounts (descriptions) is given by a formalism developed for the understanding of practical action (Garfinkel and Sacks 1986). In the present instance, the formalism expresses what is involved in following instructions as "doing [following instructions]." This formalism relates the notational particulars, "[following instructions]," with the work ("doing") that produces the embodied actions in such a rationally accountable manner. This work, produced in and by life (*Being*) itself is invisible and therefore in essence inaccessible; it only *manifests itself* in and through the descriptions. Thus, if the lab instructor in one of the examples had approached the paralyzed undergraduate and sociology student, she might have asked what they are doing; and the two students might have responded, "we are following the instructions in the laboratory manual." If asked about which of the instructions, they might have said, "pour 50 mL of 0.1 M hydrochloric acid into the base." Here, the utterance is a gloss of the work by means of which the formal instructions, written in the laboratory manual, come to be turned into a course of action. This gloss also is a *formulation* by means of which the members of the undergraduate laboratory course make available to each other just what is happening at a particular instant.

Over time and with experience and increasing expertise, the distance between plans and situated actions becomes narrower, or, rather, work and its account are compatible more often than not. However, as we see in the case of the experienced scientists, even after more than 30 years of experience in the dissection of retina, what they actually do and what they think they do and have done are of two very different orders. In many instances, what the scientists had done – as recognized after the fact – and what they had wanted to do was compatible. But this was not the case in all instances. What ought to surprise is the fact that scientists themselves did not know about the gap until many hours later. We might understand this gap if we think of life itself as being invisible to acting persons themselves and available only in and through their immanent knowing (Henry 2003). In such immanent knowing, life reveals itself to itself leading to a power that phenomenological philosophers have come to denote as "I can," a form of knowing that remains immanent in any actual performance. The empirically demonstrated invisibility of work (life) cannot be understood in representational (constructivist) approaches; but it is the starting point of the theoretical approach articulated here, according to which knowing begins with immanent knowing that emerges from the auto-affection of the flesh in primary movements.

There are serious consequences for the constructivist metaphor and the associated epistemology; these consequences derive from the essential invisibility of work (life), which reveals itself only in and as of a practical sense. Piaget and the radical constructivists suggest that much of cognitive development occurs as the individual subject abstracts from concrete actions (operations) in the world to arrive at mental actions. To do so, the actions have to be available in some (material) form from which their form can be generalized to become mental action. There have to be (sensorimotor) *schemas* that are generalized to become schemas of actions generally. However, where do these schemas come from? Especially, how can more complex schemas be evolved when all that the subject has available are the less complex schemas – which constitutes the return of the learning paradox as developmental paradox. But this already presupposes some form of cognitive actions, which is precisely what needs to be explained. That is, the constructivist metaphor draws on the explanandum, that which is to be explained, to produce the explanans, that is, the sentences that do the explaining. If actions were accessible to explicit understanding, then we would not expect the chasm between plans and the corresponding situated actions. The present investigation into the nature of actions leads us to an understanding of the passivity involved in bodily knowing and the unavoidability of the gap between planning, acting, and knowing what one has done. The first and third of these forms of events are of one order, whereas the middle form is of another order.

The problem returns when we consider the reverse direction, from mind to actions. Why should there be a problem between a plan that a person "constructs" in his or her mind and the actions that follow. According to the constructivist account, sensorimotor schemas enable actions. These schemas – for Kant as for his constructivist and embodiment successors – are mental and are called upon when the current considerations require a particular action. But if these sensorimotor schemas are representations of the person's knowing-how ("procedural knowledge," as cognitive psychologists would say), why do the actions brought about

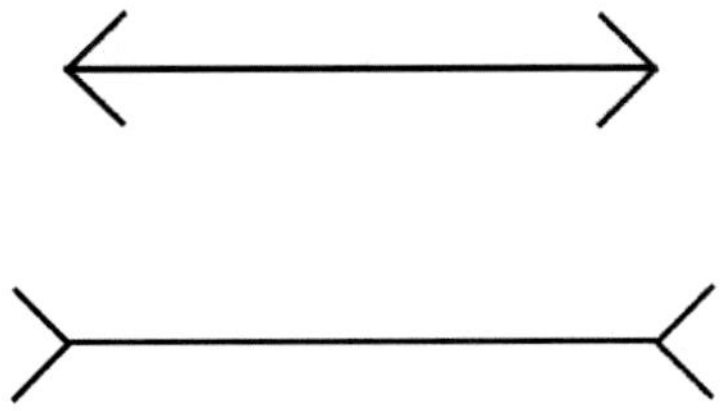

Fig. 4.1 In the Müller-Lyer illusion, we see two lines of different length even though we know and measure the lines to be of equal length.

by these schemas not correspond to the schemas themselves? Why should there be a gap between know-how encoded as sensorimotor schema and the corresponding action? As noted in the beginning of this chapter, there currently are no answers in response to these and similar questions. Any answers that do exist are based on a deficit model, according to which relevant information in the mind somehow is degraded along the efferent channel to the organs and leads to a difference between the actions as constructed in the mind and the actions as enacted.

A very different account is provided by the material phenomenology or phenomenology of life (e.g., Henry 2000) and by the phenomenology of the tactile kinesthetic body (Sheets-Johnstone 2009). In both approaches, movement is primary and constitutes a form of consciousness, a form of *thinking in movement*. Such knowing then exists in the form of kinetic-kinesthetic archetypes or kinetic melodies. This thinking in movement is possible because of an immanent knowing, that is, knowing that is not mediated by representations. It is precisely movement in the present, *Being*, which constitutes and is expression of knowing rather than the *beings*, the entities (signs, representations) that allow us to make something present again. In this sense, the account of work is such an entity, one of the beings, that denote something that in itself is invisible, the real living/lived work. But because the account is only a manifestation of the work, an externality with respect to the lived work, it cannot *determine* it. The work is done based on the immanent knowing of movement and independent of the account. The power to act itself is not the result of an intention but is the result of a pathetic autodonation of life that gives to itself the power to act (and sense). All powers therefore are affective in nature, those effected to act as much as those effected to sense. Life, in a process of auto-affection, gives itself these powers to be what they are and what we experience them to be.

That doing is different from what we want to do becomes quite clear in the case of "optical illusions," to which we are subject even if we know differently. Optical illusions therefore can be understood in terms of the difference between what we see and what we know to be there. Take the well-known Müller-Lyer illusion (Fig. 4.1). We see two lines of different lengths even though we know they are of the same length (to produce the lower line, I made an exact copy of the upper one); and we see them as different even after having measured the lengths of the two lines. A phenomenological investigation reveals what the eyes are doing without our explicit knowing. That is, the eyes are doing work and the result of this work is the apparent illusion. So what is happening here? We can investigate this easily by removing some parts of the arrows – when I had the idea for conducting this experiment, I was on a plane and used a graphics program on my laptop (Roth 2011). This involves removing the upper or lower half of the arrows in each case, which is a way of constituting variations in our perception that allow us to investigate the conditions for having different perceptual experiences. Our eyes move in the case of the upper line as if the line were in the foreground and as if the slanted lines belonging to the arrows were railroad lines moving into the distance. In the lower arrow of Fig. 4.1, the eye moves as if they followed the railroad tracks ending at a railroad tie in the background. Now two equally sized lines will be perceived differently when they are in the foreground rather than in the background: in the former instance, they appear smaller than in the latter. That is, what we see is the result of the immanent knowing of the eyes when there are lines (ties) of apparently equal size but lying at different distances from us. Normally railway ties appear smaller the further away they lie even though they are of the same size. If two ties are the same even though one is further away, the eye will perceive it as being larger than the other. The fact that the illusion is stronger than our knowing of the equality of the line lengths shows that the immanent knowing is much stronger than the explicit and linguistically mediated consciousness that is so highly celebrated in the constructivist account of what it means to know.

The upshot of these investigations is that once we know the actual work done, we always know what the experience will be and how this experience differs with changes in the contextual particulars. From knowing the work that the eyes accomplish, we always can predict how something will appear, whereas from appearances we cannot say anything about the living/lived work. That is, if we know the notational particulars of the work, we do not know what the experience will be. From knowing how the eyes move, we can predict whether a person, looking at the drawing in Fig. 3.4 will observe a hallway or a truncated wedge protruding from the page. But from knowing that a person sees a truncated wedge, we know neither the work of its constitution nor the other forms of experience that the figure enables. This feature is at the heart of an incommensurable difference between any (qualitative or quantitative) formal approach to the study of human action and the ethnomethodological account of practical work (Garfinkel 1996). The two approaches are asymmetrically alternates in that an ethnomethodological (phenomenological) description indexes the work that leads to the patterns and patterned actions that formal analysts report on in their accounts (e.g., the "queue," the "Müller-Lyer illusion").

The approach articulated here allows us to understand a number of phenomena that social scientists and educators have reported in the recent past. For example, we can now understand the difference between theory, accounts and explanations of actions, and praxis, which consists in and of the actual, living/lived work of members to the setting. Because the living/lived work is itself inaccessible and can only be felt (e.g., by means of kinesthesis) rather than known (i.e., its accountable particulars) in its full extent, there inherently is a gap between the two. However, rather than privileging theoretical accounts and explanations, we ought to give primacy to the living/lived work, which is based on immanent rather than ecstatic knowing. The approach also allows us to understand the distinction STEM educators often make between knowledge and its application. The knowing these educators refer to are the notational particulars, abstract descriptions, and words; what "application" requires is a different form of knowing that transcends the procedural knowledge thematized in the cognitive sciences: the knowledge has to be that of the living/lived body, which knows how to move prior to all representations of these movements including those representations that we have come to denote by the term "sensorimotor schema." We also come to understand the distinction between body and mind, the former characterized by its immanent knowing, the latter by ecstatic, sign-mediated knowing. The living/lived body exists in the presence, whereas consciousness requires the tools to make the presence present in the present, to re-present a present different from the one a representation denotes. Linguistic consciousness therefore always implies a dehiscence between the present and the presence of this present.

This description makes apparent the differences with a constructivist account of human knowing and acting, which are concerned with the contents of the mind. Depending on the particular version of constructivism, these contents stand in variable relations to material reality we inhabit. In the case of Piaget, the mental structures represent and are derived from the structures of the natural world – one of the reasons why he has been called a realist. For others, such as von Glasersfeld, the relation between mental structures and the world are subject to the question of viability and fit. In this case, because our bodies are of the same order as the world that the mental structures model, there ought to be a question about how mind relates to the (material) body at all. This is a non-issue in the present approach, which places primacy on immanent knowing and knowing in movement including the immanent knowing on the basis of which our mouths and vocal cords move that produce sounds that we hear as words. Because of its immanence, this knowing is the result of an unmediated auto-affection rather than the result of a construction.

The approach articulated here also allows us to understand the relation between communicative expression and thought. Both Merleau-Ponty (1945) and Vygotsky (1986) understand thought to understand itself *following* rather than preceding its expression. That is, not only the audiences but also speakers themselves discover after the fact the thought in what has been said, gestured, and otherwise marked. Although there is a sense of what we want to say when we speak in everyday situations (generally improvised), we do not select words as if we were pulling a desired plate from the kitchen cupboard or the intended writing utensil from the tray on our desks. Rather, as the words come into our mouths we realize what we are thinking in and through its concretization in material configurations (including sounds, gestures, body positions, body orientations). Our mouths, as our hands, always are ahead of themselves: Any conscious awareness of what we are saying and doing *follows* rather than precedes the living/lived work that produces them. I develop these ideas further in Chapter 7.

Chapter 5

Emergence of Duality

> Circularity speaking-hearing, seeing-being seen, perceive-being perceived (it is it that makes it appear to us as if perception produces itself *in the things themselves*) – *Activity = passivity*. (Merleau-Ponty 1964, p. 312, original emphasis)
>
> The sensations of the soul/psyche become, in the construction of the world, properties of things, or, as we can also say, the subjective qualities form the objective world. (von Uexküll 1973, p. 102)

In the course of our lives, as we become (self-) conscious, there is also a separation of the world we know from who we are. In the biblical story, it was the eating of a forbidden fruit from the tree of knowledge that led the inhabitants of Paradise to recognize themselves as objects among objects; and it led them to recognize their nakedness *as* nakedness. A duality had emerged: The children of Paradise no longer lived in harmonious unity with nature but they looked at this nature as something separate and at themselves as independent subjects. The first lies, too, appear in the biblical story with the emergence of consciousness. That is, with consciousness arose the possibility to tell events in ways that are not consistent with the records. Much later, during the 16th and 17th century, Galileo brought about another separation in which the natural world was theorized independent from the observer – shifting from the qualitative affectations of human observers in interaction with the world and worldly objects beyond the observer. Galileo categorically affirmed – in an act that would be the foundation of modern science – "that the sensible body that we take for the real body . . . is nothing but an illusion and that the real universe is not composed of this kind of body" (Henry 2000, p. 140). With this move, "the body that one can see, touch, feel, hear, which has colors, odors, tactile, sonorous and so on qualities" (p. 140) has been extirpated. The only things left are material, extended bodies without form and figure that are the true objects of knowing. René Descartes, living in the first half of the 17th century, invented the Cartesian coordinate system that allowed this ekstatic world of extended material bodies to be described entirely in geometrical terms. With these developments, a double scission had occurred: the material world to be

W.-M. Roth, *Passibility*, Classics in Science Education 3,
DOI 10.1007/978-94-007-1908-8_5, © Springer Science+Business Media B.V. 2011

described by science comes to be independent of the observer; and, in the same gesture, mind comes to be opposed to and detached from the body.

The purpose of this chapter is not to dispute the dualism of world and (sensing, knowing) subjects but to make thematic that this dualism – and the independence and permanence of objects – is the result of processes not only during a particular period of child development (ontogeny) but also during adult life. Because the process generally is too fast, it tends to escape our attention. Even when we have experiences that would call our normal understanding into question, we tend not to interrogate them to reveal a different form of understanding. This is precisely where the phenomenological method – which is in fact a non-method because it begins with the object-phenomenon as *given* tracing it to its very cultural-historical or ontogenetic origin – allows us to recover fundamental processes underlying our sensory experiences and the way in which knowledge of and about the world comes about.

Phenomenological Orientation

The process by means of which the world we perceive and our perceptual processes come to be correlated generally goes unnoticed. But there are everyday experiences that obliquely point to this dependence that then somehow disappears. For example, we often believe to have heard some noise (e.g., a bit anxious individuals alone in their home or in some other place) and therefore believe that something or someone else is present when subsequent inquiries show that there is nothing. Similarly, children and anxious people walking at night through a dark countryside often see shadows and moving things when in fact all their subsequent inquiries might turn up is nothing. In both examples, a perceptual experience appears to have been ephemeral and does not stabilize in terms of some causing agent or phenomenon in the world. The things heard and seen are then ascribed to the perceptual apparatus. The sounds and shadows might be said to have been figments of our minds. But the world surrounding us similarly depends on our perceptual experience. Somehow, and for many entities, objects and phenomena become independent of this experience. The question is this: "How does this world-out-there-independent-of-our-experience come into existence?"

The constructivist metaphor presupposes the duality of world and cognition. It is from navigating in a world that the individual organism learns and produces mental representations (schema, conceptual frameworks) that in subsequent navigational experiences are employed and submitted to tests of viability. Moreover, the mind is considered to be the seat of knowing and the living/lived body is, at best, a stepping-stone in the construction of this mind. In this chapter, I take the phenomenological approach to question this presupposition. I show that the world *comes to exist* independently but is not so a priori. From Chapter 3, we know that the perceptual processes self-organize as a result of an auto-affection during their interaction with the phenomenal world that gives itself. It is not so that mind reconstructs itself to arrive at conceptual frameworks that allow it to better navigate the world. Rather, the very perceptual processes are organized in and through perceptual experience. In Chapter 4, we find that agents do not intellectually know their actions while acting. The explicit knowledge about what we have done is, inherently and unavoidably, available only after the fact. The world given phenomenally to us, and that we take into account in our actions and decisions, emerges from this double givenness of sensation and knowledge about action. In both of these instances, sensation and action, the outside bears on what it is that we knowably have done and sensually experienced.

In Chapter 3 we already see that a change in intention allows us to go from feeling the effect of contact within ourselves (as in scratching, being burnt) or outside (as in touching, sensing the heat). Although there is a sense impression – which, as we said, is of a pathic order and therefore non-intentional – this impression is carried outside not only of itself but also of the acting individual. When I glide my fingers along the mouse pad, I do feel the surface outside of myself, though right at the surface of the skin of my fingertips. Two key points are to be retained from this observation. First, the world is not outside a priori, which we learn about and navigate by somehow constructing representations of its features. There is a genetic origin to the world experienced as independent of the observer, the world of "independent Galilean objects." Rather, the sense impressions that are produced within our bodies are carried outside to constitute a world independent of our actions and independent of our Selves. In the example of the twin silos presented in Chapter 2, I could see the towers out there, among the trees. That is, whatever sensations were created as light fell on my retinal cones and rods, the twin silos stood out there, were part of a world independent of my gaze at the instant that I became aware of them. There are therefore two planes of events; there are two orders the sensible world and consciousness, which are not only different but also heterogeneous. Pure sense impression and intentionality are external to each other. Both are sources of knowledge, but an important problem is how they work together (von Weizsäcker 1973). Kant has not provided the solution to the problem, precisely because he leaves unaddressed the question of the relation of the body and the mind, so that the solution remains to be developed. The constructivist metaphor has nothing to say about the appearance of the world outside of us from sense impressions and mental constructions that are within us.

In Chapter 4, I introduce a formalism that relates the objects of the natural (social) world as identified in the sciences and the work of producing these objects: "doing [notational particulars, account of object]." Taken together, the work and the account (notational particulars) of its result are known as a *Lebenswelt pair*. The important function of writing lies in the production of notational particulars that make possible the continual objectification of the ideal sense form, which exists virtually in the written document (Husserl 1939). The first part of the pair, the living/lived work, is responsible for producing the objects in the world, which, once they are independent objects, tend to become invisible. The objects are refered to as *independent Galilean* objects because it was Galileo who achieved the separation of nature and observer, perceived objects from perception. This separation led to

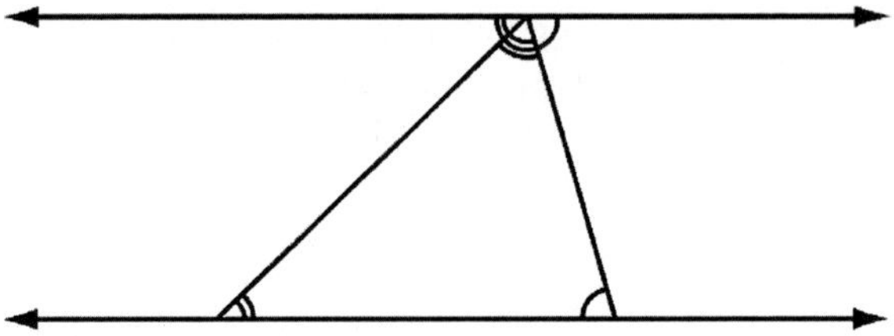

Fig. 5.1 Notational particulars with and from which the proof can be constructed that the sum of the internal angles of a triangle on the Euclidean plane is 180°.

the independence of the natural world from individual (perceptual) experience. For example, the instructions in a laboratory manual, as the recipes in a cookbook, constitute such written documents that allow us to reproduce phenomena and meals at will and anywhere we live. These results of our actions, therefore, are independent of the description and can be reproduced and observed at any time. These objects are independent as soon as it is evident that they can be produced repeatedly; and as soon as this happens, they obtain an "ideal objectivity." Something that occurs only once, something that occurs for a first time *cannot* claim an independent and objective existence for itself because we would not know how to separate signal from noise – we would not know if what we are looking at is signal or noise. We would not know whether what we perceive is a fact or an artifact of our perception.

The Lebenswelt pair "doing [notational particulars]" is easily available in the following example from geometry. The drawing in Fig. 5.1 constitutes the notational particulars from which the proof for the sum of the internal angles of a triangle on the Euclidean plane can be constructed: Because the angles marked in the same way are equal, and because the three angles within the triangle, as shown at the upper parallel, together form a half circle, the proof is provided that the sum is 180° (given that the total angle of a circle is 360°). But the diagram in itself does not constitute the whole proof. The proof is the result of the living/lived work *together with* the notational particulars. The objectivity of the geometric fact – i.e., the sum of the internal angles of a triangle is 180° – arises from the reproducibility of the proof across historical time. The geometrical fact only appears to be independent of the work of proof and the living/lived body of the proofer.

Emergence of the World in Constructivism

Piaget described duality as a phenomenon that constitutes part of children's development. Objects, which initially are dependent on the perceptual process, subsequently become independent thereof. For example, when an object disappears behind a screen, it ceases to exist for children of a certain age. Most animals stop chasing their prey if it is out of sight or if they lost the scent – though certain animals have been shown to exhibit levels of object permanence that generally do not reach those of human capacities. The specifics of this phenomenon are not of importance here other than to make clear that object permanence is not an inherent but a learned property of cognition. For Piaget (1970), object permanence emerges during the first year of a child's life. Thus, at around 7–8 months, the child will stop reaching for an interesting object if a screen is placed obstructing visual access to the object. The child will act as if the object "not only has disappeared but also is no longer accessible" (p. 43). The child will make no attempt to remove the screen and search for the object behind it, whereas it will do so near the end of its first year. At that time, the child also will follow the object through a series of operations, such as when it is placed in a box itself placed behind a screen. The child will remove the screen and look into the box to remove the object. For Piaget, this "notion of permanence of an object, then, is the sensory-motor equivalence of the notions of conservation that develop later at the operational level" (p. 44).

In "Cognition, Construction of Knowledge, and Teaching" (von Glasersfeld 1989a) we find examples of how constructivists understand the construction of the world. The example used is that of a child constructing the "rattle scheme." This construction, like the construction of all schemes, consists in and of the following three parts:

> (1) Recognition of a certain situation (e.g., the presence of a graspable item with a rounded shape at one end);
> (2) association of a specific activity with that kind of item (e.g., picking it up and shaking it);
> (3) expectation of a certain result (e.g., the rewarding noise). (p. 127)

Here, the child is said to recognize a graspable item. The item in itself is taken as available in the child's sensory experience. It is present there together with (some of) its properties – here the rounded shape at one end and the rattling that occurs upon shaking.

In constructivist theory, the scheme may be extended or modified when the child comes to sit at a table where it finds some graspable item rounded at its end.

> We call that item a spoon and may say that the infant is *assimilating* it to its rattling scheme; but from the infant's perspective at that point, the item *is* a rattle, because what the infant perceives of it is not what an adult would consider the characteristics of a spoon but just those aspects that fit the rattling scheme. (p. 127)

When a shaking of the spoon does not result in the "rewarding noise," the experience generates a perturbation, which is one of the conditions for change to occur. For example, the child might modify its schema so that there are rattling and non-rattling items in its mental framework. This modification is the accommodation of an existing scheme to new experiences.

One unresolved question is that of intentionality. Why does a child engage in the construction of cognitions? There is no reason within cognition that would allow us to understand why it develops (constructs) itself. That is, intentionality is presupposed rather than explained in constructivist theory. If it is intention that carries sense impressions outside themselves, then how does intentionality reveal itself to itself? Another question constructivist theory does not answer is this: If knowledge construction is intentional, then how is intentionality itself constructed? Postulating another intentionality involved in the construction of intentionality only would lead us to an infinite regress. Saying that it is innate does not take us any further, because it requires us to specify at which point in the evolution of *Homo sapiens sapiens* this feature emerged and came to be innate because it provided an advantage in natural selection mechanisms.

To understand how the world comes to exist independent of the subject, we need to know more about the observer. Thus, to reach object permanence the child has to be able to establish recurrences in the flow of its experience. This in turn requires (a) the ability to remember and retrieve experiences and the ability to judge and compare similarities and differences and (b) an inclination to have preferences that select between different experiences that have some elementary values. How these abilities and competencies emerge, or rather, what their requirements are can be found in my phenomenological analysis of the twin silos. This analysis shows that object permanence is associated with its standing out from the flow of my experience. Precisely at the instant when I become aware of the twin silos *as* twin silos, they also come to exist as independent object. Their *ek*sistence (existence) is tied to *re*presentation. It is at that point that the possibility emerges for the twin silos to have existed in the past and for them to exist in the future. That is, the independence of the object comes with their presence in retention, which itself is possible only when I can make them present again in another present, when I can represent them. In absorbed coping, on the other hand, such as when we drive the road to work while talking to someone else in the passenger seat we generally will not notice things as things. The twin silos do not stand out and therefore cannot be remembered – which is why they never appear in my post-trip notes taken on the first six days. That is, when we are subsequently asked about something that we should have seen along our way, we will not be able to do so, even though it might be possible to show that the light from these things must have fallen onto our retinas. That is, although we might react to particular conditions of the traffic and road, we may not be aware of these and we are therefore unable to provide an account of them when asked later about them. This is why I was unable to account for the twin silos, which, until the seventh trip when I saw them (consciously) for the first time, did not appear in my records. As a first move toward a better understanding of the apparent independence of objects from the experiencing person, let us take a look at the natural sciences, which are concerned with independent Galilean objects.

Independent Galilean Objects in the Natural Sciences

Independent Galilean objects are the result of work that not only establishes the existence of the object but also establishes their independence from the situational particulars. The discovery of such an object is a technical phenomenon that is not independent of the task of making the work itself accountable (Garfinkel et al. 1981). Using materials from the discovery of the optically discovered pulsar, these authors show how the pulsar emerged from the embodied and situated practices of a pair of astronomers, which led to the local production of order, here, the pulsar. "The possibility of [the scientists'] discovery and achievement is observable to them and is a discussable for them in this: *their night's work poses for them their tasks of its own astronomical accountability*" (p. 140). That is, the pulsar gained its independent existence precisely because the astronomical practice could be exhibited, was observable, and could be talked about in and as of the discovery work. And it is precisely when they can speak of the optical pulsar *again* that it becomes the independent object that it is. "This object, their competent-practices-evidently, was achieved *in*, it consisted *of*, and it was extracted *from* the night's work, *and in situ* it was rendered as the properties of the pulsar's Galilean independence of their practices" (p. 141). In part, the independence of the object is the result of variations in experimental conditions that changed the recordings from the telescope. Because these changes were directly associated with empirical variations, the object could be separated from the observation itself.

The credibility of scientific research requires more than the production of the phenomenon; scientists must be able to eliminate other possible and plausible causes. From the perspective of the scientists, uncertainty is decreased or removed if they can establish a causal relation between practical actions and outcomes. In Chapter 4, I show how scientists do not know with certainty that they are doing what they have set out (intended, planned) to do. This, then, introduces uncertainty into the practice of the kind that Garfinkel and colleagues have not addressed but rather referred to as being hidden and familiarly effective.

To begin responding to this problematic consider a person walking at night through a forest. There are apparent noises and shadows, which children and native peoples often take to be signs of real things, spirits, ghosts, animals, or forces. But the existence of these entities depends on the perceiving individual. They only exist independently if the perceptions can be reproduced independently by other observers and independently of the particular conditions. Adults here are in the same situation as the children, for whom the sound [kʌp] (which English speakers hear as the word "cup") and the material object that we perceive as "" are one and the same. At some later point, the sound-word and the "" come to be independent of each other. The separation between one portion of the material continuum, which we come to regard as the object, and another portion of the material continuum, which we come to regard as the sign, also has to be accomplished in the natural sciences. Thus, I note in Chapter 3 that the nature of the thing that we are looking at is a function of the ability to re/produce it repeatedly. For example, my research showed that what the biologists – in the laboratory researching the absorption of light in fish retina – are looking at under the microscope and the graph representing the amount of absorbed light is mutually constitutive. That is, they know they are looking at a "blue cone" when there is a particular curve of specific width and shape at a particular location in the graph representing the optical spectrum.

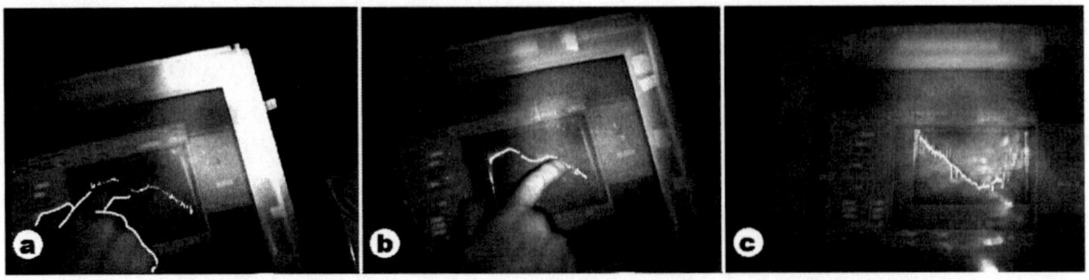

Fig. 5.2 In this situation, there are two objects. **a**. The sought for blue line above the imaginary "baseline" pointed to by Ted. **b**. The "something" "in the red" is unknown but yielding a curve. In fact, it is known retrospectively through the curve rather than through inspection of the slide image. **c**. After bleaching, the "very good blue curve" had disappeared, providing evidence for the nature of the object as a "blue cone."

But they know that this spectrum refers to a real object when they can be certain that the object under the microscope is a (blue) cone. That is, the nature of the thing under the microscope and the nature of the material perceptual process of the experiment (the curve) are intertwined. It is only when they can be separated that the curve represents a fact (object) rather than an artifact of the experimenters' practices. This separation is precisely the production of the independence of the Galilean object from the acts of perception and from the living/lived body that makes perception possible in the first place. In the following I return to the biology laboratory conducting experiments that focus on the absorption of light in retinal cells of salmonid fishes.

In this laboratory, there exist a number of methods for ascertaining that a peak, for example, is caused by absorption in a retinal cell rather than by some other entity. This is evident in the procedure of "bleaching." Sometimes the biologists notice that there is something on the slide not because they see it, but because their graph exhibits a feature that can be read as the signal of "something," as an unknown object or matter that is on the slide but has not been noticed by inspecting the visual image of the slide. In the following fragment, Ted reads a graph as it shows up on the monitor following a measurement.

Fragment 5.1

```
01   T:   this is probably the baseline here ((points as in Fig.
          5.2a)). Here ((points as in Fig. 5.2b)) in the red is
          something happening that we have to get to. If you take
          the baseline, this looks like a very good blue curve.
          ((Moves several times along its outline, as in Fig.
          5.2a)) this seems to be the best way of reading this
          graph.
```

To make certain the peak they see corresponds to a natural, that is, independent Galilean object visually available on the computer monitor (e.g., as in Fig. 5.2a), the biologists have to show that the signal disappears when the object disappears or no longer absorbs light in the way it normally does. One way of doing this with the photoreceptors studies in this laboratory is to "bleach" them. The work of bleaching involves shining light for one or two minutes onto the photoreceptor, a process that is said to transform the molecules that absorb the photons. Disconnected from the living tissue, the photoreceptor under the microscope no longer is regenerated so that during a subsequent measurement, no absorption of light should be recorded. Sometimes, however, the expected disappearance does not occur and the scientists repeat the procedure, evidence for which has come from not seeing disappearance of absorption peak. In the present instance, bleaching removes the UV peak, but not the one to the right (Fig. 5.2c). Through the positive results of bleaching, the scientists ascertain both the object under the microscope and the graph, contributing to their sense that the data and object are "real." The graph produced is not actually interesting in itself: the disappearance of the previous curve is the real object of interest – the observable *direct causation* between bleaching and perceiving the absence of a peak makes the effect a demonstrably real fact – an independent Galilean object – rather than a possible artifact. If the peak visible to the left in Fig. 5.2a, b has not disappeared (Fig. 5.2c), "not bleaching" would have been the proper gloss, as the intended action does not have an observable outcome and the peak would be considered an artifact. The entity on the slide would become again an unspecified something – just as indicated by the peak in the red (right-hand side of graph). It is through the disappearance that the biologists know not only what they have done but also what they have seen.

The lack of an observable outcome would have turned the peak into a non-peak, or perhaps a peak corresponding to something else – just as the "something" that appears to be "happening in the red," to which Ted suggests they have to get to for better understanding its origin (see Chapter 3). The missing outcome would have created uncertainty about what is on the microscope slide, or rather, about the outcome of the work involved in "seeing 'something'." When the scientists take the difference between successive absorption curves before and after bleaching, only the curve due to the absorption should remain, all other known and unknown contributions having been subtracted out. "Bleaching," too, constitutes a material action on the material at hand: if bleaching has an effect, that is, if it brings about a transformation of the material as shown by the disappearance of a peak, then the material continuity has been reified. Again, material actions and the observability of objects they create presuppose one another.

In this instance, the experimental variations that lead to the different graphs – variations that can be repeated over and over again with other, similar entities – constitute the independence of the object and therefore their status as Galilean objects. This move makes the objects independent of the work, actions and perceptions, and therefore constitutes the objectivity of the object. But initially, the graphs that serve as the account and notational particulars in the proof of the independent Galilean object (here the retinal cell in its absorption of light) do not exist independently of the work. The work (proof) and the account (proof account) themselves are not separated and therefore, the account cannot serve as an index to an independent object. It is only when the graphs as accounts are independent of the living/lived work – i.e., independent of the sensible and sensing body of the researcher – that the natural object also comes to exist independently. One and the same step, therefore, brings about both the independence of the material world from the knower and the independence of knowledge (mind) and the human living/lived body that produces the work.

In this situation, the result of the biologists' work is the independence of the material world known from the knowing subject. This radically changes the perspective on the nature of objects, which, in the constructivist approach, is considered to be independent of the knower. Piaget and the (radical, social) constructivists assume the independence of the material world – what they differ on is about what we can know of this world. For Piaget, knowledge about the world and the structures of this world are homomorphic, whereas for von Glasersfeld, there is nothing other that matters than the degree of viability of the mental structures when a person is interacting with the material and social world. For social constructivists, the mental or discursive structures are the result of collective rather than of individual constructions. In all of these approaches, the independence of the world and knowledge is presupposed, and, in the same move, also the independence of the mind from the body.

Independent Galilean Objects in Personal Experience

It's nothing until I call it. (Attributed to baseball umpire Bill Klem)

In a curious manner, Bill Klem, at best a lay philosopher, articulated a fundamental fact of human existence. Only when we can call something by its name does something (object, fact) come into existence for all of us. We can see this occurring every day, when co-workers, doctors, and psychologists denigrate an experience a person reports calling it the result of an imagination and therefore prohibit the experience to exist as fact. Individuals struck with chronic fatigue syndrome often encounter such situations, when their real and experienced pains are attributed to imaginations – leading doctors to refrain from treatment and employers to the laying off of a person.[1] Having suffered more than nine years from a chronic fatigue, I have my own share of experiences with the everyday experience of objects (facts) the factual nature of which as independent from my living/lived body is difficult to establish. So far, the doctors have refused to call the chronic fatigue by name (i.e., chronic fatigue syndrome); and as soon as they were to call it such, they would turn my condition into an objective fact allowing me to access resources or to be excused from work for long-term illness. However, as long as the doctors do not call it something, it does not exist or at best as something that potentially exists only in my mind and not as something real, a demonstrably objective fact. In Part D of this book, I focus precisely on experiences such as illness,

[1] In my wife's and my own workplaces, we have seen this happening to co-workers. The general attitude was that the persons involved were "faking" an illness, pain, or discomfort for the purpose of receiving sick leave or that the illness, pain, or discomfort was merely the result of a psychosomatic disturbance. Several individuals we know of have been forced into retirement or have been "offered" "packages" so that they would take early retirement.

their relationship to knowing and knowledge, and the problematic issues that they raise for the constructivist metaphor generally.

In the introduction to this chapter, I make reference to sensory impressions that turn out not to be attributable to things in the world. In this case, we may attribute the impression to our imagination: "I thought I saw an animal, but perhaps it was only there in my imagination." That is, in this case, the phenomenon is absorbed into and attributed to the perceptual process. On the other hand, once we see something accountably, such as the twin silos or the church in my experiences (Chapter 2), then the perceptual processes are externalized and adsorbed into the object. These Galilean objects now appear to exist independent of our perceptions and experiences, out there to discoverable for everyone. At an individual level, the genesis of independent Galilean objects is much more difficult to observe because they are so fast that they are generally hidden from sight and from our attention. But the inquiry in Chapter 2 has revealed to me the process by means of which the twin silos have become an independent object. The same investigation also shows that the absence of these objects from our lifeworld just prior to their appearance comes to withdraw itself completely from our attention. This, in fact, is what guarantees the world its firmness and durability. Object *permanence* depends on this process and is impossible without it. The common experience of thinking to see, hear, feel, smell, or taste something only to discover that there is nothing shows us that repeatability is part of the phenomenon of the stability of the world; and repeatability is tied to *re*presentation. As long as we touch an unknown entity in a black bag, it does not exist as a thing. We cannot tell someone else that I feel this or that – just my pain or my illness does not exist until the doctor calls it such.[2] Sight is so pervasive that it, more than the other senses, hides what is going on rather than revealing it.

More difficult for most people are distinctions, cognition, and recognition of odor and taste. When I first began to drink single malt Scotch whiskeys, I wanted to know more about the differences.[3] I bought a book that contained notes on taste and smell associated with different whiskeys. Whereas different individuals have differences in preference, if there is "a hint of vanilla" in the smell of a whiskey, this should be objectively available to others who know the smell of vanilla; if there is a taste of toffee, then this should also be tastable by others who know toffee apart from other tastes (repeatability of the sense impression that occurs when we eat toffee). Now it is interesting to engage in blind tasting experiences, as we have no other information available than our nose and taste buds. In this situation, the particular tastes – which became tastes precisely when we can describe them – emerge from more-or-less extended periods of tasting. The whiskey is indistinct

[2] The strangest experience I have ever had was that of sitting with pain in my arm and leg joints in the office of a rheumatologist, who, based on the X-ray images he had ordered, declared that I had nothing. That is, my pain was not real because I had nothing based on what the doctor could see on the X-ray images.

[3] I could have told the following also using my experience of becoming an expert at cheeses or at olive oils.

before the first emergence of a particular taste sensation – or it may remain indistinct (and therefore be a single malt whiskey of no interest to connoisseurs). Being indistinct is like the first experience with the perceptual pattern in Chapter 3 before the first little dog emerges from the apparently random patterned black-and-white ground. A particular taste as an independent property only exists when others also can taste it and when this particular taste can be repeated ad infinitum wherever we are in the world. It is nothing until it can be called. And it is precisely at that (infinitely short) moment that the property comes into objective existence and independent of our sense experience, though it definitely emerged in and from this experience.

What precisely is happening at that instant? Clearly it is not just a construction on my part. I do not act as a comparator in the way a computer might go through a library of known items and compare the unknown whiskey with the tastes available in the taste bank. I take the whiskey into my mouth and I swirl it, draw some air through it in case the swirling in the glass has been insufficient to aerate it. And then, all of a sudden it strikes me: I have found the sweetness of Oloroso sherry or a different sweetness that resembles that of Malaga (wine). Such taste notes are produced by the Oloroso or Malaga casks in which the whiskey is left to age for a few years. If there is mainly vanilla, I know the whiskey has spent all of its maturing time in oak casks from the US.[4] As soon as the particular taste emerges from an indistinct sensory ground – i.e., as soon as I can name it – it also comes to exist independently. But naming it means bringing a language that is not mine (see Part C) to my sensory experience that this subjective experience becomes objective. It is there to be tasted repeatedly by myself or repeated by other individuals. That is, it becomes independent when I can bring the other, the outside, to my sensory experience, allowing the objectivity of the whiskey taste come into existence as an independent Galilean feature of the world.

> The "objectification" that takes place in each of our senses when the vision is carried into the distance, when the touch touches an object, the smell smells the odor of a flower, the hearing perceives a sound that resounds in the world, this objectification signifies each time the coming outside of an outside, the excess over an intentionality that orients itself toward its horizons of transcendence. (Henry 2000, p. 203)

That is, when language, a form of exteriority and *eksistence*, is brought to something private and contingent, this something in turn becomes something exterior, public and repeatable. It is the coming outside (object, objective fact) of an outside (language) that gives objects their Galilean independence.

The sense experience is brought into existence when the liquid enters my mouth, that is, when something from the outside comes into contact with my taste sense (or smell). But this contact initially only gives an indistinct sensation, especially for the newcomer to this game of tasting. The moment of objectification occurs precisely when another feature of the outside world is brought to the sensation produced in the contact between my taste buds and the liquid. It is not that I

---

[4] The U.S. law requires sour mash whiskey makers to use new casks. They sell the once-used casks to makers of single malt whiskey in Scotland.

have to know that it is a liquid or a single malt whiskey. The game would also work when I simply open the mouth, blind folded, to accept anything that the other players were putting into my mouth – the familiar game of having someone, nose and eye closed, bite into an onion without that most persons could make a distinction with an apple. It is precisely when the twin silos came to stand against everything else as figures against the ground, with a *re*cognition of the entities *as* silos in a multitude of two, that they came to exist independently of myself.

For human beings to knowledgeably navigate the world, they do not have to be conscious of it. For example, we often tend to forget the surrounding world when we are engaged in an interesting conversation. We forget about the world even when, as noted above, we are driving somewhere only to note upon arrival that we cannot remember anything about the trip that brought us to where we are.[5] But given that we have not had an accident, had not run into the ditch, had not hit some other feature off the road, and had not bumped into another car shows that we were in the world without consciously attending to it and without consciously noting any event or object that happened on the way.

There are other experiences of this kind that (at least some) readers may have had in their own lifetimes. One of them is an experience commonly referred to as *flow*. During flow experiences, we are not detached from the world and the world does not exist independently of the subject. The individual experiences intense feelings of enjoyment and creativity (Csikszentmihalyi 1988). But the dualism of the experiencing subject and the world in which it dwells does not exist. It is in such experiences, like that of "paradisal enjoyment, without time or worries, [that] the distinction between activity and passivity is confounded in pleasure/grace" (Levinas 1971, pp. 137–138). Such moments, therefore, consist of pure presence in which the present is not made present. There are no representations necessary.

Flow experiences come in all sorts of kind. We tend to describe the instances of interest here in terms of "being absorbed." When we are absorbed in a conversation, we do not represent the surrounding world as surrounding world. We do act, sit, drive, walk, or ride the bicycle (as I know from personal experience) without attending to worldly objects qua independent Galilean objects by making them present again in thematizing consciousness. In our experience, this world, therefore, does not exist independently. It does not exist as something out there independent of our perception. Perhaps even more astonishingly, when we are absorbed, we tend to loose our sense of objective (clock) time. I tend to do garden work in this manner. I begin to weed my vegetable beds by hand only to find myself, sometimes several hours later, conscious of the objective time that has passed

[5] Cognitive scientists and psychologists have their own, generally negative ways of talking about our "lack of attention," such as those described in association with the phenomenon of the "invisible gorilla." The name of this phenomenon derives from an experiment in which cognitive psychologists ask subjects to count the number of times a basketball is passed back and forth between members of two distinct teams. The experiment shows that the subjects generally are so absorbed by the task that they do not see that a "gorilla" walks through the scene. "Illusion of attention" and "fallibility of our intuitions" are some of the deficit ways in which psychologists describe the outcomes of such experiments.

in the process. I do not remember time or place, and I have to retrace how much I have done by scanning the vegetable garden. I have had many similar experiences during long-distance hikes lasting over a week, where there are periods of hours that have disappeared from conscious memory. I have been so absorbed that objective notions of time and space have been absent for a while. During meditation, I am absorbed so that there is no consciousness of the world around or of some Self that exists independent of the experience. Even experience as such is not present in and to consciousness. Time does not stand out, does not exist outside of my existence, does not *eksist*. The existence of time as an objective aspect of experience co-implies eksistence, retention, and representation.

Différance

The absence of what happened from memory tells us that what definitely has been a present cannot be recalled and therefore is no longer available to be made present. We do not have available the representations required to make the world and its objects present again in a different present than when they originally appeared. This, too, provides us with a clue about the origin of the separation of the world – objectively given because recurrent in our own experiences and those of other people – and our presence. When we cannot recall what has happened around us while we have been absorbed then that which has been present cannot be made present again. The present precisely existed in not being (consciously) present to the Self. To be conscious of something (as something) requires representation, that is, some form of sign that allows me to make present again what has already passed. But in standing for something that is not itself present, the representation tells us about an absence, the absence of the thing. This absence lies at the core of the painting by René Magritte featuring the drawing of a smoking pipe under which is noted "Ceci n'est pas une pipe" (This is not a pipe). Thus, the sign of the smoking pipe also is a sign for its absence, which has led the philosopher Hegel to suggest that the body of the sign also is a tomb (Derrida 1974). This is so, too, when I saw the twin silos *as* twin silos. It is in and with seeing twin silos that the originary experience – the one I had prior to perceiving them the first time or while subsequently riding my bicycle past this spot being absorbed (in cycling, thinking about cognition) – is no longer present and available to me.

Many of my own "discoveries" as a social scientist and as a just plain person in the everyday world are associated with experiences of flow. In each case, what I have learned was not intended, it has emerged in moments where I am not conscious of the world, and where I have had some "sudden insight." If I had known before what I knew after the moment of insight, there would have been neither the experience of insight nor an interest in pursuing the discovery work. That is, I am "seeing" something that I have not seen before, and because I have not seen it or know of its existence, I could not foresee it, anticipate and intend it. That is, this insight could not have been intended. It is precisely because it is suddenly *given to me*, as if from nowhere, that makes an insight different from other forms of learning.

The constructivist metaphor does not provide descriptions and explanations for such experiences.[6] But there is precedence of possible forms of explanations in the philosophical literature concerned with the difference between Being (e.g., during absorption) and beings (the objective, independent Galilean objects). In Chapter 2, I already refer to Plato's notion of *khorismós*, which constitutes the difference between beings and Being. It is the space, *khôra*, that gives rise to the difference – which ~~*is*~~ this difference. We can understand the source of this difference while analyzing the experience of a sound. Each actual phase of the sound slides into an immediate past (retention) and this phase in turn slides further into the past more and more removed (Husserl 1980). And, as we see in Chapter 2, the result of this sliding is the coming-out, the *ekstasis*, of the original difference. But this difference is only possible because there already is a difference in itself, which has to be an unnoticeable difference, a *différance*.

The notion *différance* – which is pronounced precisely in the same way as the word *différence* (difference) – is a neologism designed to bring to our awareness the fact that all thought, as a living process, is based on difference as such. A point on a page is not just a point but is a point (figure) against the white page (ground). When we talk about the point, we always imply the page, because without the page, there is no point (Henry 2005). Any representation, any being different from but denoting Being, requires an inner difference, a dehiscence that produces the non-self-identity of a thing with itself. It is precisely because light is non-self-identical that it can manifest itself as wave or particle; it is precisely because value is non-self-identical that it can manifest itself as use-value or as exchange-value. *Différance* ~~is~~ – Derrida (1968) strikes through the "is" – marks such a difference though, in listening, it does not expose itself as such.

The notion of *différance* is interesting and important, because it lies at the origin of the independent Galilean world filled with objects independent of our thematizing consciousness. This consciousness, to be conscious of something, requires this something as an object, something that can be made present over and over again. Consciousness itself is tied to *différance*. This difference has to exist in such a way that it does not draw attention to itself (Heidegger 1954) – much like the point must not draw attention to the paper that allows it to exist. Flow experiences, as we see above, are characterized by the absence of a segregation of experience and world. The world regains its dominance when we return from the flow experience and become conscious of time and space in their regular ekstatic mode. In flow, knowing is immanent rather than theoretical and ekstatic. Mastery

---

[6] Actually, some constructivist educators have told me that insight is due to cognitive processes that run in the background, that is, constructive processes that occur in the unconscious parts of our mind but that are no different than the conscious processes that are normally referred to as construction. A "Freudian slip" (parapraxis) is an expression of this kind of thinking, for in this phenomenon, an unconscious wish, conflict or train of thought is said to interfere with current, actual praxis. This way of theorizing thinking, however, is typical of metaphysics and therefore lacks scientific rigor (e.g., Derrida 1996b; Henry 1985).

of the ekstatic form of knowing does not correspond to mastery of Being. Most importantly, the original movements that bring about any kind of immanent work are prerequisite to any ekstatic form of knowing, because without the non-self-identity that is the result of auto-affection and immemorial memory, no dehiscence between movement and a representation thereof would be possible.

It is precisely this *différance*, *khôra*, which gives rise to ekstatic time and, as a consequence, to ekstatic space. It is precisely in the becoming present of a past present that experience itself comes to be temporalized. Presence is replaced by a past present made present again in and through representation. This temporality *is not*, that is, it is not a being among beings, but it temporalizes itself (Heidegger 1977). That is, temporalization and standing out of the world *as* world (*ekstasis*) arrive together. It is precisely the consciousness of real time and consciousness of the world as standing over and against us that we experience as arriving together when we emerge from a flow experience. Paraphrasing Heidegger's very difficult German, we might say that the coming of the outside of itself of the outside is precisely the process of the temporalization that also externalizes the world, which it thereby makes stand out. The world "is" not "there," but becomes (temporalizes) itself in temporality. It "is" (Ger. *sein*) "there" (Ger. *Da*) simultaneously with temporalization and therefore exists only together with Being.

Part C

Otherness

> Auto-affection is a universal structure of experience. All living has the power of auto-affection. And only a being capable of symbolizing, that is, to auto-affect, can let itself be affected by the other in general. Auto-affection is the condition of experience in general. (Derrida 1967, p. 236)

Any form of experience has auto-affection as its condition. Symbolizing, use of language, thereby is the condition of experience in general, whether it is the experience of the world, the experience of others, or the experience of self. There is no experience without the auto-affection made possible by language. And, because language is always the language of the other, talk of experience also, always already is the possibility of experiences that we do share with others. Auto-affection, which we already encounter in Chapter 3 as the possibility of the touching-touched, also provides the subject with its powers. The auto-affection made possible with language allows repetition, and this repetition idealizes itself in language. As a result, this "idealization here is the movement by means of which the sensible exteriority, the one that affects me and that serves as the signifier, submits itself to my power of repetition" (Derrida 1967, p. 236). This power of repetition, "from then on appears as my spontaneity and . . . less and less escapes me" (p. 236).

In a constructivist ideology, knowledge is conceived of as something personal, subjective, the result of the cogitations of a mind concerned with self-organizing itself in such a way that it produces viable knowledge. In this Part C of the book I articulate the essential *(social) otherness* of knowledge and experience, and therefore, the essentially *passivity* that comes with agency in Being, knowing, and learning. This otherness derives from the phenomenon of non-self-identity and its origin lies in *différance*. Thus, for example, at the very instant that I refer to my innermost being using the pronoun "I," I make use of the language that is not mine, which has come to me from the other and therefore constitutes otherness itself. At the very instant that "I" tell "my" life – produce an auto/biography – "I" make use of language and its syntactic and semantic resources. I therefore tell a story that is already enabled and possible; and it is therefore not mine at all. To this otherness, therefore, I am in a relation of passivity more radical than any passive relation I might take with respect to other people in some social situation. *Différance* implies or rather explains the origin of otherness, and this otherness also means radical passivity. Who we are and who we can become, our ideas and identities, all are prefigured in and by the resources of culture, a culture whose possibilities we concretize in and through our lives, and a culture, therefore, which we reproduce and transform in Being.

Chapter 6: Talking Conceptions without Conceptions. In the constructivist literature on student knowing in science and mathematics, what students say and do in classrooms and interviews is said to be the result of their conceptions. These conceptions are thought to be "mental models" and "internal representations" that are the outcome of constructive work. But does this have to be the case? Do students have to *have* conceptions when they talk about certain topics? Or are these conceptions a property of language that the students draw upon in their explications? Moreover, if we were to accept that there are mental models (mental structures) driving what someone says at some point, how would talk look like when a learner is between two mental models? In this chapter, I draw on interview and classroom data to show that people do actively participate in conversations without ever having thought or talked about a topic before, which means, that they *could not have* previously constructed a mental model. Rather, by participating in conversations, children learn to speak – without having to learn grammar before they do so, without having to learn conceptions before talking about a phenomenon of interest to science educators – and from the way language is used, answers to questions can be derived even without the occurrence of prior knowledge-building processes. For example, a little boy – after having heard his astronomer mother talk about the beautiful sunrise and, asked about sun and earth – might say that it is the sun that rises in the morning to make the day. Constructivists have taken such talk to argue that children have misconceptions about the world, when in fact children can use these forms of talk without ever having reflected about any cosmological model. If anyone or anything "has" a misconception, then it is language. The conception is in the language and resides in its possibilities; it is therefore entirely *other* with respect to the child. It is not *his* conception, it is a conception he articulates but one that is not only possible but also understood (i.e., it is intelligible) by the misconceptions researcher. Because children always already find themselves in a language, which is not theirs, what they *actively* articulate is not (completely) their own. Moreover, because what they say is for the benefit of someone else (e.g., the interviewer, the constructivist psychologist) it is inherently assumed to be intelligible, and therefore marked as much by the other as by the speakers themselves. Any form of talk therefore is marked by heteronomy as much as by autonomy, by otherness as much as by selfhood. Conceptions, if they exist at all, are enabled by language, which is a *general system of expression*, an *ideology*, which the individual *concretely realizes* in and through talk. The upshot is that we do get by in life without conceptions, but we can always produce them on the fly from the collective resources (language) that we have access to at the moment. I develop a position grounded in the work of Mikhail Bakhtin (Valentin Vološinov),

Martin Heidegger, and Jacques Derrida. The position advanced here describes very well what and why we find the particular discourses among students concerning natural phenomena. Even philosophers and poets find themselves writing, as if overcome by a force much stronger than themselves, and, therefore, in a situation of radical passivity. Thus, Derrida suggests in an interview available online that when he writes, "there is a kind of necessity, I don't know how to say, of a force stronger than I that makes me write what I have to write whatever the consequences."[1] That is, there is something other than the author himself that makes him write and articulate the particular contents of his writing.

Chapter 7: Thought Follows Communication. According to the constructivist metaphor, what a person knows exists in the form of a mental model or framework. When people communicate, they are said to externalize the contents and "meanings" of their mental models or frameworks, thereby making the contents of their mind available to their interlocutors.[2] In Chapter 6, I already show that no mental model has to be available for a person to talk about physical or mathematical phenomena because language not only provides a resource for talking generally but also for producing and constraining contents in particular. In Chapter 7, I extend the analysis by showing and theorizing how thinking itself develops in the process of the articulation of a topic. I draw on a database with university lectures given by professors with decades of experience generally and lectures from physics particularly.[3] In the examples we can see that thought develops in articulating itself only to find itself in a contradiction almost despite the talking subject. The idea that has been communicated in the end turns out to be problematic as recognized by the speaker himself. Why would a professor have communicated a wrong idea? There is evidence in the concrete case I draw on that this was not done on purpose and for rhetorical reasons. A better way of looking at these events is that thought itself develops in communication, and communication develops with thought. Communication, however, is a fundamentally social (societal) phenomenon, so that thinking, if it is mediated by communication, inherently is a social phenomenon rather than the individual phenomenon that the constructivist metaphor makes it out to be. That is, in a radical way, what appear to be our own thoughts really are characterized by a fundamental otherness. I may be as much surprised by my thoughts as those who listen to me – a phenomenon particularly evident when I have an insight while talking to someone else so that I come to discover my thoughts in my own talk. This otherness derives not only from the fact that the language we use is the language of the other, which affords certain

---

[1] The video runs under the title "Jacques Derrida on the Problematics of Deconstruction" and is available at URL www.youtube.com/watch?v=w9YaNW7Q0c8 (viewed on August 11, 2010). The philosopher speaks in French with English subtitles. I translated the original French text rather than using the subtitles, which often do not correspond to what the philosopher actually says.

[2] If it were not like this, conceptual change researchers would be unable to claim that they extracted conceptions from interview data.

[3] In Chapter 7, I draw on the lectures in a third-year physics course. But I also have lectures where the same phenomena happened recorded in a second-year ecology course.

contents to be articulated, but also because we come to know our own thoughts only in and through our concretizations and articulations – leading us to instances where we are surprised by our own thoughts. There is a fundamental relation of otherness passivity with respect to our thoughts and insights, and, therefore, a radical passivity. *Différance* therefore may also be taken to refer to this radical alterity at the heart of ourselves. This otherness, this non-identity with our own Selves, is at the heart of the differences between any two individuals. That is, these outer, ekstatic differences are but manifestations of non-self-identity and *différance*, without which there could not be any outer difference at all.

Chapter 8: Otherness of Self. In an essay entitled "Facts and the Self from a Constructivist Point of View," Ernst von Glasersfeld (1989b) takes a Kantian position and shows "how the cognizing subject builds up the idea of fellow experiencers who help to supply stability and a sense of objectivity to the experiential world" (p. 435). The author comes to accept a Cartesian ego as the basis of the human faith in the self as an active agent. The problem in this position is – shown in the ultimately futile effort of Edmund Husserl (1950/1991) to establish an absolute egology (science of the "I," Self) – that we could never build up an understanding of others from the inside out, so to speak. Thus, "my ego, given to me apodictically . . . can be a world experiencing ego only by being in communion with others like himself: a member of a community of monads" (p. 166). That is, the Self always already has to be other – autonomous and heteronomous – to be able to see its own experiences (e.g., of emotion) as those that others can have as well. In this chapter, I use interview excerpts in which a teacher and two researcher-teachers talk about themselves, their interests, likes, and visions of future Selves to show how it is precisely *talk* about themselves that these Selves *are* generalized and inherently other. Speakers, in communion with others, use genres for telling auto/biographies,[4] which are populated with characters and plots, so that who they are and can be always already is framed in terms of the other and the general. As a result, the Self is not built up from inside out, nor is it built from outside in. The teller of a version of Self always is an acting subject, but at the same time uses a language that is not his/hers, to which s/he is passive, to which s/he submits without being consciously submitting to it. As a result, I do not merely de/construct the constructivist position on the Self to replace it with its opposite, which could be the social constructivist position. Rather, I articulate, consistent with the poet Rimbaud's expression "*JE est un autre* [I is another]," a position that philosophers such as Derrida developed. He has made this otherness thematic by emphasizing that all autobiography is *oto*-biography, biography for the ear (*oto-*) of the listener, and *allo*-biography, biography of the *allo*, the other, foreign, and strange. The phenomenological philosophers – including Didier Franck, Jean-Luc

[4] I use a slash in the term *auto/biography* to indicate that the two phenomena, autobiography and biography, essentially constitute one and the same phenomenon. The resources for telling an autobiography are the same that are used for telling a biography. Moreover, because of the fundamental otherness between the protagonist in a narrative and a living person, the narrator in the text and the narrator of the auto/biography in flesh and blood, no greater reliability should be attributed to one or the other account of a life (Bakhtin 1981).

Nancy, Michel Henry, and Bernhard Waldenfels – have further developed how the Self always is other than itself and only as such can construct the reductionist version that von Glasersfeld lays out. That is, the position articulated here can explain why von Glasersfeld says what he says, the conditions for thinking in the way he thinks; but the reverse is not possible.

Chapter 9: The Nonsense of Meaning. The construct of *meaning* is a rather ambiguous one and, as semioticians have noted, it is a term used in very different senses. In one and the same essay, von Glasersfeld (1989b) uses the term (a) in the sense of *dictionary sense* (Kant's German *Vorstellung* and *Darstellung* as two "meanings" of the term representation), which, by definition, can be found in a dictionary; and (b) in the sense of something that does "not travel through space and must under all circumstances be constructed in the heads of the language users" (p. 444). Constructivists generally do not tell us *what* they mean when they say that students "construct" "meaning" other than *that* some students have done so on this or that occasion. There is also a fundamental error in locating signification.[5] It is not that "subjective meanings get modified, honed, and adapted through the course of social interaction" (p. 444) but words are used in social interaction in ways increasingly appropriate for the situation at hand. That is, the constructivist metaphor suggests that students "construct" something that is associated with and denoted by a word, a something constructivist scholars denote by the term "meaning." But we would never be able to do this on cultural-historical grounds, because we would never be able to get from language use into the "meaning" realm unless we already knew this realm together with or prior to language. However, living does not require language and relations of signification develop prior to words and language. Words *accrue to significations* rather than the other way around. It is the social situation that is shot through with signification, which always already precedes the individual language user; and the words accrue to familiar situations. That is, *if* there were meaning at all, then it would be in the situation rather than in the head of the user. It is not surprising, therefore, to find the pragmatic philosopher Ludwig Wittgenstein (2000) declare: "meaning, is dropped from our consideration. . . . The notion of meaning *derives from a primitive philosophical conception of language*" (¶1, 7, Ts-213, emphasis added). Drawing on everyday examples, I articulate – following such philosophers as the late Wittgenstein, Martin Heidegger, but also more recent language philosophers such as Richard Rorty, Donald Davidson, and Derrida – a theoretical approach in which the difference between forms of language and the world in which these forms are used is undecidable. There is no meaning that we construct, of words or situations, there is only increasing familiarity of the places that we inhabit and in which we dwell.

[5] I tend to use the term "signification" rather than "meaning" to disambiguate a number of theoretical issues.

Chapter 6

Talking Conceptions Without Conceptions

> It is easy to show that meanings do not travel through space and *must under all circumstances be constructed in the head of the language users*. (von Glasersfeld 1989b, p. 444, emphasis added)
>
> Thus, neither theoretical cognition nor aesthetic intuition can provide an approach to the once-occurrent real Being of an event, for there is no unity and interpenetration between the content/sense (a product) and the act (an actual historical performance) in consequence of the essential and fundamental abstracting-from-myself *qua participant* in the course of establishing meaning and seeing. (Bakhtin 1993, p. 18)

In the first opening quote, Ernst von Glasersfeld excludes the role of the world outside of the organism in the constitution of meaning. It, "under all circumstances" "must" "be constructed in the head of the language users." Meaning, whatever it is (see Chapter 9), is something singular. It cannot travel through space. It must be in the head of the individual language user. Here, in the head, it is the outcome of a construction. Others, such as cultural-historical activity theorists, disagree recognizing that "the meaning of a word represents . . . a close amalgam of thought and language that it is hard to tell whether it is a phenomenon of speech or a phenomenon of thought" (Vygotsky 1986, p. 212). That is, the relationship between language and thought – which is theorized to exist in the form of or to be derived from mental structures – needs to be questioned. This is particularly the case if language never is the language of the person but always already something that comes to us from the other and in our speaking returns to the other for whose benefit we speak. Bakhtin therefore emphasizes the "once-occurrent Being of an event" as the frame from within which to understand the content and function of language in use. To understand "meaning and seeing," we need to be in the situation *qua participant*, because it is only from within participation that we can understand the commitment speakers and recipient, because each (speech) act has a consequence, which neither theoretical cognition (construction) nor aesthetic intuition grasp. Throughout his work, Bakhtin emphasizes the function of the Other (recipient) in the shaping of the utterance generally and the word specifically.

W.-M. Roth, *Passibility*, Classics in Science Education 3,
DOI 10.1007/978-94-007-1908-8_6, © Springer Science+Business Media B.V. 2011

Speaking implies the Other in a double way: We use language that is not our own as an integral part of communicative expression and we speak for the benefit of another, whom we address and take into account and to whom the expression is oriented. Any intelligible (verbal) expression, therefore, tells us little about specific speakers, little about their subjectivities, and little about what they think independently of anyone else. Rather, any verbal expression tells us a lot about the language, its possibilities, and inherently shared, collective consciousness (i.e., from Lat. *consciēre,* to know with, a composite of *sciēre*, know + *com[n]-*, with). From this position, then, conceptions are not properties of individuals; rather, individuals mobilize the resources of an inherently shared language for talking about natural phenomena and constituting the topics of science. This, in fact, allows us to understand how people can talk about a phenomenon or topic, about which they have never talked before, and therefore of which they could have never constructed the knowledge or "meaning" that constructivists claim to lie at the origin of the talk. Whence, therefore, come the ways of talking in which students engage, for example, during interviews about "their" "conceptions" or during interviews designed to find out about the "conceptual changes" that a student has undergone? A much simpler and more parsimonious explanation has speakers mobilize the resources of a more or less familiar language in and for the purposes of particular situations (activities), whether these are conversations in a science lessons or interviews about natural phenomena or some other suitable context where people talk (about) nature. These situations are like games in which we participate, which, as games, always imply others and therefore imply alterity. Because of the alterity of language and the orientation to the others of any intelligible communication, speakers always are in a situation of radical passivity with respect to the contents and process of the language in use. I begin by reviewing the received ideas about conceptions before presenting a concrete interview situation that was recorded during an interview following a conceptions/conceptual change research protocol. The analysis of this concrete situation allows me to develop a very different version of conceptions talk than the one that (radical, social) constructivist science educators present us with.

Traditional Ideas About Knowledge and Language

Received approaches to knowledge and language are Aristotelian, taking the former to be an inner structure and the latter as a system of signs standing in for and pointing to the latter. The first analysis in Chapter 3 – as the one provided in Chapter 7 – embodies a different approach according to which the difference between thinking and speaking is undecidable. To articulate how my approach differs from other received approaches, I begin by outlining just enough of the principles underlying the existing constructivist ways of thinking about knowledge and language. I do not provide this description to set up a straw-person but rather to provide an inherently incomplete sketch that allows my readers to understand my choice of a different model. In the choice for an appropriate text to review, I had to return to the early years of conceptions and conceptual change research, as later texts tend to use the term "conception" without further explication. Throughout this chapter it is evident that there are major steps to be taken to get from the interview contexts themselves to their descriptions and explanations of texts and topics in terms of (mis-) conceptions. These steps require many assumptions and presuppositions before one can speak of conceptions that individuals are said to hold in their minds. In the subsequent section, I then show how my approach leads to a different way of understanding and theorizing talk about phenomena that are of interest to scientists and science educators (e.g., models of the universe, earth, sun, day, night).

Conceptions are theorized as cognitive entities, furniture of the conscious mind, but they are unlike signs. Conceptions are human inventions that "once labeled become communicable through the use of language" (Pines 1985, p. 108); that is, conceptions are different from language, which is but the vehicle that makes them available to others. Conceptions are said to be mental/cognitive regularities that are labeled with words and, by means of these, can be "employed in thought and communication" (p. 108). "A word is like a conceptual handle, enabling one to hold on to the concept and to manipulate it" (p. 108). These definitions are consistent with a recent statement of leading conceptual change theorists and researchers, who describe conceptions as "learners'" "mental models" or as the "learner's" "internal representations" constructed from external representations of entities initially presented by teachers, textbook authors, or software. The essence of the cognitive perspective on conceptual change is characterized by "knowing as having structures of information and processes that recognize and construct patterns of symbols to understand concepts and exhibit general abilities such as reasoning, solving problems, and using and understanding language" (Greeno et al. 1996, p. 18). The mainstream approach to conceptions is captured by stating that "cognitive approaches provide analyses about the ways in which knowledge *must be structured* and about the *structures* of knowledge *in learners' minds* that will be available to support task performance and to transfer to new situations" (Anderson et al. 2000, p. 12, emphasis added).

Conceptions are organized into networks of relations, which often are represented in node-link diagrams or ontological category trees; conceptions themselves can be thought of as theoretical nodes where a multitude of meaningful relations cross. Researchers have come to use the term *misconception* to denote the use of conceptual relations in inappropriate ways and contexts. Concept maps, semantic networks, and node-link diagrams are some of the diagrammatic forms that illustrate concepts and the conceptual relations in which they are involved. These relations are expressed in propositions that may be communicated in the form of sentences. Thus, for example, node-link diagrams (concept maps) have been used to show the differences between radical and non-radical (simple, slight) conceptual change, corresponding to conceptual change across and within ontological categories (Chi 1992). Figure 6.1 presents one example of a concept map, which, being the result of a collaborative concept mapping assignment in an

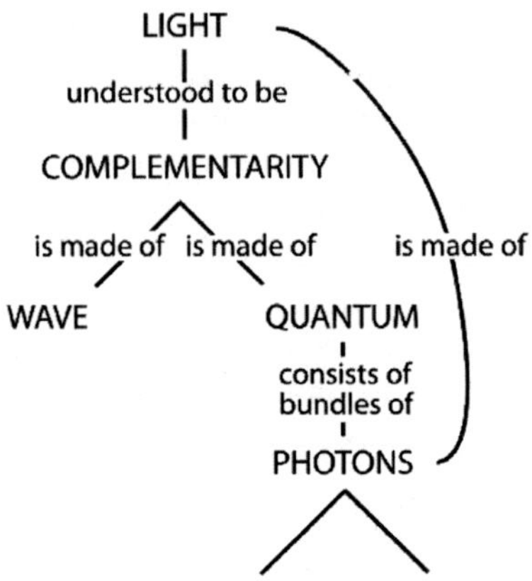

Fig. 6.1 This typical concept map, constructed by high school physics students, would be taken as a representation of their (shared) conception of the nature of light.

advanced high school physics course, might be used as evidence for the shared conceptions these students "constructed" and "have" about the nature of light. Depending on the study, conceptual change may refer to the process of change or outcome of the change process. Recent conceptual change theories integrate affective components that are said to mediate the mind such that there may be no change, superficial change/assimilation, or true conceptual change/accommodation (e.g., Treagust and Duit 2008).

Conceptual frameworks (conceptions, misconceptions, alternative and naïve conceptions) generally are inferred from interviews, clinical interviews, or tasks in which persons are asked to demonstrate/predict, observe, and explain some physical phenomenon. There is an (implicit) assumption that the (interview, task) situation itself does not mediate the conceptual organization but that the situation simply allows reading out – like a computer printout that shows the results of a calculation or the contents of computer memory – an at least temporarily stable mental organization. That is, mind is said to be structured like a concept map (Fig. 6.1)[1]; and, during the interview, the relations between the terms are read out into language. Strong arguments have been made on sociocultural grounds that contest the "reading off" of conceptions from interview texts. However, the fact that conceptions are very resistant even to intensive instruction gives legitimacy to the reasonable nature of the assumption that conceptions and mental models are independent of

[1] During the late 1980s, I conducted research on students' "conceptions" and "cognitive structure" using a protocol in which students were asked to generate as many associations to a given word (concept) within a one-minute period (e.g., Roth et al. 1992). There were between 16 and 25 words in a protocol for a given domain. Once these associations were submitted to multidimensional scaling (MDS) or individual differences in multidimensional scaling (INDSCAL) analysis, maps were produced showing proximities and relations between the chosen words (concepts). This approach, therefore not only assumes that conceptions constitute maps but also produces such maps that make us think conceptions exist in this manner. Needless to say, this book is also critical of (de/constructs) the work that I had done at that time.

the environment in which they are elicited. The resources in the setting have been treated as transparent. Thus, the method of identifying a conception is to excerpt statements that interviewees make irrespective of anything else in the setting or the nature of the setting itself. This can be seen in the following excerpt from the seminal article on the nature of conceptual change:

> (I) . . . it seems these are strange results. What attitude do you take of these results?
> (CP) I say they don't really mean all that much; it just depends on what your frame is. It's sort of like potential energy depends on the way you define zero to be?
>
> [. . .]
>
> CP's reference to potential energy is significant in pinpointing a conception which enables her to regard the values given to a variable as arbitrary, being dependent solely on the observer's point of view. She attempts to resolve some counterintuitive results of Einstein's view of time by drawing an analogy between time and potential energy. No matter that the analogy might break down with further analysis – it serves her belief in absolute time. (Posner et al. 1982, p. 219)

In this quote, the authors derive from the quoted interview excerpt that the subject CP has a conception that "enables her to regard the given to a variable as arbitrary, being dependent on the observer's point of view." There is no reference to the fact that CP has responded to a query on the part of the interviewer (I) and for the purpose of answering this query rather than primarily developing relativity theory – which is done as a matter of course *in* and *through* the interview. The authors attribute an intention to resolve "counterintuitive results," when in fact the event shows us that it is the interviewer who denotes the results as "strange." Asked what she makes of these "*strange* results [emphasis added]," CP then produces statements that complete the interviewer's utterance as a question. In responding, she addresses strangeness, the topic that the interviewer has started. There is no indication or evidence that she would have noted or talked about anything strange: It is the interviewer who contextualizes and occasions the response. As a result, the cognitive approach often focuses on what people (students) do *not* do, how they misconstrue, misconceive, or alternatively conceive of some phenomenon. It is not surprising, therefore, that "the constructivist approach to cognition has emphasized forgetting – the distortions, confabulations and general unreliability which results when memories are schematically assembled in some kind of cognitive processor" (Edwards and Potter 1992, p. 36). I show below that interview participants talk about scientific phenomena even when they have never talked or thought about them before, in which case, what they say *cannot* be driven by or be the result of a conception.

Talking Seasons

To exemplify what an actual conversation looks like and how stretches of talk emerge that constructivists (conceptual change theorists) determine to be evidence of misconceptions, I draw on an interview from a database that my graduate students conducted with people from a variety of walks of life. In the present instance, the interviewer is a doctoral student (Pei-Ling Hsu) who had come to Canada a few years earlier; Mary, an adult participant, also had immigrated to the country to pursue graduate studies. At the point where I pick up, Pei-Ling has already asked Mary to explain the phenomenon of day and night. Pei-Ling then articulates what will have been a transition into a new topic by making a statement about the season at hand at the instant of the interview (turn 01). She continues by producing what we might gloss as the question, "What is the reason for having seasons?" She also states that the seasons are different and the summer is hotter, where the comparative of a hotter summer can be heard to be in reference to the winter talked about earlier. Partially overlapping, Mary begins her response stating that "it depends on the latitude" (turn 05); but she follows with the word "right," uttered with a rising intonation characteristic of certain ways of asking questions. Pei-Ling pronounces the word latitude, which, after following a rising and falling intonation contour, rises toward the end, as if saying, "You are asking me about 'latitude'?" (turn 07).

Fragment 6.1a

01 P: okay=so now its summer; right?
02 (0.20)
03 M: uh hm:
04 P: =do you think why we have summer a::nd; (0.19) <<all>ye=know> winter; the seasons (0.14) different (0.30) in summer is HOtte:r <<all>ye=know> (1.35) why we have the [seasons]?
05 M: [i:::]s: <<len>depen:ds on your> (0.20) <<p>lAtitude; right?>
06 (0.15)
07 P: ^LAtitude?
08 (0.42)
09 M: izAT de word?
10 (0.25)
11 P: ^yea=LAtitude. [(0.16) [right?]

[
12 M: [um hm-]
13 P: uh hm
14 M: <<f>=i think> <<dim>it depends on de latitude;>
15 (0.19)
16 P: uh hm;

In her turn, Mary, too, asks a question, as can be recognized by the rising intonation (pitch) and the grammatical structure of the utterance; and we can hear her ask something like in the gloss "Is that [latitude] the right word?" (turn 09).

Pei-Ling utters the word once again (turn 11), but this time with falling intonation, which is typical of constative statements. We might gloss our hearing in this way: "Yes, the word I am asking you me about is 'latitude.'" This time, Pei-Ling also says "latitude" while making a gesture, where the right hand initially lying on her lap first moves upward together with the left hand (see drawing in turn 11). The right hand touches the back of the left hand and moves away from it upward. Someone familiar with geography can hear/see this as a reference to the geographical latitudes on a map or globe (turn 11), where the right hand moves upward/northward from the reference (back of left hand). Mary makes another statement that it depends on the latitude (turn 14), not without having flagged the statement by the expression "I think."

Mary then begins what will have been an elaboration of her answer given so far. She starts by invoking the equator stating that the farther away you are from the equator (turn 17), "the more you will get the sunshine and the more [it is] possible for you to have four different, distinctive seasons" (turn 20). As she talks about being away from the equator, Mary moves her left hand forward, which has been engaged in what we come to understand as a grooming gesture, to open up so that thumb and index finger are spread far apart, and thereby suggest a distance. The hand closes, thumb and index finger touching each other, and then opens up again (turn 17).

Fragment 6.1b

```
17 M:  =if the:re: you=are (0.16) den
       [ (0.56) de::f (0.14) āwAY from de (0.21) equator:?
```

```
       [
       ((same gesture for "away" and subsequently, "the more"))
       ((the gesture corresponds to the one Pei-Ling has made))
18     (0.14)
19 P:  uh hm
20 M:  de m:::o:re (0.27) <<dim>you will get the sunshine>. (.)
       an:d de mo:re you will (.) .hh (.) is pOSSible foryou have
       a f: (0.52) [f:our (0.71) dif:ferent s:easons;
       =distinctive seasons.
```

```
       [                                   ((beat gesture))
21     (0.16)
22 P:  uh hm::;
23     (0.29)
```

24 M: <<f>but> if you=re in de: (.) eQUAtor () <<dim>youll have summer all the time; right?>
25 (0.31)
26 P uh hm (0.15) then (0.28) ^why:::; (0.42) why is the rEA:son;

Mary then moves into what will have been a contrast to the movement away from the equator by stating that at the "equator you'll have summer all the time" (turn 24). Mary ends with an adverb that denotes agreement or correctness ("right") but with a rising pitch contour that normally marks utterances as questions. Pei-Ling's utterance, which she flags as a question by a rising intonation toward the end, requests a reason, a "why," and thereby also treats the preceding as a description that does not explain the seasons.

This episode exhibits all the features of normal everyday talk, the pauses, grammatical errors, speech particulars (e.g., "de" rather than "the"), intonations, run-ins, broken words, dangling letters, overlaps, and so on. Yet the two appear to be having a conversation where they understand what is being talked about. Here, the topic pertains to the seasons and to the fact that there is summer all of the time closer to the equator whereas farther away there are more definite seasons. In a gross reduction we may gloss the event in this way: Pei-Ling asked a question about the origin of the season and Mary stated that the latitudes are responsible for the season. At the equator, there is always summer, whereas at the higher latitudes that Pei-Ling and Mary have shown by movements away from some reference point (back of left hand, thumb). A conceptual change researcher might attribute to Mary the misconception that the seasons depend on the latitude rather than on the relative angle of the Earth's axis with respect to the incoming light rays from the sun. Before that Mary has already articulated something that conceptual change researchers might attribute to another misconception: that the sun moves around the Earth leading to the phenomenon of day and light. But such an attribution, so I suggest here, is the result of a gross characterization of a complex phenomenon to which both interlocutors have contributed. Together they have brought about this conversation. They have not just emptied their brain cases, as independent speakers, dumping the contents of their mind into the public forum of the conversation. Rather, they have been speaking *for* each other – and, given the transcript available to us, have done intelligibly so – so that each word, each idea, already has collective characteristics irreducible to the individual. But if the language, which is the language *of* and *from* the other, has been understood, this intelligibility of the topic derives from language itself. The ideas about the seasons, the conceptions, are not those of the particular speaker, here Mary; these are the conceptions made available by language. In fact, conceptual change researchers themselves need to be able to understand this language to make the assessment that it is not consistent with science speak. To be able to recognize and understand some talk as misconception talk, this talk needs to be intelligible!

Seasoned Talking

> Theoretical cognition of an object that exists by itself, independently of its actual position in the once-occurrent world from the standpoint of a participant's unique place . . . does not constitute ultimate cognition; it constitutes only an auxiliary, technical moment of such ultimate cognition. (Bakhtin 1993, p. 48)

In this quote, Bakhtin points out that theoretical cognition does not constitute ultimate cognition because it is disembodied and uncommitted. Only when realized as participative thinking does cognition bear on life, taking into account the speaker-participant's unique place in life generally and in the specific situation particularly. When a researcher interviews a research participant, for example, about the relationship between the seasons, the sun, and the earth, the participant does not merely spill the contents of mind but engages in a unique and answerable act of life.

In the first section of this chapter, we see how in conceptual change theory the content of talk is taken to be an accurate – non-mediated – reflection of the mental structures of the speaker. The interview situation is treated as an occasion for a memory dump or core dump. In computer speak a core dump refers to the recording of the entire memory used by a particular program at a particular instance in time. Core dumps are used to diagnose (or debug) errors in computer programs. But can we really treat the interview situation as something like a core dump? Are the utterances of a person reducible to him/her and independent of the interlocutor and the situation? Are the utterances a reflection of the singularity and subjectivity that we attach to what is thought to be the real Self of a person? Over the years of analyzing conversations, from interviews as from classroom situations, I have come to the conclusion – and had to abandon both radical and social constructivism as suitable explanatory frameworks – that talk is the reflection of something else, something irreducible to the individual person. Today, I understand each situation as pertaining to a particular societal (collective) activity, each of which in some way contributes to the reproduction and transformation of society.[2] As such, all forms of talk have cultural histories, and, with it, the language that constitutes an integral aspect of such talk. It is for this reason that talking seasons also constitutes seasoned talking, that is, a form of talking that has evolved and been honed in the course of the history of Western culture. The language is a reflection not only of the content of talk but also, and more importantly so, of the nature of the activity – e.g., interview, science lesson, or everyday conversation. Rather than using the term activity, Wittgenstein (1958) speaks of the games we play with

[2] The English term *activity* does not render justice to the theoretical notion as originally developed. The founders of cultural-historical activity theory understand the term *activity* in terms of the German *Tätigkeit* or the Russian *deyatel'nost'* (activity), which refer to a form of activity that contributes to the collective satisfaction of needs. The terms *Aktivität* and *aktivnost'* also translate as *activity* but are used to refer to the fact of doing something and being busy without the orientation toward need fulfillment. *Schooling* would be an activity, as would be farming or producing tools. That is, activities are characterized by collective motives, and these motives "shine through" in talk.

language. Talk therefore has to be understood as contributing to the very maintenance of the situation and activity, not just in terms of this or that content. In the following, I work out these issues drawing on the interview excerpt in the preceding section.

Any stretch of talk does more than produce a topic and its content. Talk is the very resource in making the situation that what it is. For example, an interview is a societally recognizable activity (or part thereof) designed to solicit statements that are used for publication (television, radio, journal interview) or that are used as data source for research purposes (conceptual change interview). For example, Pei-Ling has invited Mary to participate in a videotaped interview to "find out about her ideas concerning a variety of natural phenomena." Mary has agreed. When the two are together, both are oriented toward achieving the motive of the activity: completing an interview that focuses on ideas concerning natural phenomena. Although interlocutors do not necessarily make their situation the topic of the talk – it would in many cases be considered strange if the nature of the situation or activity would be topicalized – their turn-taking routine is consistent with an interview situation. The unwritten rule for this routine assigns the questioner's role to the interviewer and responder's role to the interviewee. In the present instance, we see this being the case when Pei-Ling sets up and asks a question (turns 01 & 04, turn 26); Mary provides the response. The routine is different from typical classroom talk, where the teacher would pose certain types of questions, which solicits a student response that is followed by a teacher evaluation. In interviews, the questioner asks genuine questions, that is, questions to which s/he does not yet have the response. This may also require "digging further" and requesting clarifications, such as when Pei-Ling "queries" the use of the term "latitude" (turn 07). But it also means that the interviewer does not make assessments of the responses, such as might normally be required in situations where an interlocutor asks "Right?" in the way Mary does in turn 24. Pei-Ling avoids the answer – which is the non-preferred way as a second part of a query/answer pair – which would also have been an evaluation of Mary's response. Pei-Ling does so initially by hesitating – producing the "uh hm" interjection that is also used in hesitations (turn 26) – and then moves on to ask for the reason underlying the change in seasons.

A situation may be questioned by events that emerge from talk; and talk is the resource to bring the situation back into the normal state when trouble becomes apparent. Thus, for example, when Pei-Ling repeats Mary's use of the word "latitude" but with rising rather than Mary's falling, constative-producing intonation, trouble emerges. This trouble can be seen in Mary's query whether it – i.e., latitude – is the (right) word (turn 09). We might gloss the exchange between turns 05 and 11 in this way: For whatever unstated reason, Pei-Ling puts the word in a questioning context – be it because she has not heard precisely, be it because she does not know its function. Mary treats it as if the word itself had been questioned, as if she was asking whether the term latitude is the appropriate word to talk about the seasons. Pei-Ling, in confirming (turn 11), ascertains the word to be an appropriate one: "yea, latitude, right." The troublesome issue having been removed,

Mary now returns to work on her response: "I think it depends on the latitude" (turn 14).

There is a lot more to the work that the two accomplish so that the situation recognizably comes to be an interview. For example, Pei-Ling produces an "uh hm" interjection (turn 19). This tends to be taken as a way of acknowledging listening and as a way of leaving the turn with the current speaker. That is, in producing the interjection "uh hm," Pei-Ling not only allows Mary to know that she continues to listen but also that she is providing space for the latter to elaborate. This is precisely what Mary does: she continues the work of producing an answer. Precisely the same turn taking occurs a few seconds later, when Pei-Ling produces the same interjection (turn 22) preceded and succeeded by pauses (turns 21 & 23), which gives Mary the opportunity to further elaborate. This is just what Mary does.

This brief analysis shows that from the activity perspective, the interlocutors are not entirely free to say what they want, to engage in something like a core dump. Rather, they collaborate with and for the other to recognizably produce an interview designed to elicit ideas about natural phenomena. Their talk not only constitutes content but also makes the context. Mary actively orients to Pei-Ling, speaks for her counterpart's benefit, using a language that is not her own (where I am not even referring to the fact that it is not her mother tongue). In talking, they make the situation together, which is the result of their collaboration. This result, the recorded and subsequently transcribed text, therefore cannot be reduced to the individual speakers. Because of the mutual orientation to the other, the text, which uses language from the other, for the purpose of another, belongs to both of them. Even if we were following the argument of the monadic subject constructing its own world, we would be led to the conclusion that our world is inherently and objectively the world of others as well (Husserl 1991). It would be a pure absurdity to think that communication is possible if each subject "constitutes a world of its own, so that together they constitute two worlds that are separate ad infinitum, two infinite spaces and space-times" (pp. 166–167).[3] In making the situation recognizably an interview, the two also constitute the topic together. Again, to be mutually intelligible, the two have to draw on a common resource, which they can because "there can exist only a single community of monads, the community of all coexisting monads" (p. 167). What they can say therefore is determined by its nature as a *common* resource rather than by something attributable to the singularity or subjectivity of the speaker (see Chapter 11 on the relation between private experience and its account).

The very possibility of recognizing a stretch of talk requires the intelligibility of misconception, and therefore, the possibility to talk in this way about a natural phenomenon. For example, conceptual change researchers might hypothesize Mary to have a misconception about the origin of the seasons because she responds to the question "Why [do] we have seasons?" by saying, "It depends on your latitude" (turn 05). She subsequently elaborates this answer by saying that at

[3] Surely, this is a radical indictment of the constructivist position elaborated by scholars from Kant to Piaget and von Glasersfeld.

the equator, there is summer all of the time (turn 24), whereas there are four distinct and definite seasons farther away from the equator (turn 20). There are two important aspects to this talk.

First, phenomenally Mary is right. At the equator – e.g., in Singapore or close to it – there are no seasons in the way that Europeans and North Americans know it. There are wet periods (monsoon) and dry periods, but the temperatures do not vary much between the seasons. Moreover, Mary can speak from experience: the Tropic of Cancer runs straight through her native Taiwan. The climate is tropical and temperatures are fairly constant throughout the year. In Victoria, where she does her PhD degree, which lies between the 48th and 49th parallel, there are definite seasons. What she says not only corresponds to her experience but also to the experience of anyone else who has lived in the two geographical locations. Second, for this talk to be recognized as inconsistent with the scientific standard way of talking about the relationship between latitudes and season, the talk has to be intelligible. But intelligible talk is possible talk, and possible talk is recognizable talk. It is also *seasoned* talk, as these ways of talking are cultural-historically contingent ways of talking about attendant phenomena.

Culturing Conceptions

Anything we can identify in language is cultural-historical, as we cannot ever extricate the process and contents of talking from the particular situations and motives of our culture. The attribution of conceptions to individual speakers is understandable within the constructivist paradigm, which attempts to model knowing and learning from the perspective of the individual mind – or should we say monad, the individual human subject seen as separate from society – separated not only from the living/lived body, the needs and impulses of the thinker, but also from the material world. In the previous sections, I show how each of the speakers not only contributes to producing the content of the talk by contributing to its intelligibility, but also to re/producing the situation to make it recognizably an interview and to manage those micro-situations that threaten intelligibility, recognizability, or nature of the activity. Talk, therefore, is cultural through and through. Conceptions, if it makes sense to denote stretches of talk using this term, are therefore cultural through and through. We therefore need to engage in an effort to culture conceptions, to cultivate them as shared resources in collective endeavors. This shift also requires us to move away from the intentionalist perspective on the individual that the constructivist metaphor implies. We need to take into account the radical passivity that comes with the use of language, participation in collective activity, or the responsibility toward the other in face-to-face encounters. That is, to understand what people say and do in concrete situations, we need to understand the ethical commitment underlying our concrete participation in material life, which comes with the *givenness* of Being and participative understanding

Fig. 6.2 A sunset in the Australian outback, which has attracted a group of tourists who have come to this spot to admire the scene over a glass of sparkling wine.

(Bakhtin 1993). It is only when we account for *Being-as-event* that we can understand the content and process of language-in-use.

Science educators using the constructivist metaphor often wonder why students – including those who graduate from Harvard and who have become the target of irony and jokes after the documentary that exhibited their non-scientific explanations of the seasons[4] – "have" "inappropriate conceptions" about the seasons, about day and night, or about the relative movements of sun and moon. That is, although Nicolaus Copernicus introduced the heliocentric explanation for the planetary system in the first half of the 16th century, everyday talk relating to the sun and moon continues to be in terms of a geocentric language and the heavenly spheres that the Greek used to talk about the universe or the reasons for having night and day. Rather than individual conceptions, it is this talk that continues to be reproduced, among others, in conversations between tourists in the Australian outback on a special evening trip to marvel at the most beautiful sunsets over a glass of sparkling wine (Fig. 6.2). Any child participating in such a trip, after hearing and perhaps even participating in the talk about "the beautiful sunset," may reproduce such talk in an interview with a conceptual change researcher later that night.

People inherently talk about certain phenomena for some first time and about phenomena that they have never thought about before. Thus, Mary repeatedly remarks not having thought about some phenomenon that Pei-Ling is asking her about. Nevertheless, she provides a reasonable response. That is, even though constructivist conceptual change researchers suggest that there are conceptions underlying talk, people do produce such talk even when they have not had an opportunity to construct any conception. Thus, it does not take a conception to produce such

[4] Information about this video documentary is available, as of October 30, 2010, at the URL http://www.learner.org/resources/series28.html

talk, merely the capacity to reproduce talk and the capacity to make inferences. We do not need underlying conceptions (mental structures) to re/produce certain types of talk in the same way that we do not need to know any explicit grammar to re/produce grammatically correct sentences. General competencies allow us to talk in situationally appropriate ways. Thus, it would be entirely unromantic – it would have been teacherly – if I had talked about the rotation of the earth when I gathered with other people to marvel at the sun setting behind the mountains in the Flinders Range National Park near Wilpena Pound, the colors it produced, and the emotions that were affiliated with the setting (Fig. 6.2). The emotions associated with names and concepts are important dimensions in their survival. Thus, even though scientists have banned unicorns from their language for a long time or use it to refer to something unscientific (e.g., the newspaper line "Don't waste medical research money chasing unicorns"), a simple Google search with the term "unicorn" reveals over 12 million hits, which include the term as part of company names, banks, restaurants, faith healers, http servers for Rack applications, and so on.

Talk about sunrise, where the sun is the (moving) agent, makes sense and it is consistent with talk about an "early rise" or an "early riser." Here we talk about a person rising early in the morning, and it would never come to our minds to talk about the earth falling away from the person making it only appear as if s/he were rising. Moreover, we do understand sentences such as the headline in a Scottish news paper "Star ditches early rise to become main BBC weather forecaster," even though the star is a person and the "early rise" is another way of talking about the morning shift in a radio or television studio.

Discussing a variety of questions concerning the usefulness to seek "meaning" and other things said to lie "behind" words, Wittgenstein (1958) also addresses the use of "above" and "below" with respect to the earth. In using these terms, we always know what speakers "mean" in saying that they are on top of the world/earth or that the earth is below them. Yet the directionality of above and below changes – in an absolute frame of reference – when points on the northern and southern hemispheres are considered.

> But a still better example would be that of the application of "above" and "below" to the spherical earth. Here we all have a quite clear idea of what "above" and "below" mean. I see well enough that I am on top; the earth is surely beneath me! (Do not smile at this example. We are indeed all taught at the elementary school that it is stupid to talk like that. But it is much easier to bury a problem than to solve it.) And only a reflection shows us that in this case "above" and "below" cannot be used in the usual way. (That we might, for example, say that the people at the antipodes are 'below' our part of the earth, while having to acknowledge as right for them to use the same expression about us.) (§351 [pp. 111g–112g])

The consideration of the uses of the terms becomes more complex when we add that Australians, for example, also might use the term "down under" to refer to the part of the earth that they inhabit. In this case, the antipode, as Wittgenstein says, would be "below" even when one is "down under." My research on talk about locations on a map shows that in everyday conversations, the shifting use of terms such as "down" and "up" is managed unproblematically in the here and now of the conversational setting (Roth and Lawless 2002). The question rarely is whether the talk is "correct" in some other frame of reference, whether geographically seen or with respect to a particular discourse. The very survival of such talk – i.e., its reproduction and transformation – is a sign that the question is one of situational usefulness. Phenomenally, such talk is more adequate: talk about a sunrise or sunset is more adequate than talk about the rotating earth. Such talk is reproduced and handed down precisely because it is more adequate; and children, by participating in such talk, continue to reproduce it in new generations.

More interestingly and surprisingly, perhaps, there is an inner contradiction in the recognition of misconceptions. For conceptual change researchers to recognize a stretch of talk as constituting a misconception, they need to understand this talk at least to the extent that they can determine it to be inconsistent with the ways a scientist would talk. But understanding such talk means reproducing such talk, reproducing its intelligibility as a possible form of talk, even and precisely because it does not correspond to the talk that a scientist might use when asked about the current standard conception of the universe. Constructivist science educators do not say that the talk from which they extract a misconception is non-intelligible. Quite to the contrary, as the extraction of a "misconception" requires the talk to be intelligible otherwise there would only be gibberish, talking in tongues, and not a misconception. The very cultural competency that underlies "misconception" talk also underlies the discourse about talk as deriving from a misconception.

The way of describing and explaining talk about natural phenomena in this way allows us to understand otherness as a fundamental condition for having any conversation; and the otherness of talking and in reproducing societal activities leads us to notice the essential (i.e., radical) passivity involved with respect to our contribution to the collective endeavor, what we can communicate to others, and, therefore, even of our own ideas. We can only communicate what is intelligible, which means, that we can only say what others can already hear and comprehend. This means that our talk is shaped by the understanding of the other rather than merely by our own subjective understanding, even if something like it could exist, for *any* understanding, precisely because it is *under*standing, standing in the midst (of a group, community), means reproducing the language from/for/of the other. The same holds not only for the content of our talk but also to those aspects that serve to manage the situation; in most situations, these two aspects – i.e., text and context – cannot even be separated (Roth 2010c). Thus, any stretch of talk not only says more than can be said in so many words but also it does more than what meets the eye.

It is precisely the same competency that also provides for the opportunities to change between forms of talk, that is, for example, to change from the ways of talking about the universe as Aristotle had done to the ways that have become common among scientists following the work of Nicolaus Copernicus, Johannes Kepler, Galileo Galilei, and Isaac Newton. Comprehension requires fundamental competencies that underlie making sense of talk generally. Without such competencies, differences between situationally appropriate and inappropriate talk could not be made or discerned. During the appropriation of a new form of talk, the old way serves as the condition for the new to emerge. Therefore, "misconception talk" is not something to be eradicated. It actually constitutes part of the conditions for the new forms of talk that scientists and science educators find more acceptable in context that matter to them. If it were any different, then no science could have emerged (Husserl 1939).

The approach outlined here provides us with new ways of thinking about some of the difficult problems in science education, such as the question of evolution versus creation. If we were to move away from essentializing both concepts, (theory of) evolution and creation, then we could take the attitude that for the purpose of science, it is useful to work with the theory of evolution because it allows us to make predictions under the conditions that scientists work. People may continue to think about the act of creation and may continue to read parts of the Bible in a literal manner. This is not different from my being able to marvel at a sunset in the Australian outback while relegating the rotating earth speak to a science classroom or using it to predict the phases of the moon. At least since quantum theory, (some) scientists have learned that it is pointless to ask what things *really* are, such as in the case of light, where this question is moot. Quantum theory allows us to predict observations irrespective of its "truth" and irrespective of the essence of light. What matters is that quantum talk allows us to make correct predictions and descriptions independent of whatever conceptions (beliefs, knowledge, convictions, ideas) we might have about this or that. Quantum theory is but another example of contexts that presuppose our competence in ways of talking, which are precisely those ways that children learn when growing up independent of what the particular contents and theories are that researchers attribute to them. What ultimately matters are not our inner beliefs but that we are able to participate in a variety of (language) games. This means that in our actions we have to display knowing the rules that are said to govern each game.

Chapter 7

Thought Follows Communication

> One day Michael arrived at our office and started to share his insights from a paper he had written that morning. He talked excitedly with waving hands, and I tried to follow his thinking, which sounded novel and interesting to me. During our conversation, I mainly nodded my head and produced continuers such as interjections "mm . . . yeah . . ." to show my interest and encourage him to share more. Suddenly, he said, "Oh . . . do you have a pen and paper, I need to write something down!" After finding a piece of paper, Michael then jotted down some words on the paper and said, "Just some ideas I suddenly have for my writing; and I need to write it down, otherwise I will forget!" (Hsu 2010, p. 162)

In this quote, Pei-Ling Hsu recalls an event during which I articulated for her some of the ideas that I had been working on. But there is more to this event. There is not just an account of what I have been thinking but apparently, as evident in this account, new forms of thinking emerged while talking about something that had pre/occupied me before. That is, as I articulated ideas that I have already thought through for some other person, new dimensions of ideas emerged and new forms of thought. Here, then, we see how talking leads to thinking and, clearly, earlier thinking has been articulated in speaking. Following her account of the event in the chapter, Pei-Ling asks the question how it is possible that thinking develops in speaking, especially in the light of the fact that the current canon supposes thoughts and structure of mind to predate the speaking that makes the former available to other people.

But there is still more to this episode. Upon closer inspection we see that it is from what has been said that the speaker discovers a new idea while talking about what he has been working on and thinking through. It is in the spoken that the speaker realizes to have a new idea, a form of thought that goes beyond what he has been thinking up to that instant. Here, communicating apparently precedes the realization of novel thinking; what has been newly said and realized now has to be written down, because it might be forgotten otherwise – which it would not have to be if it already were fixed thought encoded in mental structure. That is, unlike in the constructivist literature – where a mental framework, the outcome of a construction, is externalized for some audience – a close analysis of this episode

W.-M. Roth, *Passibility*, Classics in Science Education 3,
DOI 10.1007/978-94-007-1908-8_7, © Springer Science+Business Media B.V. 2011

reveals a reversal of the common idea about the relationship between thought and communication. Here, communication precedes the realization of new ideas, new forms of thought. This is possible because "language certainly has an interior, but this interior is not a thought that is closed in upon itself and self-conscious" (Merleau-Ponty 1945, p. 225). It is precisely in communicating that the new forms of thought are produced and these new ideas are recognized by the speaker in or after the fact: "Speech creates a new sense, if it is authentic speech" (p. 226). But in communication, not only the vocal cords move, it is the body as a whole that becomes expression. Therefore, it is "the body that has to – in the final analysis – become the thought or intention that it signifies for us" (p. 230). Hence it is in and through the body that our linguistic consciousness becomes aware of the thoughts that the human being as a whole is capable of. Thought, ideas, and sense are immanent and born in the living/lived body as a whole rather than in the mind alone, generally thought of as the grey matter in the brain case. In and with its movements, the living/lived body allows thought, ideas, and sense to be revealed to the recipients of the communication as well as to the communicating subjects themselves.

There is still more to this. If I discover a new idea in what I have said, that is, while or after having said it, then there is a dehiscence, an experience of otherness where my thoughts are revealed to me only during or following "externalization." I am not a master of my thoughts or thought content but a willing recipient or giftee. "Everything happens as if the intention of others inhabited my body or as if my intentions inhabited his" (p. 215). Rather than having thoughts first and revealing them to others, I discover my own thoughts in my communicative gesture, which, because it is for the other is not entirely mine. In so doing, I draw on resources that have come to me from the other, which therefore are inherently other. The sense of the communicative gesture "does not lie 'behind' it, it is confounded with the structure of the world that the gesture outlines" (pp. 216–217). That is, my "linguistic gesture, as all others, outlines its own sense" (p. 217) rather than having a thought that lies behind and creates it: My "thought and expression, therefore, constitute each other simultaneously" (pp. 213–214). In the preceding chapter, we see that the language we use has come from the other and, in my speaking, is for the other to whom the language returns. Because my thought is "nothing 'interior,' it does not exist outside of the world and words" (pp. 213), I can find it in my expressions. But if I find my thoughts in my expressions, then I am passive with respect to my own thoughts. My awareness of my thought follows the thought itself, which I find expressed in my communicative gesture. Moreover, accompanying the passivity is an otherness, whereby my own thoughts are accessible only through the expressions, which, as expressions that are intelligible for the other, bear her marks as much or more than my own. In a very deep way, therefore, autonomy and heteronomy are deeply and fundamentally intertwined.

Thought and Word – In Theory

> Thought has its own structure, and the transition from it to speech is no easy matter. (Vygotsky 1986, p. 250)

In the preceding chapter I suggest that the constructivist metaphor and associated conceptual change research treat interviews as "core dumps." In the older computer technology, a core dump was a printout of the entire memory, whereas in more recent use it is a memory image of a particular process. By reading the printout, computer scientists can figure out the present state of the computer memory or process in action at the time of the core dump. In a similar way, researchers subscribing to the constructivist metaphor read interview transcripts to extract conceptions and conceptual change.[1] To do so, constructivist (as cognitive psychological) theory requires the representation of words and concepts I use. But, a phenomenological perspective counters: "I do not have to represent to myself the word to know it and to pronounce it" (Merleau-Ponty 1945, p. 210). In Chapter 6, we see how the interview is itself a context in which particular ways of talking about the topics broached come to be produced collaboratively and where one person is speaking *for* the other as much as for him-/herself. We see that language allows persons to talk about phenomena (issues, topics) even without having talked and thought about them before. In the previous chapter, we observe how two participants to an interview actually produce the topic and produce its cohesion over a brief moment of time. What we have not considered in that chapter is the question of the constancy of thought. The point of conceptions and conceptual change research is that there are underlying structures of mind that generate the talk about phenomena. Language is but the medium by means of which the structure and what it does come to be externalized, as if in a core dump, and come to be made accessible to the conceptual change researcher.

This idea is not a new one. Already for the Greek, the voice (speech) was but a secondary phenomenon by means of which the soul expressed itself in a public manner and forum. Thus, for Aristotle, the sounds emitted by the voice are symbols of the states of the soul; and the written words are symbols for the spoken word. The soul directly corresponds to things and thoughts by means of a natural resemblance. Importantly, the states of the soul are thought to be common to all and so are all those objects that the states of the soul are copies of. The voice therefore is as close as we can get to the signified whether it is determined as sense (thought) or as thing (Derrida 1967). In the research literature today, this, by and large, still is the going approach: The voice is a reflection of thought (structures) and the transcript is but a representation of this voice. But, so the phenomenological philosopher argues,

[1] "I've been there and done it" and I have worked with and observed numerous researchers working from a "constructivist" perspective to be certain that this is a correct gloss for how interviews are treated in the field.

> if speech presupposed thought, if talking were first of all a matter of meeting the object through a conscious intention or through a representation, we would not understand why thought tends towards expression as towards its completion, why the most familiar object appears indeterminate as long as we have not recalled its name, why the thinking subject himself is in a kind of ignorance of his thoughts so long as he has not formulated them for himself, or even spoken and written them, as is shown by the example of so many writers who begin a book without knowing exactly what they will be putting into it. (Merleau-Ponty 1945, p. 206)

In his analysis of the relation of expression, language, and speech, Merleau-Ponty shows the fundamental similarities that exist among empiricist, mechanistic psychological and intellectualist (constructivist) theories. Thus, "as far as speech is concerned, intellectualism is hardly any different from empiricism . . . since the word is a passive shell. Thus we refute both intellectualism and empiricism" (p. 206).

A very different approach to the question of the relation between thought and language has been proposed in the early part of the 20th century. In this approach, thought and speech are considered to be processes that stand to each other in a dialectical, that is, mutually constitutive relationship (Vygotsky 1986). Thinking is constitutive of speaking, and speaking is constitutive of thinking. Like two branches of the kiwi tree, which interlace and in interlacing mutually shape how they grow individually and together, thinking and speaking develop together, continually exerting influence on and meeting the other process; and, in this process, they are influenced by the partner process.

We can think of each process as a one-sided expression or manifestation of an overarching process, which in the literature has come to be termed *word-meaning*. This overarching process cannot be accessed itself, in its Being.[2] Thus, the relationship between thought and language, as traditionally theorized, is that between two externalities, "as a mechanical, external connection between two distinct processes" (Vygotsky 1986, p. 3). However, "speech and thought would admit of this external relation only if they were both thematically given, whereas in fact they are enveloped one in the other" (Merleau-Ponty 1945, pp. 211–212). Sense permeates speech, and speech is the external manifestation of sense. Attempting to understand one or the other by taking them apart rather than analyzing both from the perspective of their fundamental unity comes with dangers. Thus, "the analysis of verbal thinking into two separate, basically different elements precludes any study of the intrinsic relations between language and thought" (Vygotsky 1986, p. 3).

For Vygotsky, the problem with the psychological science of his day – which has not changed in this respect until the present day – is that it does not concern itself with the unity of consciousness and the connections and relations between its different functions. Thus, to understand the relationship between thought and language, we need to think them as mere manifestations of the higher order process.

[2] For an analogy, think about light. It is not directly accessible in its nature but manifests itself as particle or wave depending on the experimental conditions. Just as thought and speech, wave and particle nature are but (one-sided) manifestations of a higher unit.

We could also say that there is a unity of the word (Gr. *logos*) and thought (Gr. *logos*), that is, there is a unity of the two in logos. Vygotsky agrees with and favorably cites the linguist Sapir, who suggests that the psychic value of language varies widely and freely, depending on selective interests and attention. Thus, "from the point of view of language, thought may be defined as the highest latent or potential content of speech, the content that is obtained by interpreting each of the elements in the flow of language as possessed of its very fullest conceptual value" (Sapir 1921/2004, p. 10). The problem of constructivist and traditional psychological approaches is that they infer from language use always the fullest possible conceptual values without making the actual psychic value of language an empirical matter.

Once the question of the relation between thought and language is approached from a unitary perspective, the relationship between intellect and affect also can be treated in a more satisfactory manner: it is more than an external relation. The traditional approach to treat affect and intellect as different subjects of study is a major weakness, "since it makes the thought process appear as an autonomous flow of 'thoughts' thinking themselves, segregated from the fullness of life, from the personal needs and interests, the inclinations and impulses of the thinker" (Vygotsky 1986, p. 10). Thought, when it is isolated in this manner, can only be a "meaningless epiphenomenon incapable of changing anything in the life or conduct of a person or else as some kind of primeval force exerting an influence on personal life in an inexplicable, mysterious way" (p. 10). Moreover, the going approach cannot provide us with explanations how thought influences, in the way it does, affect and volition.

Because of the problems with the theoretical term "meaning" (see Chapter 9), I refrain from using it and follow, instead, those who think in terms of signification, captured in the *theme* (Bakhtine [Volochinov] 1977) and *participative (unindifferent) thinking* (Bakhtin 1993). The theme is the singular and never repeatable relation of (a) the currently unfolding cultural-historical (social and material) situation, including all forms of signs in addition to speech and (b) its ideological reflection in participative thinking. Thus, both Bakhtin and Vygotsky draw on the same excerpt from a notebook that the writer Fyodor Dostoyevsky has kept. In it, the author depicts a scene of six heavily inebriated workers. The first worker utters a word that is so obscene that neither Dostoyevsky nor Bakhtin or Vygotsky dare putting it down on paper. The second utters what is recognizably the same word (sound envelope). But it is no longer the same, as its function now is radically different. The third worker repeats the word, then the next and so on until all six of them have pronounced the same word. But, and this is the conclusion of all three authors, it or rather its function is very different, not in the least because each repetition occurs against a changing background context, which includes all previous utterances. The sameness of each utterance of the obscenity is its dictionary sense, whereas its changing, contingent function is the theme. Being attuned to the situation, thinking *participatively,* the workers understand what is being said even though the word and the dictionary sense of the said remains the same. As such, the theme (also thematic unicity) is continuously unfolding in time. It is one-sidedly expressed in the signification of the word, which is its iterable dimension, the one that can be found in the dictionary. From this perspective, we can think of the word as a sign in a second manner accompanying that of expression: it is an index to a thought rather than the thought itself.

Giving Birth to Thought

Lectures are often thought of as situations in which a professor makes public his knowledge to the audience. In the common, non-constructivist literature and ways of speaking, the professor is said or might say about himself to be engaged in an effort of getting the "material (information) across." It is a process not unlike that of our analogy of a core dump, where scientific knowledge comes to be displayed and therefore made public (available) to an audience. In the course of my research on lectures in science, I have not found evidence that lecturing is analogous to a core dump. Rather, there is evidence that during lectures, thinking develops with speaking both at the micro-scale of the unfolding lecture and over longer time periods. Thus, many professors notice in the course of their careers that they not only become better at lecturing but also that they come to better understand their own subject matter in the course of teaching it. That is, speaking in lectures is associated with the sense of better understanding the subject matter. It is in speaking – in and through lecturing – that these professors come to understand the subject matter better. Sometimes this realization of a deeper understanding occurs even during the lecture itself. That is, it is not that cognitive development and new mental structures *precede* the new talk, but rather, new forms of understanding, new levels of cognitive development, are discovered in talk – just as this is made evident in the exemplary episode that opens this chapter.

For this chapter on the relation between thought and communication, I selected from my database an episode from a third-year university physics course on thermodynamics. The lecture has advanced to phenomena that exhibit adiabatic processes, one of which is the magnetocaloric effect used to cool substances to very low temperatures. At the time of the recording, the professor has had more than 30 years of experience. We may gloss what he is in the process of doing in this way: A magnetocaloric substance is placed in a magnet so that its elemental magnets are aligned when the field is turned on; in this step, the temperature is held constant. Alignment means higher order and lower entropy. When the magnetic field is then turned off at constant entropy, the temperature will drop. In an entropy-versus-temperature diagram, these two processes show up as two parts of a step: drop in entropy at constant temperature (rise of the step) and drop in temperature at constant entropy (run of the step).

We pick up the lecture just after the professor has indicated that insights about adiabatic demagnetization processes can be gained from a graphical representation in which the system is represented in a diagram where entropy (S) is plotted against temperature (T). He has begun to draw a coordinate system with the two axes labeled and has added two curves labeled "$B = 0$" and "$B \neq 0$" representing a zero and non-zero magnetic field, respectively.

Fragment 7.1a

```
22      (2.48)
        [((steps back to look at dia-
        gram))
23      when (.) the 'first ˇprocess
24      (1.41)
25      when you;
```

```
        (3.75)
        [((stares at diagram)) (as
        if thinking)) ((looks from
        down to up
```

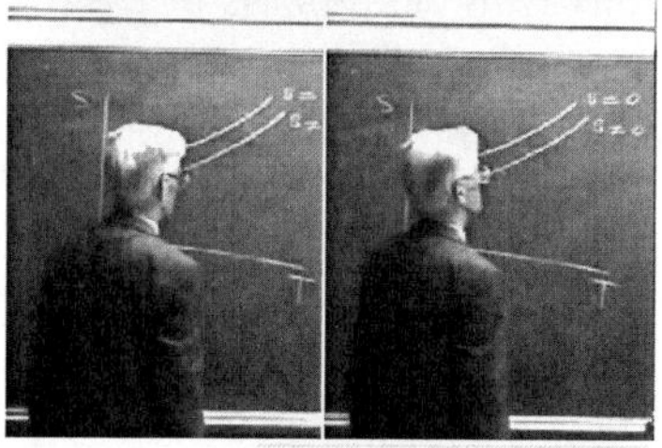

```
27      when you; (0.30) <<f>put> the material
        in a magnetic fIE:ld at a constant
28      tEMprature (0.47) <<p>[its just]> like
        'that
                                [((draws down-
        ward line beginning at current hand
        position))
```

He now has stepped back silently gazing at the diagram produced so far (offprint in turn 22), the left hand raised to the face around mouth and nose. It is as if his entire body is in the waiting, as if it were thinking about what to say next. He then begins in a stop-go manner, with long pauses, "when the first process (1.41), when you, when you put the material in magnetic field at constant temperature, it's just like that," the very last part accompanied by drawing a vertical line from the curve labeled "$B = 0$" to the one labeled "$B \neq 0$" (turns 23–28). Written as in the previous sentence, without the pauses and changes in intonation, the event easily can be thought of as a core dump. This is so because all the other signs that point to a process not only unfolding but also configuring itself have been omitted. That is, when we transcribe the lecture leaving out everything but the corrected flow of words, lecturing *does* look like a core dump. But this is an artifact of the method. Once all the pauses and mishaps of speaking are included, we have to ask for a different explanation of the production of speech.

We do not know what, if anything, is going on in the conscious mind of the professor. We see the professor pause and we see him look at the two curves near the right end of the diagram, lower his gaze toward the axis labeled "T" where it remains for a time, then raise the gaze back up to the pair of graphs. The gaze moves down to the *T*-axis, then up again, before the professor finally draws a vertical line downward between the two curves. That is, his gaze has moved already vertically in the horizontal location where he subsequently draws the line. He has also announced that the material is placed in a "magnetic field at constant temperature" (turn 27–28). It is only then that he also draws the vertical line, consistent with the vertical movement of the head and the articulation of the process as involving constant temperature. But the fact is that the production is very slow, involving a lot of staring and many head/gaze movements. This, therefore, is more consistent with a thought in the making rather than the dumping of the contents of mind. The entire person, the living/lived body as a whole with all of its dimensions is in the process of evolving thought, which includes pauses. My own phenomenological investigations of problem solving processes show that such pauses correspond to periods of waiting, where the conscious mind is empty and in a state of waiting for something to emerge from parts of the Self inaccessible to the conscious mind and the linguistic consciousness that characterizes it. It is but another aspect of the passivity with respect to our own thinking, which does not occur merely in terms of representations, explicit signs that we manipulate. Every thought is a movement – it "moves, grows and develops, fulfills a function, solves a problem" (Vygotsky 1986, p. 218) – and, as such, includes the forms of temporality proper to it, including pauses, accelerations, decelerations, and rhythms as much as periods of constant displacement. What we observe here is the process "of verbal thinking from the first dim stirring of a thought to its formulation" (p. 217), which, here, is the process from the idea of providing another picture of the magnetocaloric effect to the final articulation of the thought. This, in the present instance, once fully developed in speech, gestures, and inscriptions, turns out to be an inappropriate thought as the professor himself characterized it.

In the second part of this fragment (Fragment 7.1b), horizontal movements of the head, consistent with constant entropy, and horizontal movements of the hand, precede the actual drawing of a horizontal line from the point where the vertical has intersected the lower curve. That is, the entire living/lived body is engaged in the thinking; its movements *are* the thinking made available to the students and researcher, who can also see his hesitations. Pauses, repetitions (three times "when you"), interjections ("uh"), and signs of uncertainty (voice fades away in turn 34) mar the production of the phrase that we might gloss as "When you adiabatically demagnetize it [the material], its temperature is lowered" (turns 31–34). That is, in these expressions there is little if any evidence that we observe a core dump; rather, there is a lot of evidence not only for the speaking as an unfolding process but also for the thinking as an unfolding process.[3]

[3] Unless a psychologist attempted, as this is the case for grooming and beat gestures, to explain the hesitations and mishaps as the result of faulty wiring and pathways that interfere with the dumping of the memory contents. In Roth (2002), I also review competing theories on the role of gestures in communication, one of which theorizes them as an epiphenomenon produced during problems in the externalization of thoughts into speech.

Fragment 7.1b

29 [(1.79)
[((looks left [Fig.] and right at the graph))

30 a:nd when you; (0.95) uh:::[((a)) that
31 when you when you then uh [((b))
a:dia:^bATic[ly:: [
[((c)) - [((e))
(0.53)

32 demagnetize it, (0.44) it uh::::
33 (0.30) [y its 'that,]
[((draws horizontal line left to right))
34 <<p>[and so] its temperature is lowered.>
[((horizontal gesture))

Here, I take the entire performance as an ensemble of expressions that stands in a one-sided relation to thinking. Thought and speech are not directly linked in simultaneity. Thought and speech are two non-identical processes, "and there is no rigid correspondence between the units of thought and speech" (Vygotsky 1986, p. 249). The hand movement, which gestures a line or movement from left to right, unfolds parallel to the temperature axis (from offprint b to offprint e in turn 31). Before that the hand is held in the same position for about 3 seconds, which is an indication for the thought in development, where I understand movement as an expression of thought, as *thinking in movement* (Sheets-Johnstone 2009). Thinking in movement does not follow thought: it *constitutes* movement *as* thought. The head movement, which is from right to left near the entropy axis and back to the previously drawn vertical line is an indication of this early part of thinking – though we cannot tell from what we observe how much it is developed.

The professor continues to produce such steps but then stops when a horizontal line he gestures no longer intersects with the upper curve ($B = 0$). He steps back, as if he intended to look at his work from a remove. He then announces: "I think I said there is something wrong with that picture." In fact, he has said precisely that during the preceding lecture after having produced the same graph. Despite the fact of having already done it once before, he now has reproduced the same graph only to note again that something might be wrong with the picture of the process he has represented in and with the diagram. How could it be that he has produced something that *after the fact* he determines to be wrong or have something wrong with it? Why would anybody in his/her right mind "dump the core" knowing that what comes out is wrong, not only noted as such in the present lecture but also articulated in the same during the preceding one? From a conceptions perspective, something in his cognitive structure would have to be amiss. It is the fact that the entropy of a material at the absolute zero temperature is different from zero and has different values depending on the magnetic field (the curves intersect the S-axis at different values). That is, from a conceptual change perspective, the professor would be said to have a misconception about the entropy, even though one of the most basic facts of thermodynamics is that the entropy S at temperature $T = 0$ K is chosen to be zero ($S = 0$). Perhaps conceptual change theorists want to suggest that he has forgotten a basic fact of the theory he is in the process of teaching? How does the constructivist metaphor explain *forgetting*, when the issue is concerned with a fundamental structure of the scientific phenomenon?

After having turned away from this diagram, which he has formulated as producing because it leads to some insights, the professor continues the lecture for 15 minutes, adding a few more equations. While articulating the second of two implications from one of these (Helmholtz) equations, he all of a sudden gazes at the previously drawn diagram and makes an announcement while pointing to the diagram: "The second implication is that this is wrong." He wipes off the chalkboard what he had done before, then produces another diagram this time with the two curves intersecting at the origin ($T = 0$, $S = 0$).

Phenomenology of Thought and Expression

> Thought is not merely expressed in words; it comes into existence through them. (Vygotsky 1986, p. 218)

Vygotsky provides a lot of empirical evidence to support his contention that thought cannot be identical with speech with respect to structure and development. It is therefore not surprising that he articulates himself in the way quoted above, "thought does not express itself in words." We have to recognize that for the "speaking subject, thought is not a representation . . . The speaker does not think before speaking, not even while he speaks; his speech is his thought" (Merleau-Ponty 1945, p. 209). That is, the words – and, by extension, any other non-verbal sign produced as part of communication (gesture, prosody, body orientation, body position, body movement) – are not expressions of thought, where the verb to express signifies pushing out or expelling contents from mind in the manner toothpaste is pressed out (*ex-*) from a toothpaste tube. Rather, both the philosopher Merleau-Ponty and the social psychologist Vygotsky describe and theorize thought as coming into existence in speech. In other words, thoughts become reality while concretizing themselves in the words uttered. Before the articulation, thoughts are potentialities, like seeds are potential trees; potentiality is turned into actuality when the seed germinates, takes root, and grows in the particular location where it finds itself. This process of "actualization is . . . a display of what is one's own in the universal element whereby it becomes, and should become, a matter of everyone" (Hegel 1979a, p. 309). Any inner that we may speak of, in fact, "is the activity itself. The speaking mouth, the working hand – if you like the legs too – are the performing and actualizing organs that have within them the action *qua action*, or the inner as such" (p. 234). We therefore do not need to go into the head or mind of a person to find thought. All we need to do is analyze the activity (*Tätigkeit*) generally and the language in activity more specifically, which articulate the inner as such. It is precisely in activity that thought actualizes itself and becomes available not only to the other but also to myself.

The professor in the episode has come to his lectures with the genuine concern of teaching thermodynamics to the third-year university students. He has no intention to lead them up the garden path. From his perspective, as he told the researchers, the point of this lecture is to "get the idea across" that the magnetocaloric effect can be understood in terms of a graphical representation in which the entropy S is plotted versus the temperature T when a sample is subjected to the turning off and on of a magnetic field B. At the beginning of the lecture episode, this thought exists only generically. Like a seed that might develop into a giant tree or into a shrub depending on the climatic and geographic conditions, the thought realizes itself in the expressions so that the speaking professor finds what he thinks in what he has said.[4] Thinking about the lecture in this manner allows us to understand why the professor says "I think (I said) there is something wrong with the picture" after having articulated the picture verbally, gesturally, and diagrammatically. His own thought and the problems therein are available to him only after he has articulated them in public and only after he has articulated them for others as much as for himself. This is possible precisely because the "connection between thought and word . . . is neither preformed nor constant. It emerges in the course of development, and itself evolves" (Vygotsky 1986, p. 255). And it is in the expression that the professor discovers his own thought.

This relation between thinking and expression exists not only at the scale of the lecture segment as a whole, which has articulated a graphical expression of the magnetocaloric effect, but also at the production of aspects of the segment. Thus, for example, we see the professor move his hand back and forth across particular

[4] Biologists have recorded differences in the phenotype of plants grown from the same seeds when grown in different locations or when seeds from different regions are grown in the same location. Differences may include plant size, growth habit, leaf shapes, and fecundity. The specific phenotype in which a genotype expresses itself, therefore, is contingent upon the geographical and climatic conditions in which the seed is grown.

parts of the existing diagram prior to adding another line. The gesture here is thinking in movement, which, once expressed, can be evaluated in and by conscious awareness. Sometimes the movements appear to be searching, as the hand moves along different trajectories before finding one that becomes fixed when it is repeated with the chalk held to the chalkboard. That is, the actual drawing is but a trace of an explicit thought that was formed in expressive movement. Conscious thought thereby comes to follow thinking in movement, which itself occurs across the entire living/lived body of the person. This thinking is immanent rather than explicit, and it is only after being evaluated in consciousness that thought comes to be explicit and that the conceptual content comes to be recognized. Being immanent, this thinking is not available to me in intellectual consciousness, even though it can become the content of intellectual consciousness as the result of a process of becoming aware.

There are consequences of the fact that much of our knowing is immanent, that the constitution of all knowing is immanent knowing. First, immanent knowing allows us to do what we do without conscious, intellectual awareness. We play soccer, golf, or tennis without conscious awareness of our movements. In fact, when athletes consciously focus on what they do, hitting a tennis ball or teeing off, then they play worse than they have played before. Any improvement is observable only after they have returned to rely on their immanent knowing, thinking in movement. If this is the same for thinking, the implication is that thinking about thinking physics – metacognition – would interfere with thinking physics.

Second, because immanent knowing does not require and operates without intellectual consciousness, conscious awareness not only follows thinking in movement but also, and more importantly so, *conscious* subjects are passive with respect to what they do. We are not only subjects *of* ideas but also *subject to* ideas. We already encounter this notion in Chapter 4 as the gap between situated action and plans that are said to underlie them. In the present context, the professor finds what he thinks in expressions that he has produced himself. He finds himself to have thought in a particular way, which he evaluates after the fact to have been wrong.

Third, the fundamental otherness underlying the communicative process, the dehiscence between the person and his thoughts derives from a much more fundamental non-self-identity of any communicative sign as such. Non-self-identity means otherness at the heart of the Self. It means that the Self ~~is~~ other, as has been suggested in the formula "soi-même comme un autre [oneself as another]" (Ricœur 1990) or in Rimbaud's (1951) diction "JE est un autre [*I* is another]." No word in my mouth is mine because, as an entity in *con*sciousness, it is "impossible for one person, but . . . becomes a reality for two. The word is a direct expression of the *historical nature of human consciousness* " (Vygotsky 1986, p. 255, emphasis added). The word is not mine because it has a cultural history and because it reflects human *con*sciousness generally rather than my individual personal consciousness specifically. Consciousness is inherently shared (*con-*, *com-* = with) and therefore cannot be my own. A phenomenological analysis of the origin of this strangeness is provided in Chapter 11 in the context of an auto/biographical story where a singular experience becomes common and mundane when brought into the realm of talk. That is, Being estranges itself when it is accounted for in terms of beings – linguistic and paralinguistic signs that cannot ever make truly present the presence that has receded into the past and thereby has been lost. In other words, *re*presentations always follow that which has been present, which is also a present that cannot be brought back identically, in its original singularity, but only in terms of generalities. In this manner, all thought follows communicative movement, whether it takes the form of movements of the vocal cords and changing resonance cavities, movements of the limbs and head, or movements that change the orientation of the body as a whole.

Fourth, if thinking develops in the course of speaking, then we must not treat interview transcripts as a coherent, a-temporal text in the way we may take the printout of a core dump. Rather, because thinking changes in expression, the beginning of an interview or lecture has to be treated differently than the ending. Whereas the beginning is part of the structure and background of the ending, the reverse is not the case. We must not say that the beginning is consistent with or contains the ending. This is so because in the process of speaking, thinking changes. These changes that subsequently occur are not available in the early parts of the communication. Ultimately, therefore, the analysis of communicative events has to be a historical one, focusing on thought as a process under development rather than as the mobilization of fixed cognitive structures.

It is in the process of becoming conscious that the level of conceptualization in and of the communication comes to be shaped. Thus, it has been suggested in a passage that Vygotsky himself like a lot that "language is primarily a pre-rational function. It humbly works up to the thought that is latent in, that may eventually be read into, its classifications and its forms; it is not, as is generally but naively assumed, the final label put upon the finished thought" (Sapir 1921/2004, p. 10). Thought is read into language or, as I put it here more generally, into communicative action – not only on the part of the recipient but also on the part of the speaker. The thinker thereby becomes radically passive with respect to his thought, though he is also agential with respect to the production of the communication, to the intention of which he is again radically passive.

Chapter 8

Otherness of Self

> If it is Others from whose reactions I derive some indication as to the properties I can ascribe to myself, and if my knowledge of these Others is the result of my own construction, then there will be a lot of re-interpreting and re-constructing to be done in order to arrive at anything like a non-contradictory notion of a person that I would call "myself." (von Glasersfeld 1989b, p. 446)
>
> JE est un autre [*I* is an other]. (Rimbaud 1951, p. 254)

The two opening quotes provide us with two radically different perspectives on the nature of the Self. In the first quote, the Self is the result of a constructive process that occurs completely on the inside of an organism that can only test the viability of its construction with the outside, using its senses, without ever having any other recourse than itself and its sense of viability. This Self is in a process of regress, where some acting subject constructs itself a Self within itself. This is a Self caught within itself, trapped by discourse that makes it the center of a world that could also be a mere figment of the imagination because everything is constructed by the Self for itself. But already I when consider my own perceptions, I realize that I am not closed upon myself in the way von Glasersfeld attempts to make me believe: "When I consider my perception itself, before any objectifying reflection, at no moment am I aware of being shut in within my own sensations" (Merleau-Ponty 1945, p. 464). A finger pointing to a steeple "is not a finger-for-me that I *think* of as orientated towards a steeple-for-me, it is [the] finger which itself shows me the steeple" (p. 464). It is not that I construct for myself the steeple, the finger, the Other pointing with the finger toward the steeple: All of this is present to us without representation, including the Other and myself. It is not surprising, therefore, that other philosophers as well emphasize an architectonic division of the world into *I* and Other "that is something-*given* as well as something-*to-be-accomplished*" (Bakhtin 1993, p. 75) in and as part of *Being-as-event*.

In the second quote, the poet Arthur Rimbaud recognizes himself to be another, a Self that is derived from the Other and therefore a Self that could not exist without the Other. I am possible only because of the Other: Self is understandable and thinkable only because of an originary *with* that gave rise to both Self and Other

W.-M. Roth, *Passibility*, Classics in Science Education 3,
DOI 10.1007/978-94-007-1908-8_8, © Springer Science+Business Media B.V. 2011

(Nancy 2000). Rimbaud writes about the recognition that not only language comes to us from the other but also everything else we could use to talk about and constitute a Self is other. The upshot is that the Self inherently is non-self-identical with itself: it is both Self and Other. More recently, phenomenological philosophers concerned with understanding our condition – rather than relegating their interests exclusively to the mind at the expense of everything else – have but confirmed the intuition of the poet. Thus, "the subject is not *in itself*, at home with itself, such that it would dissimulate itself in itself or dissimulate itself in its wounds and its exile, understood as *acts* of wounding or exiling itself" (Levinas 1990, p. 83). Here we find precisely the opposite message to the intellectualist (constructivist) one available in von Glasersfeld's expressions. The subject is *not* withdrawn into itself; Being is not about me: whatever I can understand about myself, my very Self, has the Other as its condition.

In this chapter, I use interviews with a new teacher during her four-month internship in a local middle school to exhibit the otherness that functions at the heart of our Selves. At the middle school I taught a science unit with her in which our students contributed to the communal knowledge base about a creek that an environmentalist group was in the process of establishing. The teacher (Nadely Boyd) and I frequently talked about what it means to be and become a teacher. The excerpts from our conversations serve me to reflect on this essential otherness of the Self that Rimbaud remarked, which exists not only in science teaching and learning, but also is an essential condition of who we are and who we can be.

Views on Self and Other

Responding to his own question of the "subject in the I" and how the Self – here denoted by the reflexive pronoun – comes to be, von Glasersfeld (1989b) suggests that the latter is the result of an inner-individual construction. The process "probably begins with the infant's discovery that, having isolated moving shapes in its visual field, there is a way to distinguish some of them" (p. 446). Unnoticed by the author is the fact that in this beginning, the subject of action and discovery is already presupposed as well as its tools that are used in the construction. Thus, the subject that constructs an image of itself in the Self already exists prior to the constructive process to which the constructions are ascribed. To ascribe properties to "myself," there has to be something that I consider "my" "Self" to which something is ascribed – but von Glasersfeld does not tell us where this Self comes from. It lies outside and comes before all constructive work. This problem has already beleaguered the work of Kant, who never investigated the nature of the subject of the constructive process. In fact, von Glasersfeld disbands any interest in the question of the topic by suggesting that "the self as the operative agent of construction, *the self as the center of subjective awareness, seems to be a metaphysical assumption and lies, at least for this constructivist, outside the domain of empirical construction*" (p. 447, original emphasis). The very question concerning the possibility of the constructing subject is thereby evacuated from the conversation that was to ask how a Self could exist in the first place. Moreover, the author claims the existence of the Self to be a matter of metaphysics apparently unconcerned with the fact that the (Kantian) constructivism he advocates is the pinnacle of Western metaphysics.

The problem with the constructivist metaphor lies in the presuppositions it makes about the Other and the role of culture in the constitution of the Self. The constructivist metaphor posits rather than explains the cognitive subject, which is said to construct itself. However, "[i]f what a cognizing subject knows cannot be anything but what that subject has constructed, it is clear that, from the constructivist perspective, the *others* with whom the subject may interact socially cannot be posited as an ontological given" (Glasersfeld 1989a, p. 126). But if it were otherwise? What if the constructivist position was untenable and unable to provide an account of the human condition? What if the very possibility to use anything like language required to think – see Chapter 7 – necessarily comes from the Other? Then, in fact, some form of inclusive *with* exists prior to and constitutes the very condition of the Other, the individual Self, and the construction of Other and Self. According to Bakhtin (1993), for example, there is not one but there are two value centers that are correlated with each other though fundamentally and essentially different: "myself and other; and it is around these centers that all of the concrete moments of Being are distributed and arranged" (p. 74). That is, unlike presented in the constructivist metaphor, all concrete moments of Being (life) are organized around *two* centers rather than around one. Life can be understood only through this double orientation rather than through the organization around one inner self-consciousness.

Language already provides us with some hint that things are in this second way rather than in the way that the constructivist metaphor suggests. The term *consciousness* derives from a Latin composite word, the first part of which denotes "with" (Lat. *con-* or *com-*) and the second part of which denotes knowing (Lat. *sciēre*). That is, consciousness means knowing together, with others, and self-consciousness then means self-knowing-with others. Even (knowledge of) emotions, which is often reduced to the individual level, cannot be "constructed" by the individuals, because any "I" would not know, for example, what anger means and how it feels if its "I" were not to take the regard of the other upon itself. As a child, I learn when I am angry when others ask me, "Are you angry?" or tell me, following a tamper tantrum, "Go outside until your anger has gone, then you can come back in."

Philosophers concerned with understanding the nature of Self and Other tend to agree on the Other as a necessary condition for any Self – which is possible only because of the Other. Thus, my very nature is a function of the Other. "I" can have a nature only when there is an Other, "whatever or whoever he may be, whatever may be his relations with me, and without his acting upon me in any way except by the pure upsurge of his being – then I have an outside" (Sartre 1956, p. 263). This leads to the fact that "my original fall is the existence of the Other" (p. 263), the original sin, the recognition of myself and my nakedness, does not come from the apple that I have taken off the tree of knowledge. The original fall derives from the collectivity and its capacity for language – which has arisen prior to any consciousness on the part of its user to be speaking a language. Even without acting upon me, the Other is my condition. It is the existence of the Other that is my original fall. Unlike constructivists, who attempt to explain how the individual mind constructs the other as well as itself, phenomenological scholars agree that without the Other there is nothing like the Self.

Where does this perhaps strange-sounding condition come from? Why is constructivism wrong on this account? Emmanuel Levinas, concerned with establishing a first philosophy, that is, a philosophy that does not presuppose the Self in the way constructivism does, links the problem to the way in which intellectualist approaches theorize language. The original saying is not one of objectifying and of ascribing "meanings" (Levinas 1990). If speaking were understood as the giving out of signs that others interpret then we would not understand our cultural-historical condition today, our ways of thinking and speaking. Therefore, if speaking were to be understood as the giving out/off of signs, then the "'delivery of signs' would amount to a prior representation of these signs, as though speaking consisted in the translating thoughts into words and consequently in having been first *for-oneself* and *at home with oneself*, like a substantial consistency" (p. 81). In this case, the relationship with the Other would then spring forth from an intentionality of a subject that existed in and for itself. Thus, "the subject of *speech* does not give signs, it makes itself a sign, turns into an allegiance" (p. 83). And, with this framing, the issue of the relationship between Self and Other returns to that of the body as expression in communication (see Chapter 7).

As exposure, communication with the Other also means passivity. "The passivity of the exposure responds to an assignation that identifies me as the unique one, not by reducing me to myself, but by stripping me of every identical quiddity" (p. 83). The phenomenon of speech itself is an index of this passivity: "in the saying, this passivity signifies, becomes signification" (p. 84). Saying, thereby, constitutes the most radical passivity, a passive passivity that is inseparable from patience and pain – two forms of experience that I return to in Part D of this book. Another way of talking about passivity is that we are vulnerable to being affected, and we may be affected only because of a fundamental passibility that is the source of affect and affectivity. That affect is not part of a constructivist conception of knowing, learning, and identity (Self) is understandable because it does not include the dimensions that I articulate in this book, passibility and the fundamental experiences of passivity and otherness that are at the core of our constitution. We know because we are subject to being affected, to passivity, including a passivity to the very consciousness that allows us to subscribe to a constructivist ideology.

Whereas it might appear that this view of passivity is particular to phenomenology, we can actually find in speech act theory elements that are consistent with this phenomenological framing of the relationship between Self and Other. In this theory, the speech act as a whole is distributed across speaker and recipient. This is so because there are three moments to the speech act: the *locutionary* act or what is said in the saying (utterance), the *illocutionary* act or intent of (what is intended with) the saying, and the perlocutionary act or effect of (what is done with) the saying. Whereas the locutionary and illocutionary moments of the speech act reside with the speaker, the effect lies with the recipient. The effect is recognizable in the recipient's response, which, in constituting the next turn, allows all members to the setting to perceive the effect. The speech act therefore is distributed across speaker and recipient, which leads to the fact that conversation analysis – concerned as it is with understanding the constitution of society in turn-taking routines and face-to-face interaction rituals – takes the turn pair as its minimum unit of analysis. Speaking thereby comes with a double passivity: the speaker is passive with respect to the nature of the effect of speech, and the recipient is passive with respect to what affects him/her in and through the speech. But if we are recognized for who we are in and through our acts, then the fact that the effect of the speech act is grounded to a large extent in the recipient: as agents we are passive with respect to who we are. That is, "identity lies in the total patience of the one assigned, who, patient – despite himself – does not stop dying, lasts in his instant" (Levinas 1990, p. 86). Accordingly, "it is only in this way that the *for-the-other* – the passivity more passive still than any passivity, the emphasis of sense – keeps itself from being *for-oneself*" (p. 85, original emphasis).

Conversation About Being and Becoming in the Classroom

Accounts of ourselves – biographies and autobiographies – articulate that which is possible within a culture and therefore do not point to a singular individual at all. If it were not in this way, others could not understand my autobiography, which therefore also is biography. Both, biography and autobiography, make use of the same linguistic resources and, in fact, constitute one and the same genre (Bakhtin 1981). "JE est un autre," says Rimbaud in my second introductory quote to this chapter; and this saying is immediately intelligible when I, the author, say "I am a university professor." In this instance, "university professor" is not an attribute that is singularly mine so that in the phrase I merely restate in another form what Rimbaud has said: I am an other, here university professor.

There are many indications about the shared nature of auto/biographical talk, which always is such that it presumes its own intelligibility in being an account for another. Moreover, there cannot be a difference between autobiography and biography, as the participants in an interview situation, such as the one presented in this section, inherently produce the account for each other and together. Thus, the issue at the heart of the following conversation is the teaching experience of Nadely during her internship. I am in the process of teaching a science unit together with her and Michael (Mike) Bowen, being a doctoral student of mine at the time, is helping out. The three of us are talking to Nadely about the first several months at the school prior to our joining her. As we are talking about *her* experience, Nadely talks about herself (her Self), and therefore relates to the account in an autobiographical way. But the other two participants, who collaboratively produce the account (see also Chapter 6 on the co-production of conceptions) with Nadely, relate to the account produced – the one subsequently transcribed and a derivative of the conversation – in a biographical manner. From their perspectives, they contribute to the production of a biography about the first several months in a school classroom.

What one participant says has to be intelligible to the other participants – it does not make sense to speak to another person unless one can presuppose that the said is intelligible. Both form and content of the utterances there are and already have to be shared with the listener. The participants already have their language in common, especially because the interviewers also are teachers and have gone through similar experiences. But besides the words, the genres we use, too, have to be intelligible. Therefore, the ways of telling a beginning teacher's story take certain forms that are different from the ways of telling the biography of another practitioner, for example, a painter or musician. These genres, in turn, involve and are constituted by particular plots and characters. These, too, have to be intelligible and in fact constitute cultural possibilities for telling the auto/biographical story of person during a particular stage of her life.

During this interview with Nadely, everything happens in and with language – and therefore also with exposure. The users *dwell* in the language, mobilize its possibilities for the purposes at hand, which has been framed to be about the main "issues" Nadely has been facing during her internships. But neither Nadely nor her two interlocutors have constructed the language in their minds, nor have they constructed narrative genres or the plots that make up the story lines and the characters that fill and constitute them. Without the language there is no tool for construction, without language there are no story forms to construct. But this language is not constructed from the bottom up by the speakers. It has *come to them from the Other*, contrary to what (radical) constructivists claim; and because it returns to the Other, that is, because it is *for* the other, it cannot be other than what the Other already understands. therefore, language speaks in terms of what is already shared between interlocutors. We are recipients of our languages together with the narrative forms, characters and plots, grammatical possibilities and so on even without ever thinking about plot or grammar construction.

The conversation fragment begins with a question about classroom discipline and control, which I articulate as having understood to be a major "issue" for Nadely (turn 01). Nadely agrees, and then there is a brief exchange about the nature of what we "want to hear about it" (turn 02). I invite Nadely to talk about what it means to her, on this day or during any individual class.[1]

Fragment 8.1a

```
01 MR: Maybe we just start with one issue that I thought, I
       thought may have been central in your teaching and it was
```

[1] The minor exchanges and head nods have been left out, but readers should refer to the previous two chapters concerning the real-time production of talk and the pauses, stumbles, mumbles, stupidities, and so on that are produced in the process. The transcription is already cleaned up a bit to allow me to focus on the next level of issues.

the discipline. And, what, if we could just talk- what it means or how it looks like from your perspective, you know, is that correct? I seem to perceive that the discipline and control of class, yeah, sort of was a major issue for you or?

02 N: Yea, no, it definitely was with these guys. So, do you want to hear, like what do you want to hear about it?

03 MR: Talk about what it meant to you. Or let's say today or any individual class.

04 N: Well, I think, I mean it was hard for, I think I remember at the university you're hearing all these ways and methods and these idealistic ways and when you actually get out there it's different putting it into actions. And I think too, what I didn't really have was any modeling to follow, I don't know what anybody else did, I was sort of stumbling through things myself and so I know, I mean, I know that in September it was a real struggle and a real battle and then I know that my January, February, March, the three months were really good and we sort of came, I sort of work with the kids in developing expectations, I work with the kids in developing consequences and that worked for a while but it didn't continue working. I don't know, I'm, it's something that I think I'm constantly battling with and learning and trying to find new ways or new things and there're some things that I let some people get away with more than I would with others. And I don't know, I mean, I haven't found, I don't think I found, I don't feel like I figured it all out because what I see happening is those people that shouldn't be getting disciplined are getting disciplined and, like this morning, the kids were just out of control, and anyways, everybody came back at recess, but everybody that showed up weren't the problem causes. The problem causers conveniently were the ones that didn't show up for that. Therefore they're coming in at lunch and I'll be surprised to see which ones of those show up for that-

The question sets up the topic as one about discipline and control in teaching, here pertaining to a teacher during her internship experience. The narrative that the interlocutors produce, therefore, is one in which a new teacher is dealing with classroom discipline and about how to control students who might act in inappropriate ways. Already the concepts of discipline and control are characteristic of schoolteacher talk at a particular point in cultural history. How to understand particular classroom events would likely differ if they were taking place in some country school in India or on the African continent, in an alternative school in Europe or North America, in some public school in more traditional societies such as France, or in a North American inner-city school. What the present interlocutors will be producing together is an account of schooling in this geographical part of the world in a middle school of a semi-rural community at the outskirts of one of Canada's top 15 metropolitan areas. Thus, in one of the private schools in the area, the "issues" that make the topic of the present talk may not have arisen at all.

Nadely begins by talking about how hard it was to transition from the university, where she heard idealistic talk about methods, and how it was different when she actually had to put those methods into action. Even this gap between what is taught at the universities – the generic and generalizing methods of teaching, for example, to deal with disciplinary and control issues – is not something that Nadely constructs in her head. It is a way of talking that other teachers engage in as well, in the school where Nadely works at the time and in other schools. She talks about the lack of a model to follow and about how she has been stumbling through initially but then succeeded in having some really good three months at the beginning of the new calendar year. She describes having developed expectations and consequences (for students who do not meet expectations) and changing her approaches. But she also says about herself that she has not been consistent, as she "let some people get away with more than [she] would with others." She talks about kids being out of control and about disciplining the wrong students and about how those students supposed to show up during recess for disciplinary actions did not show up in the homeroom where she was waiting for them.

Up to this point there is nothing in what Nadely has said that we would have to characterize as singular. It is the kind of talk that I have found in schools all over my country, in those schools where I taught (east coast) and was a department head (central) and in those schools where I conducted my research (west coast). Of course, Nadely is talking to two other teachers, and she is not just spilling the beans – i.e., doing a "core dump" – but she is contributing to a conversation to which the others also contribute. This story that they are producing, the text that becomes a text file on a computer hard drive once it has been transcribed, requires everything to be understood by all participants and therefore cannot be singular.[2] Thus, the transcript could be given to other teachers in the country and they would understand what the present participants are talking about. Moreover, pieces of the transcript have been published in scholarly articles to be read by science educators around the world. This publication presupposes the *inherently* intelligibility of the excerpts, that is, it presupposes that a general and generic audience of international teacher educators and theorists of teaching would understand them. It is therefore not surprising that we find the interlocutors to engage with Nadely over and about what she has said.

Mike picks up on one of the issues (turn 05, Fragment 8.1b). Nadely has been talking about the first three months of the year as having been "really good," but that after initially having some success with setting expectations and establishing consequences, "it didn't continue working." Mike articulates the names of the first three months and then asks about April and its difference with other months. Nadely utters an affirmation (turn 06). That is, although she has not talked about the month of April before, there has been something in what she has said that makes it reasonable to speak about the month that followed those previously named. Looking back over the transcript in Fragment 8.1a, we see that Nadely

[2] See Chapter 11 on the relation between the singularity of Being, which is destroyed at the instance we move to accounts.

talks in it about the first three month as having been really good; she later speaks about her strategies as having "worked for a while but [that] it didn't *continue*." Even without constructing something in the head, that is, even without stopping and reflecting about what Nadely has been saying, Mike can ask the question about the events during the month of April. Anyone with a competence of English, who is familiar the order of the months in the year, and who has the required grammatical competency can hear the proper temporal relations in the utterance "it worked for a while but it didn't continue." This hearer can then ask the question about the month of April. This is precisely because of our *participative understanding* (Bakhtin 1993), which accounts for the continuously changing conversational context that is required for understanding the talk at hand. We do contribute to such conversations all of the time without stopping once to reflect and even without having talked about such a topic before (see Chapter 6). That is, in talking it is not only that speakers constitute an orientation to the world, but also they produce resources for recipients to orient in the same manner. There is nothing we do in listening that has anything to do with the activity of interpretation, such as when we painstakingly go through transcript to figure out what is happening in them.

Fragment 8.1b

```
05 MB: Do- Well, I mean, you mentioned the major changes, in
       January, February, March it was okay, April was a little
       different?
06 N:  Yea.
07 MB: I wasn't here last fall so I have no feel for what was
       going on then. But what do you think somebody's influences
       are on, not just on the long-term changes like those that
       you mentioned, but also the day-to-day kind of waxing and
       waning things that.
08 N:  Well, I think with, I think that's the consistency that
       I'm lacking with these guys and I'm letting some things;
       like I was saying earlier, I'm letting some things fine
       and others not and, I'm, I've noticed that, well, you
       know, one thing with Ken ((the regular homeroom and mathe-
       matics teacher)) being away right now, the kids know that
       and, um, they're always when he is coming back and we've
       having all these different subs and all these different
       people that are coming in with the kids and it's not if
       that's happening in my class, it's also in their other two
       classes, their teachers have been on leave for a month, an
       so they'd have all these, all these different, oops,
       sorry, all these different people working, and I think
       that that has a big effect on them. And I think, right
       now, I'm tired and I'm tired of dealing with the same
       stuff again and again and again, and I'm tired of it only
       going so far and it's never going any further, I mean.
       People, people are, I don't know, I don't know what else
       to do, I'm doing, I'm following all the steps that I've
       been told to follow, and I'm going a way I'm supposed to
       be going but things aren't changing. I mean, I can keep
```

```
sending away people down to the behavior problem room but
they need to learn how to work in a classroom. I mean that
can solve the problem temporarily but it doesn't solve the
problem long term, and, I don't know.
```

Mike then produces what the hearing constitutes as a question – Nadely does respond – but the question itself may not be inherently clear. There is something about what "somebody's influences are on . . . the day-to-day kind of waxing and waning things" (turn 07). The response treats the question as one about what Nadely is or has been doing – she talks about the consistency that she has lacked. She then articulates a narrative about the homeroom teacher (Ken) to have been away, just as some of the other teachers have, all replaced by substitute teachers. Regular teachers being away and replaced by substitute teachers "has a big effect on them [students]." So in one manner, Nadely has responded about the influences someone has, or, perhaps more appropriate, the influence the absence of teachers has on the behavior of students.

Nadely then suggests being tired of having to deal with the same issues – possibly discipline and control – over and over again and about not knowing what else to do, about doing what she is supposed to do without that much is changing (turn 08). She talks about a particular strategy, sending students to the behavior problem room, and about this strategy as working only temporarily without changing the problem in the long term. Again, she not only responds to a question Mike has asked about the long-term changes but also she has done so in a way that is inherently intelligible to others. Even if a teacher has not encountered a "behavior problem room" – there were no such rooms in the schools where I have taught between 1980 and 1992 – the very name of the room and the concept is inherently intelligible. This is so not only within this context but also within any context where behavior is "a problem" and where there are institutional arrangements for special localities where does exhibiting the problems are moved. Thus, the concept might also make sense in a prison, where prison guards use "behavior problem room" as a euphemism to denote the room where problem prisoners are placed in solitary confinement (though the slang word is "dissociation," short for the legal terms "punitive dissociation" and "administrative dissociation"). That is, without having talked about a "behavior problem room," and, therefore, without having had any opportunity to construct a concept of the behavior problem room, such an expression makes *immediately* sense and prior to all constructive efforts in the context of teacher talk.

In turn 09, I then pick up on one of the issues that Nadely has articulated in Fragment 8.1a: the relation between what is taught at the university and the actual work in classrooms. That is, if there had been no sign up to that point of the sense in what Nadely has been saying, here would have been one. That is, I am not only articulating the university-school – i.e., theory-praxis – gap as a topic, but in so doing also co-articulate it as a reasonable topic among teachers generally and among the present participants in particular. The reasonableness of the topic can be assumed especially given that I have been working at the institution that Nadely is at the time in the process of completing her degree. The response realizes the question as one about the gap, as Nadely first talks about something that she did during the first two months on the job (September, October) but that she now no longer does. She then talks about speaking to other teachers and about finding other ways of "dealing with these kids."

Fragment 8.1c

```
09 MR: So, there's a number of issues that have come out that I
       think are very interesting, one is, you're talking about
       the university and, you know, the idealistic way of deal-
       ing with things and you in the classroom. Can you elabo-
       rate on this relationship, like do you think of, when
       something happens, of the university things or how does
       that happen?
10 N:  I think I did, maybe in September and October, but I don't
       anymore and I just sort of, I'm talking to other teachers,
       I'm talking to colleagues, and finding other ways of deal-
       ing with these kids and what's working, and what seems to
       be working with these guys right now is time. I'm taking
       time away from their, I'm taking time away from their
       lunch or their recess or after school but then I get frus-
       trated with that because I don't ever see these guys, I
       don't see them before recess, I don't see them before
       lunch, I don't see them before after school anymore, and
       people take the buses, and I get frustrated with the fact,
       I guess I feel like I'm always pushed up against these
       walls so I need to find things, ways of solving it here
       and now without disrupting the rest of the class.
11 MR: To follow up on this idealistic, if we take a concrete ex-
       ample, this morning at one point you called out Stacey.
```

Nadely says that using "time" as a means to control the students is successful at the moment of the talk. But she also talks about getting frustrated implementing the process of getting students to show up in the homeroom at the scheduled time when students are to serve their punishment isolated from their peers during recess. She is frustrated by the difficulties of tracking students, who leave on buses after school, students of whom the whereabouts she does not know because of the institutional arrangements and scheduling of classes so that she does not see them at the right moment of the day to enforce the punishment. She talks about feeling pushed up against the wall and about having to find ways of dealing with solving it (we can hear "behavior problems") in the here-and-now of the lesson and to do so without interrupting the course. Again, as the next turn shows, what Nadely has been saying is treated as unproblematic as the question offers an opportunity to elaborate – by giving a concrete example – the issue about the idealistic ways, where we can hear these ways to be those of treating professional issues at the university. In all of this, Nadely not only articulates that she is battling but also, especially because her battling is a constant issue, she articulates willingness to battle.

Identity, Language, Passivity

> Man, determined first and above all as a subject, as being – subject, *finds himself interpreted* through and through according to the structure of representation. (Derrida 2007, p. 111)

In the preceding section, we see how the particular requirements of the narrative provide resources and constraints for the telling of an auto/biography. Without such narrative resources, we cannot tell auto/biography so that *precisely* when I use the most personal reference "I," which excludes *everyone* else, "*what* I say really is everyone" (Hegel 1979b, p. 74, emphasis added). These resources, therefore, not only constitute possibilities but also the constraints for establishing an identity. This leads to the fact that the drama of identity plays itself out in the tension between the real-life in its once-occurrent, never-repeatable form, on the one hand, and the intelligible, narrative accounting of someone (I, the other), on the other hand. Such narratives, drawing on language – its semantics, genres, and syntax – inherently casts life in terms of repeatable instants. But repeatable instants transcend self-identity. Identity, therefore, implies alterity, non-self-identity. "A pure identity would be an absurdity. A pure identity annuls itself, it cannot identify itself. It is only identical to itself, which is identical to itself, and therefore turns around and around and does not even arrive in existence" (Nancy 1993a, p. 14). A pure identity, one that is required by the constructivist Self, is absurd and explains nothing. Von Glasersfeld's constructivist Self ("I") that goes out to meet the Other – and, as a consequence re-constructs itself. This Self is pure, because the Other is only a construction of this Self ("I"). This Self therefore speaks an idiolect with itself, as it cannot go outside of itself. But such an idiolect, the pure idiolect that a Kantian (constructivist) theory requires, "would be *idiotic*: totally deprived of its rapports, and therefore of identity" (Nancy 1993a, p. 15). The Self ("I") can only be understood through the community that we constitute with others. This Self inherently is different from itself: it is non-self-identical, which allows it to manifest itself in different ways. This fundamental otherness, as well as the passivity at its very origin, is something that we need to develop for a better understanding of the Self and identity in science teaching and learning.

Identity and Self, these aspects of who someone is, can be, and can become, therefore, always comes with a considerable degree of passivity – as Derrida writes in the quote that introduces this section. As subject, I always already *find myself interpreted,* through and through, according to the structure of representation. When Nadely speaks about herself and when Mike and I ask questions concerning her being and becoming in the classroom, she always already is interpreted. It is not that she constructs herself: Nadely is constituted in and by the language we, the members to the setting, have available. It is not that we have to consciously engage in interpretation. Rather, in the use of language to name the things in our surroundings and the topics of talk, we constitute the world as an interpreted one.

Everyday language is a passion, interwoven with everything else not properly language of everyday life. In any account of my life, I simultaneously denote an utterly singular, never-repeatable individual, on the one hand, and a certain type of individual, always already intelligible as such. That is, Nadely is not just talking about herself. This is so because the language, concepts, narratives, plots, and characters that are employed in her discourse are the same resources that are also available to everyone else in her culture. In her language, in our language, it is not the singular experiences of Nadely that we find but ways of experiencing that *can be articulated.* In this way, too, any Self is an Other, who, in turn, is a Self in and through everyone else. Although I have considerable autonomy and agency to mobilize language and its resources to represent myself, this language always is the language of the Other, so that I am subject to this Other in a state of heteronomy. And, leading to passivity more passive than any intended passivity, the "essence of representation is not a representation, it is not representable, there is no representation of representation" (Derrida 2007, p. 111). This also deepens the fundamental condition of Being to always already find itself thrown into this world, as *being-there* (Heidegger's *Dasein*). Speaking is like a temple that imposes upon us a certain number of usages, an implicit religion, a rumor that changes in advance of anything we might want to and can say that charges our speaking with more intentions than we can ever admit to have had in and prior to uttering.

There is something easily recognizable in a human being, something that allows us to recognize her in the dark: timbre. I can recognize Nadely in the hallway even though I may not see her. There is something in her voice, there is something that makes this voice hers. It is precisely timbre that allows us to recognize a person among others even when their speech is marked by the same sound-words, speech intensity, intonation (pitch), and speech rate. That is, timbre singularizes the person. This, too, has been recognized in the phenomenological literature. Thus, timbre constitutes an important aporia of identity and passivity in an essay on the poet and philosopher Paul Valéry entitled "Qual Quelle" in both French and English versions (Derrida 1972).[3] Accordingly, the most telling feature about the individual person is her/his timbre with respect to which the content of talk is of little importance. But, in speaking, the timbre of my voice is that which I do not hear: I am deaf to the most spontaneous of my voice, and therefore, by extension, any "I" is deaf to itself. "I" can be, in the final analysis, spontaneous only under the condition of inconceivable and irremediable passivity with respect to my self-representation in and during the event. Thus, "[t]o hear oneself is the most normal and the most impossible experience," which leads to the fact that "[w]hen I speak (to myself) without moving tongue and lips, I believe that I hear myself although the source is other; or I believe that we are two, although everything is happening

---

[3] The two words in the title could be read as French expressions: the first, *qual-*, which is the stem of such words as qualification, and might be taken as some sort of neologism; and *quelle* is the female interrogative "which [one]." But both words can be read as German words, *Qual* translating as "pain," "agony," "anguish," "drudgery," and "suffering"; and *Quelle* translating as "source," "spring," and "well."

'in me'" (pp. 353–354). That is, even when Nadely sits alone in her apartment, preparing for the lessons on the next day, using language to reflect upon her experiences of the day, it is not her singularity that she confronts but always already the other: because the source of her Self is the Other although she might believe she is hearing herself.

In the scholarly discourses concerning "identity," the term is often lodged within an intentional discourse, as if the person in question actively "constructed" or "positioned" him/herself in this or that way. What such discourse fails to acknowledge is the passive dimension that comes with language use, which always is a language that has come *from* the Other, is produced *for* the Other, and, in so doing, *returns to* the Other. For example, when Nadely talks about herself as "letting some things as fine and others not," her language describes a possible form of experience. It is within this possible form of experience that her own experience is gathered in and with her description. Not only is this an intelligible ascription for the purposes at hand, the interview and debriefing sessions with me, but also it is a way any other teacher might describe her/himself. The subject is subjected and subject to the cultural-historically and discursively possible descriptions. Even the "I," of which there are many in the three fragments above, is not her own. At the very instant where the speaking subject first becomes conscious of him/herself *as a speaking* subject, s/he already is a more-or-less fluent speaker. When the person discovers him/herself as a person (i.e., becomes self-conscious) with a particular identity, the whole machinery that enables the production and co-production of identification and consciousness is already in place. Who I am, therefore, is always already in the image of the Other from whom this linguistic machinery has come to me. Not only language itself comes from the Other but also the kinds of texts people produce, the genres, the constituents and structures (e.g., narrative, plot, character), the living or dead metaphors, and so forth. The agency of the "I" always already is grounded in the passivity of the *me-I*, which finds itself thrown into a world full of significance predating all signification.

> [N]o matter what an Odyssey or Bildungsroman it might be, in whatever manner the story of a constitution of the *self*, the *autos*, the *ipse*, organizes itself, one always *imagines* that he or she who writes should know how to say *I*. In any case, the *identificatory modality* already or henceforth has to be assured: assured of language and in its language. (Derrida 1996a, p. 53)

In the culture where the speaking subject is a member, the forms of telling one's life already exist. In fact, as the intelligibility of any life narrative precedes the telling, any autobiography always already realizes existing biographical forms and contents, characters and plots.[4] Any presentation of the self always already is

[4] It is evident that intelligibility must precede and is inherently presupposed by the telling, which led Derrida (1984) to speak of oto-biography, biography for the ear. "Everything comes back to the ear" (p. 38), he writes while beginning to deal with an excerpt from Nietzsche's *Zarathustra* in which features a giant ear with a little human body attached to it. Hearing as the precondition of speaking, and therefore sharedness, is also a theme for Heidegger (1977): "Hearing is constitutive for discourse" (p. 163).

a *re*presentation. This interconnection between auto-presentation and re-presentation therefore marks the impossibility of pure presentation, or, in other words, it marks "the presence of the alter ego at the very heart of the ego" (Franck 1981, p. 157). Language, which has come to me from the Other precisely provides the required means to take the perspective of the Other on my emotions, the stirring within myself, thereby making possible empathy, transference, and counter-transference.[5] This arises from the fact that I cannot ever recognize and describe someone else's behavior as choleric unless I first take an exterior, the Other's, perspective over my own affect. In using language, I unavoidably and inherently take the perspective of the Other onto myself.

Fundamentally, we come to understand events such as those in Nadely's classroom better when we make radical passivity and otherness the starting point of our considerations. That is, Nadely could then understand herself as but one agent among many others that make the lived curriculum. In this case, she would no longer understand the events as having to be "under her control," but she would be able to think about how she contributes to the situation so that the lived curriculum emerges from the collective effort. As an aspect of collective agency, she could understand that she is not just the subject of the teaching/learning activity, but also that she is subjected and subject to this activity. That is, there is always a level of agency, too, but this agency itself grounds out in the passivity of givenness, the experience of a world and language always already precedes myself. Even in the case where some *I* were to have formed itself, it would have to have situated itself in a non-situatable experience of language:

> The *I* in question *formed* itself without a doubt, as one can believe, if it managed to do at least that, and if the trouble/confusion of identity [*trouble d'identité*] of which we were speaking a while ago does not, precisely, affect the very constitution of the *I*, the formation of the *speak-I*, the *me-I*, or the appearance, as such, of a pre-egological ipseity. It would have *formed* itself, this *I*, at the site of an unfindable *situation*, always referring elsewhere, to another thing, to another language, to the other in general. It would have *situated* itself in a non-situatable experience of *language*, that is, language in the broad sense of this word. (Derrida 1996a, p. 55)

The phenomena of identity, identification and dis-identification, or the experience of who I am and with whom I affiliate all have to be theorized in terms of the experience of language available to the speaking subject who is subjected and subject to the language it uses. When we began to teach together, Nadely suggested that science was not her strongest area even though she liked teaching it. In so saying, the English language and the way it allows to speak of oneself – the way in which it constitutes our Selves – articulates her as a particular *type* of person rather than as a *singular* individual. This language, by the very act of communication, presupposes its possibilities as preceding my speaking and the listener's

---

[5] In psychoanalysis, *transference* is the transfer of previously experienced, often-forgotten or repressed emotions, feelings, and sentiments from the patient to his/her psychoanalyst; *counter-transference* is the transference of feelings and sentiments, often deriving from experiences in early life, from the analyst toward the patient.

hearing. The upshot of this is that the highly agential theories of and presuppositions about identity currently in vogue need to be revised – as suggested in Chapter 4 – to take into account the inherent passivity and otherness that exist when we speak a language that is never our own. Because it is never my own, language constrains me in my telling/constitution of myself, of who I am with respect to myself and to the Other. Thus, it is not merely the case that a student who agentially "positions" herself does some form of identity work when, "unsatisfied with her progress in mathematical discourse [, she] is likely to call herself a 'terrible mathematician' or a 'slow thinker'" (Sfard 2008, p. 290). Rather, the very designations "terrible mathematician" and "slow thinker" are inherently intelligible ways that have come from the Other to account for and denote certain experiences. In fact, the language enables the existence of such experiences. The student therefore *is given,* by using language, as much as she *gives herself*, a particular form of identity. This identity inherently is *from* the Other, to whom, in the articulation of oneself as a "terrible mathematician," this formulation returns. This passivity, which I have called radical – because it is inherent and unavoidable in and with language-in-use – is an aspect of language, learning, and context *completely* unattended to in current scholarly endeavors. This passivity and the radical otherness that derives from the use of language has yet to be investigated and applied to its fullest in the discipline of science education.

In the interviews, Mike Bowen and I listen to Nadely. In reading the transcripts, readers listen to Nadely. But it is not really Nadely that we are listening to. It is language itself that we hear, concretely realized in and through Nadely's voice, singularized by its timbre that makes *this* voice *uniquely* hers. But timbre is difficult if not impossible to represent, that is, impossible to put into the pages of this or any other book. Ultimately, then, when we refer to the interview texts attributed to Nadely we might say – with Martin Heidegger, Maurice Blanchot, or Jacques Derrida – that language speaks (Ger., *Die Sprache spricht*). Paraphrasing Blanchot (1959, p. 286) we might say that language is not deprived of sense but deprived of a center. It does not begin or end and it never stops. It is something that we suffer only if it stops, at which point we would find out that it continues to speak, that it perseveres. Human beings, then, speak primarily and only in as far as they correspond to language, when they listen to its appeal. Language perseveres but not silently, because it is in language that silence speaks to itself eternally. This does not resolve the problem of the experience of the "I" in its utter singularity, its once-occurrent nature. Here, emotion plays an important role, because it is precisely affectivity that constitutes the ipseity – self-hood – of the "I." As a consequence, "the ipseity of the I consists in staying outside of the distinction between the individual and the general" (Levinas 1971, p. 90). Affect preserves, suppresses, and supersedes the individual and the general, a topic that I address in Chapter 10.

Fundamentally, constructivism specifically and all intentionalist approaches to cognition more generally misplace the emphasis with respect to the nature of who we are. In these discourses, the emphasis is on some conscious Self that constructs itself for-itself. When we take the perspective of life, which is at the heart of the phenomenological position that I articulate here, then our self-understanding changes. It is no longer any particular "me" that is of importance; there is no reason for asking the questions about the "meaning of (my) life," for each human life is but a concretization of life as a possible form. In Nadely, Mike, and myself, life and society sustain and reproduce themselves. But to be alive, they also have to change, which means, transform themselves. Such transformation therefore also means death – any proper theory of a living system such as culture, language, or life involves the non-self-identical unit of life and death (Bakhtin 1984). I find myself in this situation into which I am thrown, as a form of life that realizes itself in and through me. Any self-understanding that I might develop also derives from life, the collectivity we form with others, a form of life that evolved initially because of biological principles and later because of cultural-historical possibilities for the purpose of reproducing life itself.

With the coming of humans, the question of the sense of life may be posed. Sense is the *for-the-other* required in the collective control over conditions. In this situation, I am not only the subject of action but also subject and subjected to life and the collectivity, which reproduce and transform themselves in and through me. The *despite myself*, which denotes the fact that everything happens whatever I do, that is, the essential passivity and otherness of Being, "marks this life in its living itself. Life is life despite life" (Levinas 1990, p. 86). Thus, in contrast to the constructivist and all other intentional approaches, the subject of action cannot be properly described and theorized on the basis of "intentionality, representational activity, objectification, liberty, and will" (p. 90). Rather, the subject has to be theorized on the basis of passivity, the passivity of and with respect to time, which itself is the result of the emergence of representation without which there is no recognition of time.

This line of thought also leads us to a new way of understanding difference. Heretofore, the tradition in science education has been to think difference in terms of self-identity and difference of externalities. Thus, for example, differences are constituted in terms of different culture, gender, race, or language. But these differences imply self-identity within the term, one culture that is different from another, one gender that is different from another, one race that is different from another, or one language that is different from another. In this, each of these terms is treated as pure, which has lead, among others, to the theoretical term of "third space." In this third space, science students are said to cobble together their first space, which is "the" culture in their home, and "the" school science culture, which is what schools and teachers offer up in their classrooms. But the analysis of culture shows that there *cannot be* something like a pure culture: any culture always already constitutes a mixture, a mêlée. Yet a mêlée ~~is~~ not: "it arrives, it overcomes. There is mêlée, crossing, weaving, exchange, sharing, and it is never only one thing, not the same" (Nancy 1993a, p. 12). That is, the only thing that any set of cultures, any set of races, any group of Canadians, any cluster of students, or any group of teachers have in common is that they are different from any other culture, race, Canadian, student, or teacher. Moreover, the very possibility of difference presupposes the non-self-identity of culture, race, being Canadian, student, or teacher: only something that is different in itself can manifest itself differently.

Chapter 9

The Nonsense of *Meaning*

> The concept of meaning originates in a primitive philosophical conception of language. (Wittgenstein 2000, TS-123 ¶7, p. 25)
>
> The term [literal meaning] is too incrusted with philosophical and other extras to do much work. (Davidson 1986, p. 434)
>
> "I know what the word 'toothache' means, it produces one particular image in my mind." But what image? "That can't be explained." – But if it can't be explained what was the meaning of saying that it produced one particular image? You could say the same about the words "image in your mind." And all that it comes to is that you are using certain words without an explanation. "But can't I explain them to myself? or understand them myself without giving an explanation? Can't I give a private explanation?" But is this anything you can call an explanation? Is staring a private explanation? (Wittgenstein 1968, p. 315)

In the only text published during his lifetime, Ludwig Wittgenstein (1921) developed a language theory that was also a theory of meaning (Ger. *Bedeutung*). It fundamentally is a picture theory, according to which language mirrors (pictures) the world as it really is: "The logical propositions describe the structure of the world, or rather, they represent it. They are not 'concerned' with anything. They presuppose that names have meaning and elementary propositions sense: And this is their connection with the world" (¶6.124, p. 253). The author of this statement then spent the remainder of his life rescinding what he had written in his twenties. The introductory quotes derive from his later years.

In the first opening quote, Wittgenstein notes that the concept of meaning has originated in a primitive philosophical conception of language; in the same typescript, he also entitles an entire section with a statement that in his way of seeing things, there is no place for "meaning" (Wittgenstein 2000, TS-213, ¶1, p. 1). In the second introductory quote, he discusses the attempt to locate meaning in private thinking and experience, such as the meaning behind the expression "I have a toothache." Wittgenstein shows in the third introductory quote that the search for "meaning" is fraught with so many conceptual problems that it is better to abandon the traditional use of the term. It is better to ask how a word or aspect of language is used in concrete situations and what allows people to achieve rather than

W.-M. Roth, *Passibility*, Classics in Science Education 3,
DOI 10.1007/978-94-007-1908-8_9, © Springer Science+Business Media B.V. 2011

asking for something behind words, something that comes to be "constructed" for and attached to words. In other words, it is meaningless to talk about the meaning of a word somehow lying behind it. Thus, when we hear the sound [kʌp] ("cup") or [giv mi ðə kʌp] ("give me the cup"), we ought to be concerned only with the response, which might be that someone (the addressee) hands us the object "." It is not that there is some "meaning" behind the word "cup"; the "meaning" is not some image of the object or the object itself, as Wittgenstein suggests. The only question is one of use, the production of a particular sound and what it entrains as its effects in collective situations. Words are intimately related to and emerge from societal, living/lived praxis, which, in turn, comes to be crystallized in these words; and it is only in and through living/lived practice – in concrete, collectively motivated activity – that human beings discover their objective reality (Leontjew 1982). That is, language is one aspect of practical consciousness, which is a reflection of real, living/lived practical activity. The cultural-historical approach, therefore, is consistent with the pragmatic philosophy of language that Wittgenstein developed in the second part of his scholarly life, in which he radically overturned the results of his initial research.

The fundamental problem with the concept of meaning is that it attributes something to words other than their use. The concept fits into a metaphysical theory of knowing that is based on representations, which stand for something other than themselves. In the second introductory quote, Davidson tells us that "the term is too incrusted with philosophical and other extras" to be of any use. Wittgenstein exhorts us to shift our attention to the use of words, that is, to the words themselves rather than to something else behind and denoted by them. The concept also fits the view that separates mind and mental structures from the world that is modeled by the mind (mental structures).

Other philosophers are in agreement with this stance, thereby putting the relationship between words and significations on their head. Everyday being-in-the world means being *attuned* to the intelligibility of social situations (Heidegger 1977).[1] This intelligibility expresses itself as discourse. That is, the everyday mundane situations as such are intelligible so that we recognize without having to reflect, interpret, or construct whether in the room to which the door we just opened there is a science lesson, a staff meeting of the science department, a face-to-face meeting of the department head with an intern, or a tutoring session. We know this is the case because of the totality of things that together make the situation intelligible. The relation of words to this intelligibility is such that the "totality of significations *is put into words*" (p. 161). As a result, it is not that we "construct" the "meanings" of words used and thereby provide word-things with

[1] Heidegger uses the word *Befindlichkeit*, which tends to be translated as "attunement." The German term is unusual and cannot be found in dictionaries, but it is a noun construction of a (reflexive) verb. The verb *befinden* translates as "to judge"; *sich befinden* translates as "to be in …" a (health) condition, position, or location. The noun *Befinden* (neuter) pertains to someone's state (health) or translates "judgment" and judicial "opinion." Therefore, when Heidegger uses the term, this entire semantic field is made accessible by the German language.

significations. Rather, any new words we come to hear when we join new situations, those words that science educators have become familiar with as having their meaning constructed, those "words accrue to significations" (p. 161). For Heidegger, language is but one of the things that we encounter in the everyday world that we use without interpreting them or constructing their meaning. We use a knife and fork or spoon to eat without having to give them the slightest thought: these, as all familiar tools, are something ready to hand. This does not prevent someone to look at a word, make it objectively present, and then talk about it, the various senses in which it is used. For example, in a dinner conversation, the word *jackknife* might be used, following which those present talk about the uses of this word: to denote a pocket knife, a resampling technique in statistics, a jackknife hold in wrestling, a dive in swimming competition, a configuration in a truck or train accident, a kicking move in Taekwondo, a clam, a sit-up, or the names of a movie, movie characters, or episode of a television series.

In the literature so far discussed, then, it is not that words receive something by means of a conscious and intentional "construction" that we may denote as their "meaning," whatever it may be, which tends to be hard to determine in scholarly articles because it is not clearly specified itself. Rather, new words tend to find a place in situations that users are already familiar with, where things have their place, and which therefore bear significant relations to each other. Thus, when I need my Swiss army knife, a jackknife, I tend to look first in the pockets of my garden work pants, for it tends to be handy – i.e., ready to hand – in all sorts of situations. New words, precisely because they are foreign/strange, having come from the Other, do not lend themselves easily to use; they tend to find, take, or fall into their place almost despite ourselves. It is in and with increasing familiarity in a situation that new, unfamiliar, and strange words find their place without any construction whatsoever. The situations in which we find ourselves are part of cultural activities, which have evolved relations of significations in the course of their histories. Just as we come to life, being thrown into the world, we come to societal activities in which words have their place. The purpose of this chapter is to contribute to the deconstruction of constructivist claims concerning the relationship between words and their use. I begin by articulating some of the ways in which the word *meaning* is (has been) used and discussed in the literature before articulating more pragmatic approaches and providing a specific example from my own ethnographic research concerning words and their use.

"Meaning" in the Literature

The concept of *meaning* is particular to the English language and is used to translate very different words from other languages, frequently in very confusing ways. It comes with little surprise to find the linguistic Todorov (1984) suggest that English translations of Mikhail Bakhtin's work – including the work entitled *Marxism and the Philosophy of Language* attributed to Valentin Vološinov in the

Anglo-Saxon world – tend to be very poor. Among others, English translations employ the word "meaning" where there are very different words in other languages. Other specialists of language, such as the semiotician Winfried Nöth (1990), note that the term is used inconsistently. In this section, I articulate some of the ways in which the term appears in the literature generally and in the (social) constructivist science education literature specifically. I begin with a playful conversation about meaning, in which I confront different philosophers using excerpts from their work. I then exemplify the use of the term in science education discussing one specific study to exhibit in greater detail the problems with this theoretical term. Finally, I use children's learning of language as a paradigm for presenting different positions on the notion of "meaning."

Played Playful Conversation About Meaning

Somewhere in the same Zurich where playwright Tom Stoppard (1975) has James Joyce meet up with Vladimir Lenin and Tristan Tzara (the founder of Dadaism) as part of the play "Travesties," four philosophers meet up to talk about language and meaning, which one of them has characterized as one of the most difficult problems of linguistics (Vološinov 1973). The four are, in alphabetical order, Jacques Derrida, a French "postmodern" philosopher, Valentin Vološinov, a Marxian philosopher of language (it may actually be Mikhail Bakhtin having come with a carnival mask), Ernst von Glasersfeld, an epistemologist, and Ludwig Wittgenstein, a language philosopher.

Wittgenstein (1958, 3e): But what is the meaning of the word "five"?

von Glasersfeld (1989a, p. 133): The child must first isolate the sensory signals that constitute this auditory experience from the background (the manifold auditory signals that are available at the moment); and the parent's repetition of the word obviously enhances the process of isolating the auditory pattern as well as its association with the moving visual pattern.

Wittgenstein (1958, p. 3e): No such thing was in question here, only how the word "five" is used.

von Glasersfeld 1989a, p. 134): [No.] To be considered proficient in a given language requires two things among others: to have available a large enough vocabulary, and to have constructed and sufficiently accommodated and adapted the meanings associated with the words of that vocabulary so that no conceptual discrepancies become apparent in ordinary linguistic interactions.

Derrida (1995, p. 49): [I disagree.] What engenders all these meanings and links them, internally and necessarily, is a logic that at bottom (that is why it can still, up to a certain point, be called a "logic") has no need of the event of a revelation or the revelation of an event. It needs to think the possibility of such an event but not the event itself.

von Glasersfeld (1989a, p. 132; 1989b, p. 444): [But l]anguage users must individually *construct* the meaning of words, phrases, sentences, and texts. There is no doubt that these subjective meanings get modified, honed, and adapted throughout the course of social interaction.

Vološinov (1973, p. 101): [I disagree with you.] Meaning, in essence, means nothing; it only possesses potentiality – the possibility of having a meaning within a concrete theme.

von Glasersfeld (1989b, p. 444): [But i]t is easy to show that meanings do not travel through space and must under all circumstances be constructed in the heads of the language users.

Wittgenstein (1958, p. 3e): [Well, t]hat philosophical concept of meaning has its place in a primitive idea of the way language functions. But one can also say that it is the idea of a language more primitive than ours.

In this playful conversation, the four eminent scholars disagree about the nature of meaning. It almost appears as if Derrida, Vološinov, and Wittgenstein – to whom meaning, if it is anything at all, is a possibility inherently public and shared ways of employing words – are ganging up on von Glasersfeld – to whom meanings are individual constructions in the heads of people. For the latter, "meaning" constitutes something like a direct correspondence between items in an individually constructed list of entities abstracted from experiences that are matched up with items in another list made of word-things, the sounds that travel through the air. The listener "select[s] particular meanings from a list which, together with the list of agreed signals, constitutes the 'code' of the particular communication system" (von Glasersfeld 1989a, p. 132). But, as I point out in the introduction to this chapter, analyzing everyday use of words, Heidegger shows us that word-things do not receive significations ("meanings"). There is nothing people construct *in their heads* because "what is expressed [in communication] is precisely this being outside, that is, the actual mode of attunement (mood) that was shown to pertain to the full disclosedness of being-in" (Heidegger 1977, p. 162). Communication never is the articulation of something that derives from the inside of one person. Thus, von Glasersfeld and other constructivist scholars make direct correspondences between words and what they stand for, their "meanings." This position, however, is the one that Wittgenstein, Vološinov, and Derrida critique. It is, in the blunt words of Wittgenstein, a concept that "has its place in a primitive idea of the way in which language functions"; and for Vološinov, *meaning* means nothing. We should therefore abandon this nonsense called "meaning."

The concept of meaning is ambiguous at best. In the Anglo-Saxon literature (or literature translated into English), scholars use the term meaning, though often in very different ways and contexts, and equally often in incompatible ways. The Wikipedia entry on sense and reference equally suggests the presence of confusion because the words sense, reference, and meaning are used differently by different authors and sometimes are used synonymously. In part this is due to the translations of philosophical works written in some other language into English. Such other languages – Wittgenstein's German, Vološinov's Russian, or Derrida's French – do not have direct equivalents for the term "meaning." Thus, in similar contexts discussed in German texts, the words *Sinn* (sense) and *Bedeutung* are used, where the latter is sometimes translated as meaning, but the former also in some instances. *Bedeutung* has as the stem the verb *deuten*, to point, to give an indication of, to interpret. A better translation, which sometimes is used in the semiotics literature, therefore is reference. For example, one key work in the field entitled *Über Sinn und Bedeutung* is translated as *On Sense and Reference*. In French, too, the equivalent words would be *sens* (sense) and *référence* (reference). But more often, the discussion in French would be employing the terms *signifiant* (signifier) and *signifié* (signified), which, in semiology, are understood as constituting a relation denoted by the term sign (*signe*).

Literature Example of Meaning Construction

In a pragmatist approach, there is no place for "meaning." Yet in the science education literature, uses of the idea of the "construction" or "making" of "meaning" and "understanding" abound. It is never quite clear, though, in which sense authors use the term "meaning." Thus, what do the authors want to say when they write something like this: "an intelligible conception is sensible if it is non-contradictory and its meaning is understood by the student" (Treagust and Duit 2008, p. 299)? Here, there is an intelligible, non-contradictory conception on the one hand and a meaning thereof on the other hand. What is this meaning if it is not the discourse in use that expresses the conception? To further explore the uses of "meaning" and "construction," I provide a brief analysis of one excerpt from the literature by a research team that uses the adjectives "sociocultural" and "social constructivist" (Varelas et al. 2008, p. 67) to characterize its theoretical position.[2]

The curiously indirect and confusing deployment of the term "meaning" found in the literature generally permeates this work as well. For example, the authors suggest that students "communicate and share meanings and ideas" (p. 68) and that "meaning making shapes and is shaped by the ways individual children make sense of ideas" (p. 68). They are interested (a) in the ways in which "second graders make sense and meaning of the ideas of evaporation, boiling, and condensation" (p. 69), (b) how "their discursive practices influence their meaning making in particular ways" (p. 70), and (c) in "children's ways of meaning making in the midst of negotiation and argumentation with their peers and their teacher" (p. 70). In such framings, meaning making is something that shapes sense making; sense is something different from meaning and ideas; discursive practices – in a position clearly contradicting Wittgenstein – are different from meanings, constituting something like forces that "influence" meaning making; and argumentation and negotiation are situations in which meaning making occurs, again contradicting Wittgenstein's exhortations to abandon "meanings" and to focus instead on the uses of words. Looking for meanings somewhere else rather than in concrete use is part of a metaphysical project the pinnacle of which Kant has established.[3] It

[2] Any other article from this particular theoretical position could have been chosen, but as core members of the pertinent community of practice, these authors are more typical of its stance in the same way as core examples of a concept are more typical than peripheral examples.

[3] Kant (1956a) uses the term meaning (Ger. *Bedeutung*) to denote the relation between a mental representation (Ger. *Vorstellung*) and an object rather than between two representations: "objective

treats in a non-physical, idealist world of ideas "segregated from the fullness of life, from the personal needs and interests, the inclinations and impulses, of the thinker" (Vygotsky 1986, p. 10), that is, in a world apart from the sensuously sensible world that we inhabit with and qua our sensuous, living/lived bodies. To support their metaphysical position, the authors present excerpts from a unit on the states of matter.

In this unit, second-grade children listen to the reading from a book about the states of matter and classify, among others, a series of substances by placing them on construction paper labeled by one of the terms "liquids," "solids," and "gases." The children are also asked to come up with reasons for their classifications and subsequently discuss what they have come up with. At the instant quoted below, the children are in the process of discussing the classification of shaving cream. Having been placed on a plate, the cream was moist and smooth on the first day of the task but, drying up, had changed consistency by the next day. The discussion about the shaving cream begins with the attempt to explain the changes. There is a lot of talk about gas being in the shaving cream, its "poofy" nature, and the presence of air that would give the sample its shape and consistency.

113	Indira:	Maybe it had more air at first when it was poofy and it was on the plate and maybe it lost some of its air and it went down.
114	Caitlyn:	A little is still left.
115	Ms. Cowan:	Could be. I am not sure.
116	David:	There is air all around us so how could it just disappear?
117	Ms. Cowan:	I don't think it did disappear.
118	Sean:	It is kinda like, you know how|
119	David:	Evaporate?
120	Sean:	No. You know how when you put a potato to a timer it melts down, it gets less and less.
121	David:	At the bottom it was the same shape. There is air all around us. None of the air could have been just away. The gas couldn't have gone away from the shaving cream.

. . .

David cannot make sense of this (unit 121). He is probably thinking: If air is all around us, how can the air inside the shaving cream disappear?

Apparently there are two lines of reasoning on the table that are not differentiated. One line has to do with evaporation and drying out so that what it remains is less similar to a liquid and more similar to a solid. Sean's analogy with a baked potato (unit 120) is an example of <u>making meaning along this line</u>. Although liquids do not "meltdown" (unit 87) to gases, but they "evaporate into" gases, children who start constructing their understanding of changes of states of matter usually use words that they are more familiar with and that describe changes via the same principles. (pp. 88–89, underline added)

In this situation, the authors suggest what David "is probably thinking" without staying right with the text of what David is actually saying. Thus, although von Glasersfeld (1989a) explicitly points to the fact that we cannot derive from the fact that someone understands what we say that the person "must necessarily have conceptual structures that are identical with ours," researchers attempt to get into

---

meaning cannot consist in the relation to another representation . . . only because a particular temporal relation is necessary between our representations can we confer them objective meaning" (pp. 232–233). Wittgenstein (1921) gives this metaphysical position its ultimate expression.

the heads of the researched telling their readers what there (likely) is to be found. Moreover, unlike Heidegger who tells us that what is expressed in language *is precisely the outside*, the present authors are trying to get inside the person by guessing what David is thinking, which they would not have to do if they were to take language in the way Heidegger and other language philosophers articulate for us. Thus, David is said to be unable to make sense, when in fact he is using language to contribute to the situation. What he says and how he says it has consequences. For example, in saying "I don't think it did disappear," Ms. Cowan produces the completion of a statement/comment pair (turns 116/117); and she could do so only because everything that matters was made available in the first part of the turn. Yet David is in the process of articulating sense, and what he says is perfectly understandable. He articulates a way of talking about and explaining a particular experimental situation – and this is the case even if scientists and science educators decide that this manner of speaking is inappropriate.

This analysis is not surprising when we take the individual who dumps the contents of his/her subjective mind into public space as the starting point of an analysis. If, however, we take the position that a conversation is a societal phenomenon sui generis – here part of the activity schooling that is realized by means of a science lesson – then we look for turn pairs in which utterances mutually constitute each other. Indira articulates the issue of the disappearing air, which itself occurs against an earlier comment by the teacher, who summed up the preceding discussion: "There was air in it and that was what made it poofy?" David utters a question: How could the air disappear if there is air all around? The teacher formulates to be thinking that the air does not disappear.[4] David utters the verb "evaporate" (turn 119), apparently with a rising intonation, and then provides a more extensive explanatory description for why the gas in the shaving cream could not just have disappeared given that there is air all around. Why should it leave the shaving cream given that there is air around to take the space that the leaving air lets behind?

In contrast, Sean is said to be "making meaning" along some line, which these authors specify as having "to do with evaporation and drying out so that what remains is less similar to a liquid and more similar to a solid." But what precisely is the meaning he makes? Where is this meaning? What does it consist of? Does Sean not articulate sense in a manner similar to David by producing what we can hear as an analogy for explaining what is happening in the phenomenon that the class investigates at the time? In this, these authors perfectly exemplify the common practice of talking about meaning making without ever specifying what precisely students make. We find out that it is a process that takes place within social action, that students can share meanings, and that meanings emerge. Meanings are

[4] In conversation analysis, *formulating* is a technical term used for those instances where a speaker articulates what s/he is doing at the time. Thus, when Ms. Cowan says "I don't think …," she formulates to be thinking. When she says in turn 107 (not reproduced here), "Let me sum this up," then she formulates what she subsequently utters to be a summary of the preceding talk ("this").

something separate of ideas ("students share meanings and ideas") and something that ideas can have ("meaning of the ideas of evaporation, boiling, and condensation").

The transcription of turn 119 is actually interesting in the present discussion. There is a question mark following the verb "evaporate," which suggests that the transcriber has heard a question even though there is only one word that is not an interrogative term itself. Whatever David has done to make his utterance appear as a question, it was available publicly so that the transcriber could hear the utterance as a question. But this something is not itself part of the utterance. What is at issue is not the "meaning" of the verb but the fact that David has offered up a question. This question shapes how the event evolves, which is a function of whether the offer is accepted – by means of a completion as question/response pair – or whether something else happens. What matters to understanding how the classroom discussion unfolds is not some "meaning" behind the word "evaporate" but what David has done in uttering it. Here, the transcriber, in placing a question mark after the verb, exhibits hearing the utterance as a question. The performative explodes the concept of meaning: "The performative is a 'communication' that does not essentially limit itself to transport a semantic content already constituted and guarded by an aiming at truth" (Derrida 1972, p. 383). The concept of meaning – the entity that is denoted by the word – does not advance us here, because what is important is the fact that a question has been asked rather than some entity that is denoted by the word "evaporate."

The use non-theoreticians and other laypersons make of terms is often excused precisely by making the distinction between formal technical and everyday applied interests. Some may argue that to better understand the meaning of meaning, we ought to take a look at the way in which philosophers employ this concept – though Kant (1960) notes that philosophers take the "meaning" of a word from the way it is used generally. In the next section, therefore, I turn to the way in which von Glasersfeld and Wittgenstein use children's learning of a language as a paradigm. In learning language generally and new words more specifically, children are often said to "make meaning" of the new and unfamiliar (foreign/strange) sounds that they inherently understand to be words of the language in which they are immersed.

Children's Learning as Paradigm

In the constructivist approach, learning a language is used to indicate that individuals "construct" something, some entity in their minds; and this something is different from the word, concept, sense, or reference. If it were not different, there would be no use for the term, because it would suffice to talk about the word itself. The most fundamental way of learning a word is by means of deixis and association. For example, a parent points to an object such as "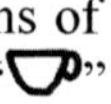" and produces the sound [kʌp], [tas], or ['tasə] depending on whether the language is English,

French, or German.[5] To make such an association according to the radical constructivist theory, three steps are essential. In the first step, the child has to learn to isolate particular structures in totality of the sensory signals. The parents pointing constitutes but an ambiguous hint as to the association. In the second step, the child perceptually isolates the object "." In the third step, the child learns to isolate the sound [kʌp] from other ambient sounds and signals.

> If this sequence of steps provides an adequate analysis of the initial acquisition of the meaning of the word "cup," it is clear that the child's meaning of that word is made up exclusively of elements which the child abstracts from its own experience. Indeed, anyone who has more or less methodically watched children acquire the use of new words will have noticed that what they isolate as meanings from their experiences in conjunction with words is often only partially compatible with the meanings of the adult speakers of the language take for granted. Thus the child's initial concept of cup often includes the activity of drinking, and sometimes even what is being drunk, e.g., milk. . . .
>
> The process of accommodating and tuning the meaning of words and linguistic expressions actually continues for each of us throughout our lives. No matter how long we have spoken a language, there will still be the occasions when we realize that, up to that point, we have been using a word in a way that now turns out to be idiosyncratic in some particular respect.
>
> Once we come to see this essential and inescapable subjectivity of linguistic meaning, we can no longer maintain the preconceived notion that words *convey* ideas or knowledge; nor can we believe that a listener who apparently "understands" what we say must necessarily have conceptual structures that are identical with ours. Instead, we come to realize that "understanding" is a matter of fit rather than match. Put in the simplest way, to understand what someone has said or written means no less but also no more than to have built up a conceptual structure that, in the given context, appears to be *compatible* with the structure the speaker had in mind – and this compatibility, as a rule, manifests itself in no other way than the receiver says and does nothing that contravenes the speaker's expectations.
>
> Among proficient speakers of a language, the individual's conceptual idiosyncrasies rarely surface when the topics of conversation are everyday objects and events. To be considered proficient in a given language requires two things among others: to have available a large enough vocabulary, and to have constructed and sufficiently accommodated the meanings associated with the words of that vocabulary so that no conceptual discrepancies become apparent in ordinary linguistic interactions. (von Glasersfeld 1989a, pp. 133–134, underline added)

In this excerpt, it is quite clear that from the perspective of the radical constructivist perspective outlined, learning a language means "constructing" associations between two kinds of continua, one existing of objects (e.g., "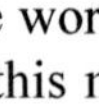"), the other one of sounds (e.g., [kʌp]). The object, "what [children] isolate as meanings from their experiences in conjunction with words," *is* the meaning of the word in this way of thinking about the issues. The associations "constructed" in this manner between "the meaning of words and linguistic expressions" may undergo development throughout the life of an individual. This is so in particular because the individual

[5] The purpose of the conventions of the International Phonetics Association is to render the way in which phonemes sound independent of the particular language in which it is used. This allows foreign language students to learn the ways in which words are pronounced or linguists to communicate about other languages and how they sound.

might discover having used a word in an "idiosyncratic" manner. Most importantly, the individual constructs a structure in the mind, a structure that drives subsequent use of the language, which becomes increasingly consistent with the way others use the language.

It is evident that the emphasis here is on the individual who constructs a cognitive structure in such a way that the latter is viable in the world. Everything in this theory is viewed by focusing on an isolated individual. What the individual constructs is independent of particular situations, the purposes, the collectively motivated activities, contexts, and so forth.[6] Not only does the individual "abstract" meanings from experiences, meanings and experiences apparently exist independently of anything that orients human practices. We might ask, "Why would a child do anything of the nature von Glasersfeld claims it to do?" In fact, the very thing children are said to construct ("understanding") or do ("understand") etymologically derives its name from the Proto-Indo-Germanic root *stā-*, to stand, paired with the preposition under-, in the sense of below, among, and amidst. To understand is equally primordial with being attuned to the place where we dwell (Heidegger 1977). "'Understanding' in the sense of a possible kind of cognition among others, let us say distinguished from 'explaining,' . . . must be interpreted as an existential derivative of the primary understanding" (p. 143). This primary understanding denotes nothing other than the fact of being attuned, which comes with having the sense of the (language) game currently played. As with all games, this does not require knowledge of explicit rules.

Very different approaches to the problem of language learning have emerged from a line of psychological theories that started with Lev Vygotsky (e.g. 1986) in the Soviet Union. Grounded in dialectical materialism, according to which reflective consciousness follows labor, real work in a real world, cultural-historical approaches emphasize the societal origin of higher-order cognitive functions. Accordingly, any higher-order cognitive function is the result of interactions in societally specific situations, is a sort of trace that social interaction leaves in the individual. Thus, anything we might call "cognitive" can actually be found in human-human interactions in the course of purposeful collectively motivated activity. This is consistent with Heidegger's approach, in which everything needed for intelligibility is found right in the situation, in the here and now accessible to all interlocutors, onlookers, by-standers, and overhearers. This approach also underlies the example another language philosopher provides about learning a language from the perspective of children born into some culture.

Wittgenstein (1958) presents a situation in which a primitive language is used for communication between a builder A and a helper B. In their work world, there are blocks, pillars, slabs, and beams. As he works along, A makes sounds that in the conventions of the International Phonetics Alphabet would be transcribed as

---

[6] As noted in Chapter 6, I clearly distinguish between collectively motivated activity (*Tätigkeit, deyatel'nost'*), such as producing food or manufacturing tools, and activity as a form of being busy (*Aktivität, aktivnost'*). Von Glasersfeld's activities lack the collective motive as an integral dimension of human activities.

[blɔk], [ˈpilə], [slæb], or [biːm] ("block," "pillar," "slab," and "beam") and the helper walks away to the piles, gets a kind of stone that he has learned to pick up following a certain call, and brings it to the builder. If this were the language of a tribe, children would learn it. What they would in fact learn is picking up and bringing specific stones following particular sounds, here transcribed as "block," "pillar," "slab," or "beam." That is, the children are brought up to perform certain actions following certain sounds in certain contexts, in which people orient to whole situations in particular ways. In these, entities and processes are connected, constituting the whole and taking their sense from this whole that they constitute. In part, the training therefore will consist in a teacher's uttering some sound while pointing to the corresponding objects, directing the child's attention to them, and at the time uttering a word.

> This ostensive teaching of words can be said to establish an association between the word and the thing.
>
> But what does this mean? Well, it may mean various things; but one possibly thinks first of all that the picture of the object comes before the child's mind when it hears the word. But now, if this does happen – is it the purpose of the word? – Yes, it *may* be the purpose. I can imagine such a use of words (sound series). . . . But in the language of [the builder] it is *not* the purpose of the words to evoke images. . . .
>
> In instruction in the language *this* process will occur: the learner *names* the objects; that is, he utters the word when the teacher points to the stone. – And there will be this still simpler exercise: the pupil repeats the words that the teacher articulates before – both are language-like events. We can also think of the whole process of using words in [the builder example] as one of those games by means of which children learn their mother tongue. I will call these games "*language-games*" and will sometimes speak of a primitive language as a language-game.
>
> And the processes of naming the stones and of repeating the words after someone could also be called language-games. Think of much of the use of words in games like ring-a-ring-a-roses.
>
> I shall also call the whole, consisting of language and the activity into which it is woven, the "language-game." (Wittgenstein 1958, ¶6–¶7, pp. 4d–5d, underlines added)

The example Wittgenstein chooses is an everyday activity (*Tätigkeit*). A mason engages in building a house or temple and is assisted by a helper. The purpose of the sound is communication so that they can do the job together. The question about the "meaning" of words is irrelevant, for what really matters is this: to get the collaboration function well. That the helper or children associate images with the words, however, *is not* the primary purpose of the word. Initially it is for the social interaction and anything we might find in the builder, helper, or child is a derivative of this interaction in a specific societal activity (e.g., constructing a house). The real question is not whether someone can picture or figure something in the head but whether and where the deployment of patterned sounds (i.e., words) is useful. That the participants, when requested, might create and describe images and use the sounds on their own is a fact that derives from the social interactions. In a series of expansions, Wittgenstein then exemplifies what he denotes by the term language-game. Initially it is the game in which children learn their language. But the process of naming stones, repeating the names, and calling out the names in the process of building a house or temple is also a game of language.

Finally, "the whole consisting of the language *and* the actions into which it is woven" can be denoted by the term. In fact, the official English translation does not do justice to the German, which uses the term *Tätigkeit* (activity), the sense of which is the collectively motivated production of cultural-historical products that are exchanged within society. Thus, for Wittgenstein, being a competent language user is indistinguishable from knowing one's way around the world in which this language is spoken.

This last position is the one taken by many other language philosophers. For example, Richard Rorty and Donald Davidson make this precise point, which is that language and the world are irreducibly tied up such that being able to navigate a language and being able to navigate the world are indistinguishable. That is, the difference between coping in the everyday world and competently speaking everyday language is undecidable. Following a line of thought initially articulated by Heidegger, Derrida (e.g. 1996a) also makes the same point using the former philosopher's German term *Geflecht* (mesh, tissue) by emphasizing the interlacing of the familiar world and the word. In fact, it is the familiarity of the world that takes priority over the word. The familiar world constitutes a relational totality of signification, which we may denote by the term *significance* (Heidegger 1977). It is significance that is discovered before the things that are contained in a situation. This is why in learning, words *accrue to* significations rather than *acquire* significations ("meanings") that come to be associated with or attached to sounds. This is also the point that Wittgenstein makes when he suggests that the lever in a brake is a lever only because it is part of the brake. On its own, the piece of metal would be nothing but a piece of metal.

In my own experience of having apprenticed to many different forms of activities – hatching fish, doing scientific research (science laboratories and field research), building homes, and so on – language learning is better characterized by the second theoretical approach than by the first (constructivist). In each case, familiarity with the work world and familiarity with the language that is being used as part of work have arisen simultaneously. It is not that I learned a list of words, which I then matched to situations. In fact, learning the language at work is like learning a foreign language, which we cannot learn by using a dictionary, because this most likely leads to inappropriate use of words. This I can see daily while working with graduate students for whom English is the second language. They often use words in inappropriate situations and places in texts, based on dictionary definitions; it is through everyday use that they learn to correctly employ words – even if they do not have a mental framework, as I show in Chapters 6 and 7. In the following section, I draw on one of the experiences where I immersed myself in a new world heretofore unfamiliar to me, where people used words that I did not know, and which have had their place at work and in their workplace. Learning these words was equivalent to becoming familiar with the work and situations; and with this familiarity also came the proper use of the formerly unknown, foreign/strange words.

Knowing Our Way Around in the World

> We may say that linguistic ability is the ability to converge on a passing theory from time to time – this is what I have suggested, and I have no better proposal. But if we do say this, then we should realize that we have abandoned not only the ordinary notion of a language, but we have erased the boundary between knowing a language and knowing our way around in the world generally. (Davidson 1986, pp. 445–446)

To better understand the use of language and the lacking need for the concept of meaning, we might investigate the experiences of what happens when we learn a language. Here I do not refer to situations where we learn a second language that maps more or less well onto an existing first language – such as when individuals from non-English-speaking countries arrive in Canada and learn to speak English or French, which allows them to go shopping for groceries in ways similar to what they have done in their first language. Rather, I am interested in situations where we come to learn a new language, that is, a way of getting around in and talking about a world heretofore unknown to us. My experience as a researcher investigating workplaces that I did not know before provided me with many opportunities to experience first hand how we come to know and learn a language. For example, when I first arrived in the fish hatchery that I would be studying from 2000 to 2005, there were many situations in which I wondered what the fish culturists and their managers were talking about. Similarly, when I arrived in the biology laboratory described in Chapters 3, 4, and 5, both the world of their work and the language that goes with it were unfamiliar to me. But, because in both cases, I used apprenticeship as an ethnographic method, I learned not only *about* these different worlds that I had not known before but also learned to *participate in* producing what the people produced: salmon smolt ready to be released into the river and ocean, in the first instance, and scientific data and papers, in the second instance.

To exemplify what it means to learn a new language, which is not really new, take the following conversation (Fragment 9.1). It has been recorded as part of the research project from which I also feature excerpts in Chapters 3, 4, and 5. Although readers will be familiar with (nearly) all words – i.e., understand their dictionary sense – they are not likely going to follow what is being said and why. This has also been the case during the first week of my stay in the biology laboratory, where I took part in the data collection as part of my paradigm, apprenticeship as ethnographic research method. I became a member of the research team and participated in all aspects including the publication of papers. The fact that I had a Masters degree in physics allowed me to become familiar much more quickly with the events, especially the conceptual underpinnings of the method and the tools and equipment themselves.[7] The point is that understanding the conversation does not require "constructing" "meaning" or anything else that the constructivist

[7] As part of my Masters thesis, I measured how much energy protons lost when traveling through beryllium. The energy lost was determined by taking the difference in X-ray radiation the protons created in a carbon backing with (scan) and without (reference) a beryllium film in the proton beam.

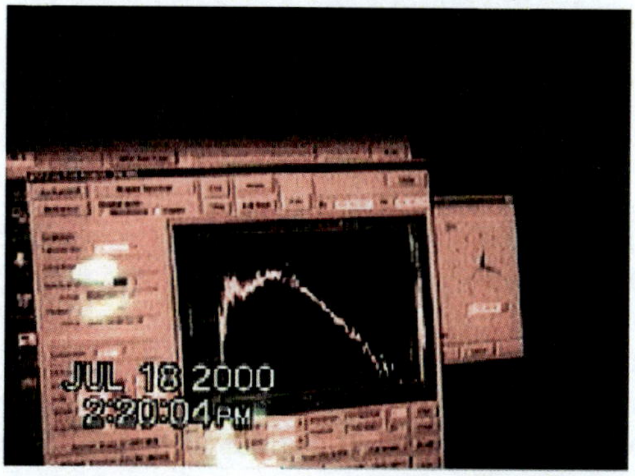

Fig. 9.1 Ted and the ethnographer observe the computer monitor on which the results of the measurements are displayed; starting turn 10, Carl also orients to this screen, the contents of which become the topic of the conversation.

metaphor makes us describe to be happening. It required me to become familiar with this world. When I had begun to know my way around this laboratory, conversation such as the one featured in Fragment 9.1 were inherently familiar, immediately understood, and required no interpretation or construction whatsoever.

Fragment 9.1

```
01      ((noise: click, clack, clack))
02 C:   scan.
03 T:   reference?
04 C:   already have done the reference.
05 T:   reference done.
06      ((noise: click clack))
07      ((pause))
08 C:   ((noise: click clack, clack) scan ... single cone
09 T:   okay ((modifies graph, Fig. 9.1)) see it is very broad
        thats a problem.
10 C:   but i mean i think that this is a detrending problem
        ((points to monitor, Fig. 9.1, rotates hand with two fin-
        ger in front of image))
11 T:   yea.
12 C:   and that ((points to peak in Fig. 9.1)) i think is . . .
13 T:   a bit of low.
14 C:   put the . . . whats the lambda max? on that . . . approx-
        imately?
15 T:   ((puts cursor on peak, Fig. 9.1)) can you read the number
        here?
16 C:   fourtwentysix
17 T:   so that would be a blue.
18 C:   okay.
19 T:   okay.
20 C:   so save this; its a blue.
21 T:   ((Fiddles with computer to save data.))
```

Fundamentally, the research in this laboratory concerned the absorption of light in the retina of salmonid fishes – Pacific salmon and trout – composed of rods and cones. The three different kinds of cones are sensitive to ultraviolet, blue, and green and red light ("double cone"). But the absorption changes during the life cycle of the fish: not only does the peak of the absorption spectrum (i.e., λ_{max} or "lambda max") shift with respect to the wavelength spectrum but also the peak changes in width. The measurement of the absorption is based on a comparison between the amount of light that goes through the microscope slide and medium next to a photoreceptor, the reference beam ("reference"), and the amount of light that goes through the photoreceptor ("scan"). There is, of course, more to the experimental background, the biology, the equipment, but this information suffices for the moment to understand the opening part of the fragment. Carl operates the microscope, where he lines up the microscope slight for referent and sample measurements. Ted operates the computer, which drives the source of the light, collects the data from the light sensors, and calculates the difference between the amount of light through the source and next to it to arrive at the absorption spectrum.

The newcomer to the laboratory will soon understand that the noises – onomatopoetically transcribed as "click, clack, clack" – are due to the opening and closing of a gate that permits a beam of light to fall onto the sample. The light for the searching process comes from a different source than that for the measuring process. In turn 02, Carl indicates readiness for and demands a measurement of the amount of light that goes through a cell ("scan"). In uttering "reference?" with a rising intonation, Ted asks whether this would not be a reference measurement to be taken, but Carl responds that this measurement has already been taken. Ted confirms and gets the software ready for the measurement. Carl prepares the sample and, in uttering "scan" (turn 08), announces that it is ready for the measurement. He also marks the presence of a "single cone," that is, a cone that should be absorbing in the ultraviolet or blue region.

As the absorption spectrum appears (Fig. 9.1), Ted announces a problem, "it's very broad" (turn 09). Carl, however, suggests that the problem really is one of "detrending," which, as those present know, is easily fixed. To understand this stretch of talk, we need to be in tune with the events. This allows us to know that the graph visible in Fig. 9.1 is not the "real" absorption spectrum but the spectrum "sitting" on a steeply sloped line from the top left to the bottom right. Carl makes a rotating gesture in front of the computer monitor as if he were turning the image counterclockwise: as if turning the imaginary sloped line (in lab lingo the "baseline") until it is parallel with the horizontal. Doing this removes the line, that is, the "trend," and therefore corresponds to the action that Ted would engage in if asked to "detrend."

Ted raises another issue: "a bit of low" (turn 13), which is a commentary on the height of the peak over and above the trend line. Pursuing whether this is a real signal or some artifact, Carl asks for the wavelength at which the maximum of the curve lies ("what's the lambda max on that, approximately"). After Ted rearranges the display, he asks Carl, who sits closer to the monitor, to read off the value at the cursor. Carl says "four-twenty-six." He thereby indicates that the curve peaks at a wavelength (λ) of 426 nanometer – a fact that others in the laboratory could verify. Ted comments that this would correspond to a blue cone, one of the two single cones that exist in the retina. They agree, and Carl then asks Ted to save the data as pertaining to a blue cone.

In the course of my ethnographic work, I never felt to be engaged in a process of "constructing" "meaning." Rather, I became increasingly familiar with what people where doing and the implications of what they were saying. Thus, for example, "reference," "scan," and "detrend" ask for particular actions much in the same way that Wittgenstein's builder and helper employ "pillar" and "slab" to call for certain actions. Nothing of what happens is somehow insight the head of people: everything is outside, available for newcomers such as myself to observe, listen to, and learn from. Within a few days of the instant when this fragment was recorded, I was sitting in Carl's chair, aligning the sampling light beam next to a cell and on the cell and asking for a "reference" and a "scan" precisely in the same way that Carl is doing it in the fragment presented here. I had learned pointing to the "baseline" and reading the display to see whether the curve was consistent with an ultraviolet or blue single cone or with one of the members of a red-green double cone. I had learned to see the difference between a "single cone" and a "double cone," and the difference between a "broken rod" and a "single cone." I could learn this because it was there "for the taking," publically available and not some meaning hidden in the heads of Ted and Carl, who were not only doing their work but also, precisely in doing their work, were allowing me to learn how to do it. When I had a question, neither Ted nor Carl had to get into my head or interpret "what I meant," but rather, they provided descriptions much in the way I do in the preceding paragraphs.

In a very strong sense, therefore, learning the language used in this laboratory was equivalent to knowing my way around the laboratory, learning to do what the more senior members do, including the production of specific sounds, which entailed actions that I also learned to accomplish. This did not prevent me to do some reflecting, envisioning what had happened during the day when I wrote my ethnographic field notes at night. For example, on July 17, 2000, I included the following paragraphs in the field notes:

Notes.000717

"As a *sign*, as a *name*, a word has nothing in common with what it is the sign of. What is 'common' is only discovered in the act of transforming the word into a deed, and through the deed into a thing (and then again in the reverse process), in practice and the mastering of its results." (Il'enkov, 1977)

The meaning of words is not recoverable from the word, but depends on the practice that relates words to "deeds" (language as action), that is, through the material aspect of being and interacting in a material and social world.

Readers can see that already at the time, my concerns lie with "the meaning of words," which, as I state, cannot be recovered from the word itself. There are practices in the laboratory and the words go with these practices in the same way as objects, materials, and tools. Once I was using objects, materials, and tools in the way the other laboratory members did, the sounds, which we hear as words, were no longer strange, no longer constituted "lingo" or "jargon," but articulated precisely what needed articulation. What is said is both sufficient and required to keep the laboratory working as a whole, producing the data required for the next scientific paper to emerge as the result of the work.

In a book suitably entitled *Mind in Society* (Vygotsky 1978), the author shows how every higher psychological process has been a societal relation. That is, mind first and foremost resides in societal relations. This, too, is the case for the laboratory. If there is mind, it exists in the relations between people who, in face-to-face interactions, make available to and for one another *everything you need to know to be able to participate and understand.* Vygotsky suggests that every psychological function appears twice: "first, on the social level, and later, on the individual level, first, *between* people (*interpsychological*), and then *inside* the child (*intrapsychological*)" (p. 57).[8] Whereas I agree with this statement to some extent, it really requires reformulation, because the child or novice would not be able to participate in interchanges with others if there were not already something happening on the inside. Rather, precisely those inner processes that children or novices eventually do on their own already occur when they participate in interactions. Therefore, many Russian psychologists nowadays working in the area ask for the abandonment of the inner/outer distinction because every action involves the inner and outer so that the distinction does not explain anything. When I do the preparation of a sample or locate and prepare a cone for a measurement, there are always inner and outer processes whether I am familiar with this part of laboratory work, as I was later, or whether I am unfamiliar and work under the direction of Ted or Carl. Knowing my way around this laboratory also means being attuned, feeling at home. Such attunement is a fundamental mode of Being in the mundane world of our everyday (rather than foreign/strange) surroundings. Attunement refers to primary understanding, literally, to Being (standing) amidst the things, people, and their doings that constitute the setting at hand.

Tuning and Being Tuned

The process of learning to find one's way around some place in the world means becoming attuned or being tuned, for "understanding is always attuned" (Heidegger 1977, p. 142). But attunement is attunement to something, which in my ethnographic study was attunement to a working research laboratory. This world was shot through with significations prior to my arrival, and attunement meant getting in tune with this world. Whereas "getting in tune" is an agential expression, it

---

[8] I do not have access to the original Russian version to see whether Vygotsky uses the adjective *social'no* (social) or *obščestvennyj* (societal). English translations – unlike German translations (e.g., Vygotsky 2002) – of Russian social psychologists habitually and falsely translate the latter in terms of the former.

leaves silent the attunement to something, a tune already playing. That is, attunement involves otherness, an otherness of the foreign/strange with which I get in tune. Attunement means tuning and being tuned, an agential as much as a passive process – such as when one clock comes to be entrained by another. What was initially foreign/ strange became familiar to me in a process of "de-distancing," where things that initially stand out – i.e., things that are *ekstatic* – disappear from consciousness when we are absorbed in the world of work. That means, attunement also means abandoning oneself to a world of which we, even when we are most familiar with it, never become the master.

This process of learning, becoming familiar with what is originally foreign/strange, goes into the opposite direction but involves the same dimensions as when ethnographers research their own familiar worlds. To find out about their world, the latter attempt to make the familiar strange, of distancing what is initially close, so that in a process of de-distancing they can come closer to their own culture while moving away. Heidegger's German language captures this dual process, because the theoretical term he created can be heard both as "distancing" (*Entfernung*) and as "de-distancing" (*Ent-fernung*). The fact that I have been familiar with the world, which the living/lived body remembers in and through its actions, leads to the fact that any "factically existing Being-there [*Dasein*] in a way always already knows its way around, even in a strange 'world'" (Heidegger 1977, p. 356).

Learning the way around the world involves a substantially passive moment. This is so because my living/lived body is the condition of my knowing. This living/lived body also is a material thing among material things that together make this world. My understanding arises precisely because of this material inclusion: "The world encompasses me, comprehends me as a thing among things, but I, as a thing, for which there are things, comprehend this world . . . *because* it encompasses and comprehends me" (Bourdieu 1997, p. 157). It is precisely this material inclusion, which extends to the social and dispositional structures, that allows me to acquire practical mastery, knowledge, and control. This material inclusion and my openness to the world, my exposure to the world, my passibility and therefore my susceptibility to be fashioned by the world, led to a process of socialization into the world of the laboratory. Bourdieu suggests that all forms of intellectualism are wrong precisely because they do not recognize the role of the living/lived body, which entails both agential and passive moments. Intellectualism – of which constructivism is but a particular variant – cannot understand the relation between the dualistically opposed dimensions of body and mind. This difficulty is most salient and exasperating in the phenomenon of language: "each act of language, as an incorporeal sense ("meaning") constitutes a veritable miracle, a sort of transubstantiation" (p. 160).

To really understand the practical comprehension that I acquired in the laboratory, including the practical comprehension of language, "we have to situate ourselves above the alternative of the thing and consciousness, of mechanical materialism, and constructivist idealism" (p. 163). That is, we need to "get rid of mentalism and intellectualism, which lead us to conceive the practical relation to the world as a 'perception' and of that perception as a 'mental synthesis'" (p. 164). Bourdieu proposes the notion of *habitus* to take into account the way in which we think about practical competence in everyday affairs. It constitutes the agential and passive dimensions of practical knowing. On the one hand, habitus, a set of structured structuring dispositions, gives human beings "generative and unifying, constructive and classifying powers" (p. 164) without making them transcendental subjects of their own activity. This theoretical concept also makes thematic that the body is a socialized one, that is, a body subject to the concrete influences of an inherently social-material world. That is, precisely those dispositions that are structured in and by the individual's relation with/to the world also make the human being agential. In this statement, we have come full circle to Vygotsky, who, as suggested above, ascribes the source of all higher mental functions to the societal relations in which an individual does engage and has engaged. How I think and what I can think does not depend on me, is not the result of *my* constructions, but is the result of interactions that have shaped me as much as I have been shaping them.

The essence of this chapter therefore can be summed up in this way: *Let us stop the nonsense of meaning.* We must stop searching for meaning separate from the words people use, for it is based on the same metaphysical ideology that separates bodies from minds and minds from the world. We need to stop worrying about what is behind the words, their meanings, and be concerned with what people do and how they do it, including how they use language (in generally non-reflective ways) to get things done. We get things done with our living/lived bodies, which therefore cannot be separate from our knowing. There must not be a difference between knowledge and its application – or, equivalently, between words and meaning – for then we would be back into the intellectualists' separation of body and mind, mind and the world.

The task ahead of us will not come easy. How difficult it is to abandon the use of the concept of meaning may have been anticipated in the foreword of the definitive work on meaning: "And if I am not wrong therein, then the value of my work secondly consists in showing how little has been done by having solved the problems" (Wittgenstein 1921, p. 186). Little had been done indeed, for the constructivist metaphor upholds this manner of theorizing language even though the author of the "definitive work" on meaning himself characterized it as the problem and spent the remainder of his life to exhibit the nonsense of talk about *meaning*.

Part D

Passion

> Every relation which does not stand out against the horizon of the world and which does not render the world manifest in its way draws its possibility from pathos. Pathos designates the mode of phenomenalization according to which life phenomenalizes in its originary self-revelation; it designates the phenomenological material out of which this self-donation is made, its flesh: a transcendental and pure affectivity in which everything experiencing itself finds its concrete, phenomenological actualization. (Henry 1999, p. 353)
>
> But this relation to oneself – this access to ourselves – precedes us, it is that from which we result. (Henry 2000, p. 123)

The constructivist metaphor is an integral part of intellectualist (i.e., metaphysical) epistemological theory. The only role the body has in knowing is to produce the experiences from which the mind abstracts the regularities that give rise to cognitive structures. Thus, for example, Piaget's (1970) child scientist moves about blocks and realizes that however these are arranged, their number remains the same. The movement itself is only the material from which Piaget allows mind to abstract a concept or schema. In fact, it is not from the movement that understanding is abstracted, but from the schemas on which movement is based; these schemas already have the structures typical of mind. Even in the so-called embodiment and enactivist theories, the not the living/lived body proper is theorized but rather image schemas are introduced, which already have linguistic structure that subsequently are further metaphorized. As a result, all of this work on sensori-motor schema and embodiment does little more than raise the specter of Cartesianism (Sheets-Johnstone 2009). The bodily experiences, therefore, are relevant only in so far as they are of intellectual nature, that is, in as far as they are and can be made subject to the intellectual (linguistic) consciousness.

In recent years, there have been a number of attempts to complete the gaps in the intellectualist (constructivist) approaches by adding emotion, motivation, and other forms of experiences. But I consider these additions to be epicycles to the existing theories, for they are not *integral* to the phenomenon but external influences that tend to diminish cognition. In these approaches, cognition and emotion are externalities to each other. The problem therefore is that these added phenomena are external to cognition; and the way in which these phenomena come to affect conceptions is, again, in an intellectualist manner. (Not surprising because the mental can be affected only by mental entities.) Externalities, however, cannot change the phenomenon at heart: "As externalities, they are indifferent toward each other and lack the necessity for one another that ought to lie in the relation of an *outer* to an *inner*" (Hegel 1979a, p. 236). Already in the 1930s, Vygotsky had reproached psychology for approaching intellect and affect in this manner, as externalities one in respect to the other.[1] But the very fact that cognition can be *affect*ed should indicate to us that there is another dimension to it, its capacity to be affected: its *passibility*. Mind, therefore, is more than a metaphysical entity. Surprisingly, the first clues about the undecidability of the mental and bodily come from the very same Immanuel Kant – the *Logodedalus*, the ultimate, most-accomplished artisan of the mind – whom constructivists claim as their intellectual origin and key reference.[2] At the very end of his life, Kant (1964), as part of his writings on anthropology, thought about jokes and realized that the difference between the mental and bodily effects of jokes is undecidable: "getting" the joke and laughing about it are simultaneous and an explained joke is not funny at all.

In this fourth part of the book, I present three types of phenomena that cover three groups of senses all falling under the concept of passion: emotion and motivation, suffering and pain, and passivity. These phenomena are unnoticed in intellectualist approaches because they fall on their perceptual blind spots. Intellectualist theories, focused on logos and the logical, completely fail to understand all those forms of experiences that derive from *pathos*, a Greek word denoting suffering, feeling, emotion, disease, and passion. This leads to two problems: first, "the door is closed on the issue of the causation and origin of our thoughts, since deterministic analysis would require clarification of the motive forces that direct thought into this or that channel" (Vygotsky 1986, p. 10); and second, the "approach precludes any fruitful study of . . . the influence of thought on affect and volition" (p. 10). In fact, there is a logic arising from pathos, *patho*logical in a positive sense. That we can learn about knowing, learning, and human competencies precisely when the normal ways of going about everyday life no longer work is a realization that Heidegger conceptualized in *breakdown*, an understanding that ethnomethodologists make use of while inquiring into the methods that everyday folk use to produce the very structures of society that researchers investigate. A contributor to a faculty seminar on Piaget comments on the most crucial problems in Piaget's work:

---

[1] As a keen reader of Karl Marx, whom he frequently cites in his works, Vygotsky was familiar with the way in which dialectical philosophers conceive of inner relations and contradictions and the manner in which these manifest themselves in one-sided ways.

[2] Jean Piaget, Ernst von Glasersfeld, and Mark Johnson all make explicit and favorable reference to Kant, whose ideas about the self-constructing mind they have taken on, including the ideas about the image schema(ta)s that "need to be specifically packaged corporeally – to be *embodied* – in order not to remain embedded in a purely mental sphere" (Sheets-Johnstone 2009, p. 220, original emphasis).

> For Piaget, structure and development are synonymous. The cognitive structure is that which develops. For Piaget, there is a structural or cognitive component to any behavior. Now the crucial problem is that what Piaget seems to mean when he talks of development is cognitive development and everything else is a consequence of cognitive development. When we talk about emotional development for Piaget, we really mean cognitive development applied to emotions. And when Piaget studies cognitive development, he really thinks he is studying the development of the child. Whatever the type of development whether it be musical, affective, social or societal, the development component is the cognitive component. And that is what Piaget studies. *When Piaget talks about affective schemes he is just paying lip service. He has not really studied it. He just says: "Sure enough, I know they are there."* (H. G. Furth, in Silverman 1980/1997, p. 56, emphasis added)

It is quite clear from this quote that Piaget, as the (social, radical) constructivists following him, is concerned with cognitive development. He pays lip service to emotions; he does not incorporate them in the way that Vygotsky asked psychologist to do for understanding the mutually constitutive role of emotion and cognition. Piaget theorizes cognitive development independently of affect, as if it could be understood in this manner, opening the door to the epicyclic addition of emotion. This has not changed in recent formulation of learning in terms of conceptual change, which asked that cognitive and emotional dimensions be considered "as variables of equal status which usually also have to undergo conceptual changes" (Treagust and Duit 2008, p. 301). There are at least two problems with such a view: (a) variables, emotion and cognitive are already external to each other and (b) an emotion is theorized as something that undergoes conceptual change and therefore is of the same rather than a different kind. The unity Vygotsky has in mind is the unity of differences rather than the unity of the same. In this manner, nothing but an epicycle is added to the central theory (see Chapter 1).

The question has to be framed in a completely different way: How would we theorize cognitive development if passibility were more important than cognition to human development? What if the primary experience that founds all other experiences is to be the capacity to be affected? Passion (Gr. *pathos*) – a term that is etymologically derived from the Proto-Indo-European root *k^{w}enth-*, to suffer, endure, by way of the Greek *páskein* and the Latin *patī*, to suffer – rather than actions and agency would be the originary experience on which knowing is founded!

In this fourth part of the book, I focus on experiences that lead to forms of learning that intellectualist approaches cannot explain, in fact, that lie completely outside the explanatory possibilities of intellectualist thinking. Passion not only describes the experiences of hardship but also those of joy and pleasure. Here hardship, joy, and their many modalities derive form a much more fundamental unity: the essence of life in its ipseity (Henry 2003). In this fourth part of the book, I present and discuss multiple forms of experiences and learning that derive from different forms of passion. I thereby articulate dimensions of human experience and therefore dimensions of learning that comprehensive learning and developmental theories generally and learning theories in the sciences particularly have to address. Because passibility and affectivity are primary experiences that precede any (intellectualist, intentionalist) cognition, theories of cognitive development that are founded on these phenomena will inherently include the dimension of affect, emotion, and passion rather than having to add them as afterthoughts to non-viable theories.

Chapter 10: Emotion, Motives, Motivations. In their radical formulations, intellectualist theories do not have a place for emotion and motivation. Subsequent revisions initially included these as variables external to cognition that somehow were supposed to modify cognition (despite Hegel's advice that factors and aspects do not change a phenomenon when they are *external* to it). In more recent formulations of constructivism in science education, affect and cognition continue to be variables of equal status. It is evident that in this formulation, we are not one iota closer to the kind of relationship between affect and cognition that Vygotsky formulated nearly 80 years ago. To provide an analogy from the sciences: Think of a physicist intending to add up two manifestations of light, the particle and the wave character. Of course, readers will say, this does not lead to anything useful. But precisely the same is true for trying to add up cognition and emotion "as equal factors."

In this chapter, I draw on a data set that I have collected over the years to show how in concrete situations, the difference between emotion and cognition becomes undecidable. It is only in this way that our everyday experience of emotions that mediate cognition (e.g., how I feel mediates how I respond on a mathematics exam, how I make purchases) and cognition that mediates emotion (e.g., knowing that my grandmother has died changes how I feel) can be explained in a logically non-contradictory way. Emotions are the result of bodily states not merely states of the physical body but of the lived body; and the living/lived body with its sensory organs and auto-sensory mechanisms is also the origin of thought. Thus, an integration of emotion and thought becomes possible when we consider both cognition and affect as different manifestations of the living/lived body in its interactions with the world. But emotion is not to be thought of something exclusively pertaining to the living/lived body: It is a collective phenomenon in important ways, leading to an integration of emotion and motives of activities. Motivations, rather than being a force that makes people do what they would not normally do on their own, equally come to be understood from the holistic approach to emotion and cognition.

Chapter 11: Crises, Suffering, Joy. Suffering and pain are part of a second dictionary sense of the category of passion. Suffering and pain would be considered external to cognition in constructivism. But this is not so, as I could personally experience while having been subject to a chronic illness. This illness does not diminish my cognition generally, though it may do so too at times. Rather, it has given me access to new forms of understanding. These forms of understanding are not the product of conscious construction but, if they were indeed of the same nature as conceptions, they have emerged. In this chapter – drawing on everyday human experiences I show how illness, suffering, pain, and crises can give rise to new forms of consciousness. Such forms of consciousness are possible precisely because of the undecidability concerning the difference between the living/lived body and the mind. Here I do not understand the mind as an *embodied* phenomenon.

Rather, the bodily-kinetic forms constitute mind: The difference between the living/lived body and the mind is undecidable – that is, constitutes an instant of *différance*. In other words, the mind is but a modality of the living/lived body.

Chapter 12: From Incarnation to Responsibility. In this chapter, I provide a further account of the theme of radical passivity that traverses the preceding chapters as a guiding thread. In radical constructivism, there is an acknowledgement that the knowledge constructed is tested for viability in the world. The experiences that the cognizing organism can have in the world derive from its physical senses. However, there are numerous questions concerning the manifold nature of the senses and how they are stimulated, which the constructivist position *cannot* resolve because, a result of its Kantian heritage, it is entirely unclear how any cognitive agent would be able to coordinate the inputs coming from the different senses. How do visual experiences come to be coordinated with tactile, auditory, olfactory, or gustatory experiences? Mind and rationality need to be so powerful in the Kantian framework because otherwise it would not be able to integrate the different sense impressions. This only changes when the order of the senses comes from the sensory-kinetic events themselves. In a phenomenology of the living/lived body, the organization of the different sense impressions derives from a differentiation rather than from an identification and unification of disconnected sense impressions. There is therefore a level of passivity even more passive than the one that is associated with learning from sense impressions (see Chapters 2,3, and 4). It is from an original (auto-) affectivity that the affectivity of the senses and therefore the sense of the living body emerges. This kind of affectivity is characteristic of the living/lived body, referred to in the phenomenological literature as *the flesh*, which has led to the idea of incarnation as the organizing principle for knowing in mathematics and science (Roth 2010b). Order in and from the senses precedes cognitive order. It is from this passivity also that intentionality emerges, a problem to which constructivists have no answer: We do not intend our intentions but rather, they are given to us. We fly with our intentions without ever asking ourselves who or what gave rise to them for doing this or that. This form of passivity is also involved in decision making, which requires us to make leaps of faith because decisions, to warrant this name, inherently cannot be computed, calculated, or entirely grounded in reason (see various works by Derrida on decision making, forgiveness, gift-giving).

Chapter 10

Emotion, Motives, Motivations

> The word does not merely designate an object as a present-on-hand entity, but also expresses by its intonation my valuative attitude toward the object, toward what is desirable or undesirable in it, and, in doing so, sets it in motion toward that which is yet-to-be-determined about it, turns it into a constituent moment of the living, ongoing event. … No content would be actualized, no thought would be actually thought, if an essential interconnection were not established between a content and its emotional-volitional tone, i.e., *its actually affirmed value for the one thinking*. The active experiencing of an experience, the active thinking of a thought, means not being absolutely indifferent to it, means an affirming of it in an emotional-volitional manner. (Bakhtin 1993, pp. 33–34, emphasis added)

In the introductory quote, Bakhtin points out that each word not only designates some object but also expresses a valuative attitude toward the object, the affect with respect to the object as desirable or undesirable. Without this valuative aspect, we cannot understand the living, ongoing event. Without this affect, the actually affirmed *value* for the one thinking, we would not understand any content of the person's speech. The constructivist metaphor makes us theorize events in the mind of the thinker, the only place where something such as "meaning" can be found. However, it is precisely affect that allows us to explode this retreat into the subjectivist conception of mind because the "emotional-volitional tone opens up the self-seclusion and self-sufficiency of the possible content of a thought, makes it a participant in unitary and once-occurrent Being-as-event" (p. 36). The emotional-volitional tone, the valuative dimension of thought, not only relates to the "*whole* concrete and once-occurrent unity in its entirety" (p. 36), but also makes any thought relevant to collective life as such.

This framing of the relationship between cognition and emotion is astonishing, as affect seemingly has received a bad reputation in intellectualist philosophy and psychology. Thus, Kant (1956b), for example, suggests that affect "belongs to sentiment (emotion), in so far as it, preceding deliberation, makes the latter impossible or more difficult" (p. 539). In fact, Bakhtin's *Philosophy of the Act* is a rebuke of the constructivist approach that separates reason from historical, irrevocably once-occurrent life. Reason – driven by virtues – is supposed to control

W.-M. Roth, *Passibility*, Classics in Science Education 3,
DOI 10.1007/978-94-007-1908-8_10, © Springer Science+Business Media B.V. 2011

affect. Affect is considered to be a weakness, even when it expresses itself as enthusiasm (Kant 1964). Even love in friendship must not be grounded in affect, because the latter is fundamentally blind. Affect comes with surprise as sensation, which sublates the powers of the mind. Affect is overly quick, "jumps to conclusions," and therefore makes thinking impossible – it is imprudent. On the other hand, affectlessness, when accompanied with the strengths to act, is phlegm in a positive sense: character of a brave person, who does not give in to the passions, including anger, hate, lust, or love. Affect is, according to Kant, like a stroke, like water that breaks through a dam, or like a state of intoxication, which one sleeps off only to wake up with a headache. Affect, considered on its own, is without reason, imprudent; it is characterized not by the strength of a certain feeling but by the lack of reason. In fact, my living/lived body becomes in this approach "an affective object, whereas external things are from my point of view merely represented" (Merleau-Ponty 1945, p. 109) a way of thinking that both contains and supports the body–mind distinctions typical of metaphysics. Thus, "the union of soul and body . . . was understood, in Cartesian fashion, as a *de facto union* whose *de jure* possibility need not be established" (pp. 111–112). As a result, affectivity is not recognized as a form of consciousness that parallels intellectual consciousness in reflecting the collective activity at hand.

In science education, the Cartesian split that separates mind from the world and the body from the mind continues to reign. Affect generally and emotion, motives, and motivations specifically have not received attention as *inner aspects of* a phenomenon that has cognition as one of its other manifestations. That is, affect, when it is considered at all by a theory of cognition, then it is framed as a factor external to cognition, as but another factor in the consideration of and mediating learning. This is quite evident in the following quote, which articulates the need in conceptual change-oriented science education to "also develop" "affective dimensions . . . in science teaching":

> [A]ffective dimensions involving student interests or motivation also need to be developed in science teaching because they play a significant role in supporting conceptual change on the level of science content knowledge . . . students' self-efficacy and control beliefs, the classroom social context, and the individual's goals, intentions, purposes, expectations and needs are as important as cognitive strategies in concept learning. (Treagust and Duit 2008, pp. 300–301)

These affective dimensions are said to "play a significant role in supporting conceptual change." These dimensions are the contexts surrounding cognition and conceptual change rather than manifestations of something else that also expresses itself as cognition. This is clearly seen in the classification of emotional dimensions – self-efficacy, control beliefs, social context, goals, intentions, purposes, expectations, and needs – as "*cognitive strategies* in concept learning." Both the adjective "cognitive" and the noun "strategies" point us to the unrelenting focus on mental representation as the constructivist approach to affect.

Because affect is not theorized as an *inner* aspect of a more general phenomenon, related phenomena – including motivation – are said to occur externally, such as when teachers are made responsible for motivating students:

> [The teacher] overtly used the analogy as a motivator and a morale booster. . . . By having students predict what was going to happen to the blood and then praising students, he established in the students' minds confidence to use the analogy as well as their own self-efficacy. (Treagust and Duit 2008, p. 315)

Readers notice without doubt that the constructivist science educators use a discourse almost entirely behaviorist, where the teacher in question motivated students and boosted their morale, allowing *him* to establish "*in the students' minds* confidence . . . and self-efficacy." Here, then, the "affective dimensions" are used to manipulate students; affective dimensions are addressed by external control mechanisms for making students do what they might not have done on their own. That is, much as the Kantian discourse, where thinking and reason are agential but affect is something to be subdued and controlled by thought, the constructivist discourse theorizes affect as something to be controlled and manipulated to change cognition from the outside. Vygotsky, certainly, would be turning over in his grave if he had to read or hear about such ways of thinking about and theorizing the relationship between affect and cognition.

In the following I make a case for an integrated approach to emotion, motives, and motivations and how they come to be shared in activity inherently oriented toward the collective provisions for meeting human needs. The approach is fundamentally materialist, grounded in an account consistent with the evolution of *Homo sapiens sapiens*. I present a case study from a science classroom, where affect is expressed in the passion with which a student presents her point of view. This case study provides me with the materials for exhibiting the role of expression as a "carrier" of affect, which "spreads" in and through the attunement by others. Rhythm will be an important intermodal dimension of human expression and communication that not only demands an integrated approach to the living/lived body as the seat and possibility of knowledge but also as the means for the coordination of social interaction by means of affect. I conclude with comments on the place of affect in the general orientation of a person to his/her world.

Materialist Approaches

In materialist approaches – whether they have been developed in philosophy or social psychology – affect plays a central role in cognition because it enables cognition. Thus, dialectical materialist approaches, concerned with a categorical reconstruction of the human psyche, show the centrality of emotion as an evaluative mechanism. In this mechanism, different existing capacities – needs (e.g., lack of food) that bring about negative valuation, arbitrary motions that lead to need satisfaction (e.g., because of higher food concentrations in the primal environment), and the positive valuation that arises from them – come to form an integral *dynamic* whole (Leont'ev 1981). Because the initially arbitrary movement comes to be associated after several repetitions with the change from negative to positive valuation, the very capacity to move leads to the emergence of intentionality.

Movement no longer is arbitrary but directed to bring about a change from negative to positive valuation within the organism; but the non-arbitrary nature of movement comes about as a result of the role that the valuation makes. A very different, phenomenological reconstruction of the relation of affect, movement, and cognition comes to the same conclusion:

> If having a feeling in an emotional sense depends on a certain postural set, a certain tactile-kinesthetic attitude and thus a certain tactile-kinesthetic feel, and if one must get out of this tactile-kinesthetic attitude and feel in order to have a different emotion, then clearly, definitions and distinctions are less important than the recognition and descriptive analysis of a basic corporeal matter of fact: *affective feelings and tactile-kinesthetic feelings are experientially intertwined.* (Sheets-Johnstone 2009, pp. 201–202)

Affective expressions are based on and develop with movements: movement and affect are irremediably intertwined. Thus, various researchers who have worked with chimpanzees recognize their repertoires of affective expressions, which include facial displays, vocalizations, gestures, movements expressing social emotions, greetings, expressions of desires, and so on. They produce threat displays including glaring, head tipping, arm raising, hitting, flapping, stamping, arm waving, swaggering, quadrupedal hunch, and bipedal swagger. In virtue of this "congruency, motion and emotion – kinetic and affective bodies – are of a dynamic piece" (Sheets-Johnstone 2009, p. 196). Rituals, that is, rhythmic occurrences of movements and of entire events constitute an important means of "spreading" affect.

Once we begin to reflect about the term *emotion*, its relationship to motion should not come as a surprise. The term emotion already contains the word "motion" in its entirety. Etymologically, emotion derives from the Proto-Indo-Germanic root *meu̯-, meu̯ə-,* from where it made it into Greek and Latin. The Latin verb *ē-movēre*, from *e(x)-*, out, and *movēre*, to move (by means of pushing) was used to refer to individual and collective stirring, agitation, and tumult. In psychology, this movement tends to be associated with feeling and affection. The term motive has the same etymological origin and has been used to denote that (the force) which causes the motion. The same is the case for motivation, which denotes the fact of having a motive, intention, itself driving the movement of the person. As a result of their initially spontaneous motion, organisms come to be affected by their environment. Because of their senses, which also underlie sense – like affect, a reflection of the environment but on an ideal plane – they are able to *e*valuate the changes that occur by attributing to them positive or negative valence. As Sheets-Johnstone suggests, affective feelings and tactile-kinesthetic feelings are experientially intertwined; and they are so to such an extent that the difference between them is undecidable. That is, the organism does not know the difference between affective and tactile-kinesthetic feelings because they are of the same cloth. Any instance of affective feeling involves tactile-kinesthetic feeling; and every instance of tactile-kinesthetic feeling involves affective feeling.

As the example of the chimpanzees shows, during evolution, the intimate connection between needs, affect, and movement came to be generalized. Affect-laden displays are observable in other animals of "lower order." During the evolution of what eventually led to the human species, group behaviors and characteristics came to selected and, in a sudden change over, the capacity to have a culture became dominant for the *Homo sapiens* species. In this species, initially collective hunting behaviors came to be subdivided into hunting and chasing (beating), followed by further divisions of labor whereby tool-making, already and still observable among chimpanzees, separated from hunting activity. But this division of labor was possible only because the needs now were generalized and addressed at the collective level. By participating in the collective control over life conditions, individuals expanded the control over their individual life conditions (Holzkamp 1979). Thus, the individuals who built tools no longer had to contribute to the hunt, because they could exchange tools for food rather than having to hunt themselves. Hunters could focus on hunting without being concerned about making tools, because they knew that they could trade a piece of meat for a hunting tool. Such trading already exists in the animal world, as both chimpanzees have been observed to trade food for subsequent sexual favors or birds have been observed to trade help in accessing food with subsequent sharing of food. In fact, as long as the individual contributes in some way to the collective provision of needs, his/her individual control over life conditions is enhanced; this enhancement of control comes with a positive valence. That is, in the human species tool-making and other forms of activity contributed to the collective satisfaction of dietary needs. A *collective* motive had developed that oriented the activity of tool making, hunting, and chasing. Affect (emotions) became a way of monitoring the state of the activity; and, because affective expressions were already communicable, individual and collective expressions came to be integrally tied. As has been shown, the categories of affect and motive, which is a collective form of orientation toward the world, make the construct of motivation unnecessary. Learning, once understood in terms of an increase of control over life conditions – i.e., the increase of agency – does not have to be motivated because it is inherently in the interest of the individual, inherently associated with a positive affective value (Holzkamp 1993).

Intentions, too, are connected to affect, as shown by the example of the single-cellular organism living in an originary soup. Intentionality is integral to an orientation to change the condition from negative to positive value. For Vygotsky (1986) intentionality develops from initial indicatory gestures and words. He uses a child's utterance "mama" as an example, to show that intention does not lie in and is expressed by the word. Rather, "*the child's whole behavior at that moment* (his reaching out toward the chair, trying to hold on to it, etc.)" (p. 65) articulates the child's intention to/for the mother. In this instance, "the 'affective-conative' directedness toward an object . . . is as yet inseparable from the 'intentional tendency' of speech" (p. 65). Object-orientation and intentional tendencies are manifestations of a n integral whole. Beginning his analyses very differently, Merleau-Ponty (1945) comes to the same conclusions. Language is but a part of a more holistic taking up of a position in the world shot through with signification. Thus, the "phonetic 'gesture' brings about, both for the speaking subject and for his hearers, a certain structural coordination of experience" (p. 225). This structure thereby constitutes "a certain modulation of existence, exactly as a pattern of my bodily behavior endows the objects around me with a certain significance both for me and for others" (p. 225). Communicative expression is an orientation, and such orientation *always* is generated by a certain affective tonality, associated with an affective value.

A Passionate Argument

Generally intending to be involved, checking on homework, wanting to pass the chemistry course, and avoiding detentions that would keep her off the basketball team, Monica has an idea about how to figure out the valence of a chemical element given its position in the periodic table. She asks for a turn and attempts to explain her idea. Her science teacher challenges the explanation and Monica begins to argue in a manner similar to how she might argue if challenged by someone outside of the classroom. She sits in the back of the room (Fig. 10.1) where most of her classmates can see her only when they turn around – which happens at least once during this discussion just preceding the major part of Monica's argument. The teacher stands at the whiteboard, where she sometimes notes something down during the episode, whereas at other times she turns her back to it without moving from this position. The teacher has explained a way of deriving the valence of a chemical element given its position in the period table. Monica has an idea, which she presents publicly once called upon. The fragment of interest here begins just after she has completed the articulation of her own method.

Monica has just completed explaining to the teacher and her class how one can arrive at the valence of a chemical element by taking two away from the atomic number. She ends saying, "I lost my numbers" and produces a big smile, her right arm and hand resting on her desk (Fig. 10.2a). There is a clear expression of a positive emotional valence in the tone of voice, comfortable body position, and facial expression. The teacher (T) utters what we may hear as a query, and which is realized both in Natasha's taking up the last word of the teacher's phrase ("from") and Monica's own turn at talk. At this point, there is an explosive expression radically changing the emotional tone that characterized the preceding instants. In apparent frustration, Monica raises her voice and lifts her arm/hand, uttering "I'm just saying, just do the number two" (turn 15). The raise in the voice is constituted by both an increase in speech intensity, which is four times (i.e., 6 dB) that of the normal upper boundary (Fig. 10.3), and by a movement of the pitch to nearly 700 Hz, which is almost 500 Hz above the normal. These speech parameters are expressions of what we hear as a shrillness of the voice. We also see that the speech rate is much faster than normal, which is attained again near the end of the utterance ("take the two from the") prior to another peak in speech intensity and pitch (Fig. 10.3).

Fragment 10.1

```
11 T:  where is the 'twO coming from.
12     (0.52)
```

Fig. 10.1 This seating arrangement shows that Monica is placed in the back of the classroom where most of her classroom cannot perceive her unless they turn around, which, at one point during the discussion, actually happens.

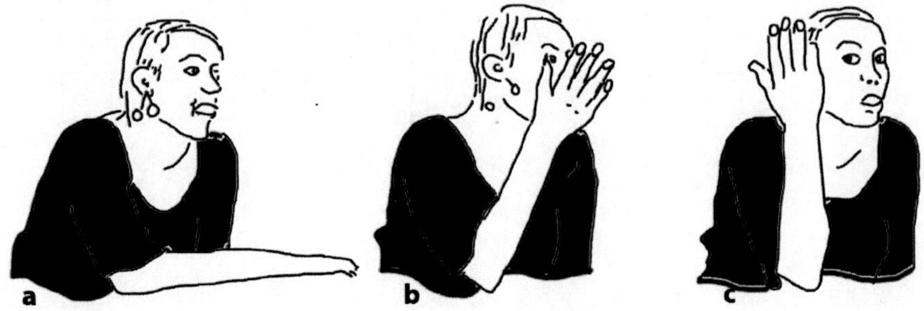

Fig 10.2 **a.** Monica rests after having made an attempt at explaining her way of determining chemical valence, smiling over having "lost my numbers". **b.** Monica takes a stance in apparent frustration raises voice and hand. **c.** Monica glances at the researcher who operates the camera.

```
13 N:  <<pp>from>
14     (0.19)
15 M:  <<f>I=M JUST [SAY>in just [do the number 'TWO: (0.14) ↓i:
                    [((Arm/hand as in Fig. 10.2b))
                                 [((Arm/hand as in Fig. 10.2c))
       ↑youdontwanmeda copy text from noWHERe, or
       [you can take t=the two  ]
       [((Gesture in Fig. 10.4))]         come from the (0.16)
       the 'rOW,> i mean
       (0.42)
16 P:  <<p>that wont wORK fr (0.57) forALLof them.>
17     (0.42)
18 V:  itdoesntwork forALL of them.
19 M:  'well' what. you <<f>want> then the <<f>↑TWO>-
```

Here, just as has suggested above, the phonetic gesture realizes for the speaker and her listeners a certain structuration of experience, a modulation of their existence in this classroom, which, in this manner, endows the room, the objects and individuals it contains, and the issue at hand with a certain signification. Thus, all the measurable speech parameters (Fig. 10.3) are consistent with those identified in psychological research as correlates of hot anger (rage). The pitch range is way up and the pitch perturbation is up (jitter), and the F1 means are up (top). Speech intensity is way up – considerably and consistently above the normal range (Fig. 10.3), and so is the energy, which is three to four times the normal amount of energy, and high frequency energy (see dark upper regions of spectrogram in Fig. 10.3, especially at the beginning of the outburst). The speech intensity is high, repeatedly peaking above 80 dB, that is, about 8 times the normal intensity of Monica's speech. The speech rate also is way up (at beginning of the outburst, 10 syllables in 1.8 seconds, whereas 11 syllables in 3 seconds).

In this situation, we see how the articulation of a conceptual position goes hand in hand with taking a position more generally involving the entire living/lived body, and especially modes that are not conscious to the speaker. In all the expressive features that she makes available in and through her living/lived body,

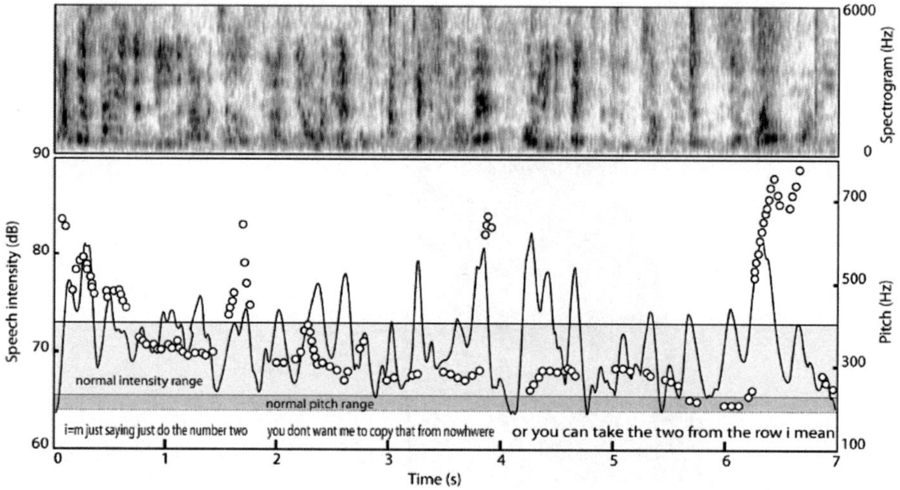

Fig. 10.3 For almost the entire sequence, Monica's pitch lies above her normal pitch range and her speech intensity frequently peaks above the normal range (doubling and even tripling normal volumes). The spectrogram shows energy contributions to the high frequencies especially early in the utterance.

Monica expresses a passionate engagement and defense of her (conceptual) position. Observationally, we might say that "she makes a passionate argument." But this passion is not something we have to interpret given certain signs. This passionate nature of her argument is available to anyone else in the room, especially because of the intermodal form that this passion expresses itself. As Fig. 10.4 shows, while Monica speaks her body has moved upward out of the seat and she is leaning forward. Just as she articulates an answer to the question "where is the two coming from" (turn 11), Monica thrusts her arm and hand forward toward the teacher, the hand changing from an open-palm to a finger pointing back to an open-palm configuration. Her entire face expresses anger, her mouth is wide open, and her irises are in the corner of the eyes so that there is a lot of the white eyeball visible. The force of Monica's presentation, and the level of emotional engagement, is available precisely from those modes not conceptualized when we think about classroom conversations in terms of the constructivist metaphor. It does not surprise, therefore, that nowhere in the article on third-grade children's discussion of the "poofy" substance do the authors write about emotion, affect, or passion (see Chapter 9). There the children appear to engage in a dispassionate manner. We can also see that Monica's bodily expression is not external and another factor but is itself expression of the stance that she is taking. Only a minor aspect of the present instance is the conceptual position, articulated by verbal means, which is but one of the (one-sided) expression of taking a position.

We should not think, however, that the various expressions are but outward signs of an emotion, here anger, which is seated somewhere else, perhaps in the mind. All the movements *are* the anger. In erecting the body, Monica gets herself

Fig 10.4 Monica exhibits affective engagement with the topic in and through her entire body.

into position for the argument. "The postural attitude is thus coincident with what might be designed the onset of emotion: *with a felt urge to do something*" (Sheets-Johnstone 2009, p. 202). Here it is to react to the opposition toward her previously presented method of calculating the chemical valence of elements and to provide an explanation for her attempt to understand rather than merely copy some text from "nowhere." Emotion then is not *identical* to the taking of a position, change in prosodic parameters, thrusting of the hand and arm, and the facial expression. Rather, emotion is "from the beginning by way of the postural attitude, the motivational-affective source of such actions" (p. 202). The same affective essence is made available and expresses itself in the orientation and position of the body, the hand-arm gesture, and the sound of voice *in addition and prior to* any conceptual content that might be expressed. Here, then, "affectivity is 'impregnated with intelligence'" (Merleau-Ponty 1945, p. 180) and, in extension, intelligence is impregnated with affectivity. The difference between affectivity and intelligence is undecidable so that "bodily feelings and feelings of emotion are divisible only reflectively, after the experience" (Sheets-Johnstone 2009, p. 202).

In this fragment, the affective dimension is apparent not as something internal to Monica but, in and through her expression, it is visibly and hearably manifested in the situation as a whole. She is not just presenting an argument, as it might appear on a piece of paper, "in cold blood" so to speak, without emotion, without engagement, without force. Rather, it is affect that is at the origin of the movement that we can observe in a variety of ways, as body movement, hand-arm gesture, sound, sound modulations, and so on. In intellectualist (constructivist) traditions, affect and emotion are external to cognition. Everything is the result of social and individual construction. It is as if we had access to emotions only through constructions, as if we had to interpret the affective feelings in and of our bodies through intellectual consciousness. In contrast, in cultural-historical and phenomenological materialist approaches, both emotion and cognition are but moments that constitute one-sided manifestations of the situation as a whole. We would not be able to understand that emotion mediates cognition and cognition mediates emotion if the two were externally related as one factor to another. Rather, to get a mutual mediation – thoughts can get us emotionally aroused and calmed, and emotions can arouse and calm thoughts – the two have to be moments of the same, inherently non-self-identical phenomenon. This phenomenon, activity theorists would say, mediates between cognition and emotion. It is only if this phenomenon is different within itself that it can manifest itself in and as external differences.

In this instance, we observe the relationship between collective motive and the motivated goal-orientation that Monica displays for us. Thus, the collective activity is schooling, which is oriented toward the reproduction of society generally and its culture more specifically. The handing down of cultural knowledge to be reproduced by new generations is thematized as "learning." In the episode, Monica articulates the motivation for understanding: "You don't want me to copy text from nowhere!" She wants to understand; and she expresses having developed an understanding that differs from what the teacher has presented so far. Oriented to going beyond copying texts "from nowhere" opens for Monica a way to understanding, that is, an expansion of her control over the present context, that is, her way of figuring out the chemical valence of an element. If she were able to figure this valence out on her own, then her power to act would have increased and, with it, her control over the subject matter of chemistry. Her emotion is a reflection of the state of the activity, which, in the present instance, appears to be a resistance to the idea that she is pursuing and presenting. Thus, for example, she has just presented her own method for figuring out how to extract the chemical valence of an element. But she meets with resistance in the form of a question where a major aspect of her method comes from: the number 2 that is subtracted in her method from the atomic number of the elements ("Where is the two coming from?"). This resistance impedes with the goal – which is written all over the episode – of having her method accepted as a legitimate alternative. The negative emotional value expressed in and through the entire expression of the living/lived body is associated with the continued distance between goal and actual state of the situation. That is, affect parallels intellectual consciousness in its reflection of material reality. It provides both an orientation for action, because of the promise of a positive emotional value associated with increased control, and an evaluation of action in direction of the realization of the goal, here negative because of the resistance and impedance with reaching the goal.

Mood and Attunement

Discussions concerning the epistemology of science or the epistemologies students, teachers, or scientists espouse rarely if ever concern themselves with the emotional and emotive aspects of Being, knowing, learning, and so forth. Yet emotion, too, needs to be understood as an integral aspect of any situation we study so that speaking and emoting no longer are treated as separate entities of research. The question of knowing generally tends to be treated independently of emotions, though the latter are an integral part of our daily lives, both at the individual and collective levels. Already at the beginning of the 20th century, Vygotsky (1986) notes the major shortcoming of psychology to be that it did not attempt to understand thinking | speaking without also considering affect. Vygotsky considers the major weakness of psychology to be the separation of theories of knowledge from theories of emotion. Emoting is theorized as something external to thinking. But, thereby realizing an idea that Hegel (1979a) first articulated at the beginning of the 19th century, Vygotsky holds that externalities *cannot* influence each other. If emotions were external to thought, then thoughts would appear to be "thinking themselves, segregated from the fullness of life" (p. 10). That is, without emotion we cannot understand thinking let alone the continually unfolding speaking | thinking unit. This also is the position Bakhtin (1993) takes when he suggests that we can only understand participative thinking – i.e., thinking integrated in the once-occurrent Being-as-act – when we approach it from within its unity. This unity also includes the emotional-volitional dimension that reflects my commitment to and responsibility for my act "in the unitary and once-occurrent context of life in which I participate" (p. 36).

Why do we speak | think? It is precisely because it moves us ahead in realizing our goals, which themselves are oriented to realizing the collective activity at hand. Both action-oriented goals and activity-oriented motives are reflected and monitored not only in intellectual consciousness but also in affective consciousness. Although the latter can be made the topic of intellectualizing reflection, in the pursuit of normal everyday activities, affect is not made thematic. We do what we do because we have to, even if our current affective state will mediate in decided ways the process and product of our actions. For example, when we are sad because a friend or family member has deceased, we act differently and with different levels of competency than under normal circumstances, and again differently when, for some reason, our affective state has a very positive tonality. The terms mood and attunement generally are used to denote the role and impact of emotions not only on Being but also on the way we interact with our world. "He is not in a good mood today" relates the affective state of a person as perceived by the speaker, and also explains why the person acts and interacts differently than on other days. We do not have to "interpret" these states, but these are available to us in the person's orientation to the world, his positions and orientations, gestures, voice, and so on. These expressions mark the entire situation so that the mood of the person may actually spreads to others in the setting.

In the approach I advocate here, emotion comes to take the place it deserves in the study of communication, which includes both the production of the particular situation as such and the reproduction of this type of situation according to the society in its current cultural-historical state and condition. Emoting and thinking | speaking interpenetrate each other; they are, in fact, two moments of the same phenomenon. Thus, affect, as mood and attunement, makes itself known in speaking by means of "intonation, modulation, in the tempo of talk, 'in the way of speaking'" (Heidegger 1977, p. 162). This is quite evident in the passionate argument between Monica and her chemistry teacher. The difference between emoting and thinking | speaking | acting is undecidable – only in this way can they mutually effect each other in the way Vygotsky suggests.

Monica and her teacher, as other interlocutors, communicate affect to others through speech parameters such as pitch, intensity (volume, loudness), and rate; they also communicate emotions (and empathy) through their orientation toward each other, for example, placing their hands on, or arms around, another person's shoulder. But this communication is not likely conscious. These are the ways in which children, prior to having developed self-consciousness, come to know and know about emotions. These speech parameters are indicators of emotions and mood/attunement (*Befindlichkeit*), just as the other physical characteristics that we may determine in a situation, including body position and orientation, gesture, and facial expression. Thus, for example, my research shows that conversation participants who are in agreement with one another (unconsciously) tend to align their pitch levels and they tend to diverge in pitch levels in the case of conflict of the form we see in the above example. When we look at the turn sequences between Monica and her teacher in the parts of the episode not represented here, we note how, in this episode of conflict, seemingly "trump" each other, as each next speaker's pitch level is higher than that of the preceding speaker. In the fragmentary episode provided here, we see how Monica has sped up and is further speeding up. It is not that her mind has to say, "speed up to become more credible or more convincing." Rather, the speech speeds up as part of an overall orientation to the situation, which is one of passionate engagement with the current topic, Monica's ways of figuring out how to derive the valence of a chemical element. From the physical placement and orientation as well as from the topic of the talk, any culturally competent speaker understands that the two are not in alignment. We may gloss the teacher by saying that she has the right way of figuring the valence, and that it is time to go on. Her impatience is communicated in the manner that she orients to Monica and suggests, "THIS is the way to do it." Monica, however, in and through her entire living/lived body, expresses a different orientation, and, in so doing, marks and endows the objects in the room with different signification. Everybody present experiences this, as we can see from the videotape, which shows how the rhythm Monica produces is the same rhythm that many of her classmates in the room also take up and produce.

Rhythm, Affect, and Sociality

Expression is not just characterized by the prosodic parameters discussed so far. We already note a change in the rates of speech and a running together of the sounds that normally are separated to produce distinct words. Another important feature related to the temporality of movement is rhythm, which can be expressed in and across all of the forms of expression discussed so far. Thus, speech may be rhythmic, changes in pitch and speech intensity may be rhythmic, and so may be changes in body position, body orientation, and gestural configurations. Rhythm therefore is cross- and intermodal. From an intellectualist perspective, rhythm is completely unnecessary. But in real life, it is a pervasive feature not only in expression but also across the turns that different speakers take. In fact, rhythm, although it can be made the content of a reflective analysis, requires a form of consciousness that is very different from intellectual consciousness (Abraham 1995). As a phenomenon, it can be understood only as the bringing together of outer (objective, natural) and inner (subjective, performative) dimensions. Thus, for example, when two or more individuals are aligned in rhythmical terms, this alignment cannot be explained as the result of an interpretation of the signals someone else produces so that the hearer can actively align with the producer of the rhythmic signal. This interpretive loop would take much too long for it to be a feasible manner for aligning rhythmically with others. The recurrent features would have to be anticipated to appear at the correct time, which is made difficult by the fact that precise repeat frequencies would no longer be perceived as rhythm. Periodicities have to fall out of tact to be recognized as a pattern of movement (Brüstle et al. 2005). Rhythm is interesting, above all, because of its well-known character as medium of affectivity and emotion across cultures.

In this classroom, there are signs that show how others are in synchrony with Monica, even when students are seated and oriented such that they cannot see her (Fig. 10.1). The rhythmic "beat" of her communicative performance is available to others. In one part of her performance, Monica presents how to calculate the chemical valence for the elements in the second row (Li, Be, B, C, N, and O). With each atomic number in the series three (Li) to six (C), she briefly glances to the periodic table, as if verifying what the next number is. The heads and gazes of other students (e.g., Gabriel, Natasha, Shawn, Fig. 10.1) also move their regard from left to right on the periodic table of elements. They move their gaze simultaneously with Monica although they do not see her and although there is no indication in the speech content itself that suggests others ought to look at the table of elements. Because of the development of thought in speaking, they also do not know what Monica will say but, simultaneously, they have no time to stop and interpret what she is saying. That is, the resource for producing this synchrony in orienting gaze is made available through the scanting rhythm in her vocal production rather than through seeing her body move. This synchrony and *participative thinking/understanding* are much better concepts to theorize these events.

There are other signs of synchrony. For example, Gabriel rocks his head slightly back and forth. Even though he cannot see Monica, his rhythm perfectly produces and reproduces the forward position of Monica's hand, which itself is aligned with the rhythm of her vocal production (intensity peaks, pitch maxima). Gabriel also produces the same rhythmic pattern with his right leg, which swings in a left-to-right motion matching in its extreme left position the foremost position of his rocking head. Thus, when his chin reaches its most forward position, Mirabelle's hand also reaches its most forward position, and the peaks in her speech intensity and pitch are at a maximum. When Mirabelle arrives at the end of her presentation, Gabriel stops rocking his head, the last coincident movement having been a slightest movement with a coincident closing of eyes following an upward movement of head to direct the gaze to the periodic table.

In the preceding sections, we observe the integral relation of knowing – which is essentially expressed in the form of movement (vocal apparatus, body orientation, hand/arm gestures) – and affect. It is precisely because of this integration that emotion can be communicated and becomes a social fact rather than one that can be theorized in terms of hormones that flood the individual thematized (*ek*static) body. It is not surprising, therefore, to find that "intercorporeal meanings are thus etched along the lines of kinetic/tactile-kinesthetic bodies" (Sheets-Johnstone 2009, p. 231). Through our living/lived bodies, significations are expressed and highlighted in the setting; through attunement between living/lived bodies, significations are reproduced and (thereby) affirmed. Such an analysis of language from an evolutionary perspective comes to the conclusion that "a common sentient-kinetic body is thus the foundation of a common body of significations that are solidified in archetypal corporeal-kinetic forms and relations" (p. 231).

Being in synchrony and harmony is an important condition and outcome of interactions and interaction ritual chains that reproduce cultural-historical activities (Collins 2004). Synchrony tends to positively influence current affective states and has a positive valence, whereas being out of synchrony has negative valence, that its, it is experienced as discordance and otherness. Speech parameters therefore are an important way in which orientation toward another is made available to others; and, because of the situation of pathos involved while attending and listening, these parameters are part of the way in which listeners literally are *affect*-ed. In opening up while actively listening, Monica's classmates create the opportunity to become attuned to her, and this tuning has occurred – like a pendulum mounted closely to another on the same wall that has become entrained to the other. The attunement to Monica is observable (and therefore objectively available) in the alignments of both pitch and speech rate (and the rhythms that the two lead to).

There is another important emotional aspect that is more elusive and never appears to figure in social analyses: Monica and her teacher are not just talking, not just participating in a lesson. They are reproducing and transforming a form of societal activity, schooling in the form of a discussion of a particular chemical fact. There is a motive embedded and embodied in such activities, which, in and through the participants' work (labor) also comes to be reproduced and transformed. Collective motives serve to orient collective need satisfaction of society. At the level of the individual, need satisfaction is regulated by emotions such that we act to achieve goals that have positive valence (need satisfaction) and avoid threats to our well-being, which have negative valence. The participation (actions) of both Monica and her teacher has to be understood in terms of their personal long-term goals, which always are oriented toward anticipated outcomes that have positive affective value – even at the cost of actions with negative affective value in the short-term – and toward the avoidance of outcomes that have negative affective value. This participation is understood not as a phenomenon sui generis but as an event by means of which a part of society is reproduced and transformed. Getting in synch and harmonizing prepares the present (local) affective conditions and context that support the positive valence in the anticipated outcomes.

Emotions are central to the ways in which human beings orient to situations; and they are modified by the articulation of emotions of other participants in face-to-face meetings. Besides facial expression, prosody is a major pragmatic resource for displaying and experiencing emotions. There have been suggestions that participants with less power and status converge in their prosodic parameters to align with those of more power and status; and differences and conflict between conversationalists having roughly equivalent power and status (e.g., children participating in hopscotch games) are characterized by very high values and differences in the pitch levels of participants. Differences in participants' institutional positions do not determine the way in which conversations unfold or the contents that they cover. In the present chapter, prosody is an expression co-extensive with taking up a position in the world that is shaped in and through the taking of a position. Prosody is but one of the moments of communication, all subordinated to the same task at hand and, therefore, being different expressions of the same societal-psychological or ideological unit. That is, sound-words, prosody, body position, hand gestures, and other communicative resources articulated at some point in time are different manifestations of the same underlying orientation, emotional valence, and signification. The sense marked out by vocal gestures therefore does not lie behind but is "intermingled with the structure of the world outlined by the gesture" so that "the smile, the relaxed face, gaiety of gesture really embody the rhythm of action, the mode of being in the world which are joy itself" (Merleau-Ponty 1945, p. 217). Sound-words are but one part of a texture that the members to a setting contribute to its constitution.

Affect and the Orientation to the World

This chapter makes evident that we need to theorize the reproduction and transformation of prosody not only as part of a larger communicative unit but as part of taking up | expressing a position in and on the world. Intonation is an integral aspect of communication, as it reflects "my valuative attitude toward the object," clearly exhibited in the manner that Monica takes a stance and defends her contribution to the class discussion. Rather than viewing power and status as factors that *determine* pitch levels and convergence, the production of pitch levels, pitch continuation, pitch level repetitions, and so on are co-expressions of difference/resistance and accommodation. The concept of social alignment denotes high levels of unity or agreement. Convergence in prosodic parameters among two or more participants is manifestation of – and serves as a resource for – the further production of alignment, itself an expression of the emotional synchrony of participants. In cases of conflict, some participants may, through the production of lower pitch levels and speech intensities, change the context and, in so doing, cool the situation and calm down an angered, excited, or animated member.

We find social alignment – as a phenomenon that participants continuously reproduce | transform – within the student body. This alignment occurs in a variety of ways, including prosody. Such alignments may well be sources for solidarity that the students experience within their peer group. Thus, the synchronous rhythmic features simultaneously found at various places in the classroom suggest that the students are affectively aligned, which expressed in their being "in tune" and "in sync" with one another. They also express anticipation of particular events, such as when numerous students turn around to face Monica as if they had literally "seen something coming" while Monica was readying herself – audibly and visibly – to articulate her response. Anticipation inherently reflects *being attuned to* and *participatively understanding* the situation in a phenomenological sense. That is, it reflects an intuitive, lived sense of what is happening and what dangers might loom ahead rather than a reflective understanding. I note above the sense of anticipation that three students express when Monica starts her explanation. This sense is available directly to the viewer/listener of the videotape in Monica's performance as well. At the same time, these various expressions, observable from Monica's position in and her take on the world, may have been resources encouraging her to take up the challenge and propose her alternative description of remembering and recalling valences.

Prosody and rhythm are important phenomena for producing affective entrainment, a driving force within the set of interaction ritual ingredients that have positive emotional value and, consequently, solidarity as part of their ritual outcomes. We can understand such outcomes also as context, where feedback intensification through rhythmic (i.e., affective) entrainment plays a crucial role in the reproduction and transformation of mutual focus and transient emotions. Entrainment is not describable in terms of a causal link, for the production of synchrony, other than in a mechanical system, requires anticipation. Thus, in one instance, Monica produces a particular rhythm, the same that her classmate Gabriel also displays. It is not that Monica's rhythm *causes* Gabriel to rock in the same rhythm, because Gabriel, looking toward the front of the classroom, has no other resource than Monica's voice. If he had to consciously attend to making his rocking coincide with her activity peaks (intensity, pitch), he would be out of synchrony by something on the order of a second or two. Even if there were non-conscious ways of causing synchrony, he would still be behind her, for he could only know when Monica's pitch peaks after having heard it in its entirety. This means that Gabriel has to anticipate peaks that he does not know when they are coming, which is a production of his own, requiring that he already be in tune. That is, there are two significant dimensions to the appearance of rhythm. On the one hand, Gabriel is attuned to the rhythm made available in Monica's verbal expressions; these expressions come to him from the outside, *affect*ing him. At the same time, both in hearing as in his own external movements, he has to actively produce the rhythm, and, with it, the affective tone that characterizes his own living/lived body now in synch with Monica's bodily self. Being in tune is a phenomenon distended in time rather than a punctual momentary one. The synchrony of Gabriel's movement with Monica's prosody is a consequence of being in tune. Such an anticipation clearly is observable at the instance when her classmate Natasha utters "three," just as their teacher hits the board with the chalk but prior to latter's utterance of the same number word, and inconsistent with all other instances where number word and the hitting of the board fall together. That is, Natasha anticipates the correct placement of the count with respect to the noise from the chalk, but the chemistry teacher, who produces the chalk noise, is out of alignment with her own speech, which follows the rhythm of the beats produced by the piece of chalk hitting the chalkboard surface.

Prosody and rhythm, as emotion, motive, and emotion, constitute challenges to the constructivist metaphor of cognition because they escape intellectualization by their very nature. To understand this entire complex of human experience, we require a better metaphor than the constructivist one, which reduces everything to intellectual cogitation. In the next chapter, I turn to additional phenomena that constitute even greater challenges to the constructivist metaphor because they arise from passion and passivity.

The passions are integrally tied to kinetic melodies and emerge with movement. Sheets-Johnstone (2009), a phenomenological philosopher, who first was an accomplished dancer and subsequently completed a second PhD in evolutionary biology, provides interesting insights on affect in her discussion of the relationship between emotion and movement. Mounting evidence from a variety of disciplines, she shows that motion and emotion – the experiences of our kinetic and affective bodies – constitute one dynamic phenomenon. Emotion is not external to acting but neuromuscular acts themselves shape emotions and anything we might denote by mental. In turn, emotions shape neuromuscular actions. Emotions *affect* us: they are integral aspects of what we call affective experience. Emotions are passions. To passions, as the etymology of the word teaches us, we are subject and subjected as much as we are the subjects of passion. Emotions teach us that we are both moved and actively moving. This leads the philosopher to the conclusion that "emotions are from this perspective *possible kinetic forms of the tactile-kinesthetic body*" (p. 205). This nature of emotions also requires researchers to find appropriate methods that lend themselves to the capture of the kinetic forms that co-express emotions. Thus, when

> serious attention is turned to kinetic form and to the qualitative complexities of movement, emotions are properly recognized as dynamic forms of feeling, kinesthesia is properly recognized a dimension of cognition, cognition is properly recognized a dimension of animation, and animation is no longer regarded mere output but the proper point of departure for the study of life. (p. 214)

Sheets-Johnstone uses her insights from the study of motion and emotion as the basis for her critique not only of intellectualist approaches but also of those approaches that characterize themselves by means of such adjectives as "embodied" and "enactivist." These adjectives suggest that there is something else that needs to be placed in a body or that needs to be done. This something else tends to be theorized as image schemata of the mind, which are used for and expressed in body movement. But such terms deflect our attention from the originary sensory-kinetic movements and the archetypal corporeal-kinetic forms that they give rise to – existing as a form of immemorial memory that makes intention possible (Henry 2000). It also raises the specter of the Galilean-Cartesian split between mind and the world and body and mind.

Chapter 11

Crisis, Suffering, Joy

> I cannot include my actual self and my life (*qua* moment) in the world constituted by the constructions of theoretical consciousness in abstraction from the answerable and individual historical act. (Bakhtin 1993, p. 9)
>
> These are the powers of life: the power to will, to love, and to suffer. (Henry 2005, p. 202)

In the first opening quote, Bakhtin notes that my actual self and life are not covered by "the constructions of theoretical consciousness"; that is, Bakhtin questions the basis of all constructivist claims. The second introductory quotes states that the fundamental powers of life are the power to will, love, and suffer. These are precisely the powers that the constructions of theoretical consciousness miss. The purpose of this chapter is to provide some evidence that suffering, joy, and crises are experiences that (a) happen to us rather than being constructed and (b) lead to forms of knowing not otherwise attainable. Suffering, joy, and crises rarely, if ever, constitute the topic of research in science education specifically and in the constructivist (conceptual change) literature more generally. This chapter provides evidence that counters Piaget's (1981) position on the relationship between affect and cognition, which "specifically den[ies] that affectivity can create new structures" (p. 15). Moreover, this chapter mobilizes evidence that contradicts the claim that "affective" structures are "organizational forms of mental activity" (Piaget 1967, p. 5).

The three terms in the title of this chapter point us in the direction of passion and the passive moments that characterize our participation in life. The purpose of this chapter is to exhibit dimensions of learning and knowing that we rarely see made thematic in intellectualist theories (including radical and social constructivism). These dimensions concern understandings that directly arise from passion without mediation by a conscious agent engaged in the "construction of knowledge." Rather, in the three sections that follow I take account of understandings that directly arise from – are given in – experiences that are not wanted but undergone and suffered. Passivity therefore characterizes the emergence of understanding in a double manner: the experiences are those of passion to which I am

W.-M. Roth, *Passibility*, Classics in Science Education 3,
DOI 10.1007/978-94-007-1908-8_11, © Springer Science+Business Media B.V. 2011

subjected and the new understanding happened ("dawned") upon me rather than being the outcome of intentional construction. Such experiences are important to understand because it is precisely in pathic states, in the experience of passion, that the world as we know is changed. In these states, the unknown and novel happens upon us, that is, the ontic character (being) of an anterior life is sublated (von Weizsäcker 1973).[1] In those instances of life characterized by passion (Gr. *pathos*), what is it that decides and who makes the decisions?

The experience of passion teaches us that life *cannot* be explained in terms of cause and effect, because the pathic confronts us with the future as the unknown. The pathic more than other experiences show that any decision, any experience, rests on the dialectic of *willing* (agency) and *having to* (passivity). The pathic can be defined as the origin and source of the dialectic of willing and having to such that no decision worthy of this name can be predicted and computed. If something can be computed, if it is the outcome of a clear causal relation between antecedents and conclusion, then it is not a decision, a choice between alternatives, qualitatively different positions. This is so because the "only decision possible is the impossible decision. It is when it is impossible to know what has to be done, when knowledge does not determine and does not have to be determinant, that a decision as such becomes possible" (Derrida 1992, p. 157). Decisions are important aspects of life, for they imply commitment to real actions, which have consequences not only to personal but also to collective life – they obtain "historical factuality" (Bakhtin 1993) – and therefore are inherently associated with affective-volitional and deontological tonalities. Therefore, the non-computable, that which cannot be anticipated and therefore constructed intentionally, constitutes an essential aspect of human life generally and knowing and learning more specifically. This means that any epistemological theory aspiring to an explanation of learning needs to incorporate passion and passivity as a central and irreducible dimension. Passion cannot be just add-ons to an existing theory. If passion is constitutive of life, then the theory of knowing and learning needs to be *categorically* reconstructed, that is, our theory needs to be built on passivity as well and as much as on agency. All the types of passion that I reflect upon below have led me to understand aspects of learning in science; they have contributed to understandings that I subsequently published in relation to my books and refereed journal articles in science and mathematics education.

Life has Pathic Dimensions

Our world is governed by an ideology of the agential subject – the doer, the maker of his/her own mind – who pursues its intentions, sometimes driven by affect. But, being transitive verbs, *doing* and *making* presuppose consciousness: doing/making

[1] The verb "to sublate" translates Hegel's German *aufheben*, which has two opposing dictionary senses of "to retain/preserve" and "to suspend/destroy."

always is the doing/making of something from the things at hand. There is a vision of something that does not yet exist, a vision that therefore goes beyond the present conditions. To go beyond the present condition by intentionally doing/making something requires anticipating a future present. Such a presence of something non-present requires a *re*presentation, which is a condition for the conscious anticipation of future events (Husserl 1991). Such "representation does not contain any passivity" (Levinas 1971, p. 98). To construct knowledge, therefore, implies an intentional, willed engagement in and with the world for the purpose of better understanding (and controlling) it. It requires the anticipated object to be present in consciousness, for otherwise it cannot be aimed at. However, on cultural-historical grounds, we could not have made consciousness itself in an agential manner if consciousness and intentionality were required in agency. Rather, human beings always already find (and have found) themselves thrown into this world, with life and consciousness given to them; and they always already find (and have found) themselves speaking without having intentionally developed language to communicate. Language, as life, is given to us prior to our intention to acquire it. Language and intentional signification come together. This is why the figure of the missing father, the missing origin, returns so frequently in the writings of Jacques Derrida and is easily recognizable as part of one book title: "the prosthesis of origin." If, for the moment, we considered the possibility that intentions themselves had been intended, we would immediately face a problem: How can we intend intentions? Of course, innatist approaches have their convenient answers. But this would not explain why evolution selected for innate intentions. Thus, "intentionality, where thought remains *adequation* to the object, therefore does not define consciousness at its fundamental level" (Levinas 1971, p. xv). Intentions constitute a gift, they are given to us – as our personal lives in the making of which we had no hand – without our having intended them; we receive our own intentions in passivity. Ultimately, therefore, subjectivity means hospitality.

Life generally and human life specifically cannot be thought without the pathic and the different forms of passion. We only arrive at partial understandings if we presuppose current psychological understandings as causes of the origin of cultural and psychological dimensions of humanity. But if the pathic lies at the origin of life as given – language, psychology, culture, and so forth – then it has to be possible to exhibit its pertinence to phenomena such as knowing, learning, and experiencing. Throughout this book, I emphasize those aspects that confront us with the givenness of consciousness, language, identity, and knowing. However, the pathic is not to be thought as a frame, as a box, within which we suffer from and endure life. It is not like the Piagetian gasoline (affect) that drives or interferes with intelligence (the engine). Suffering and enduring is not the central aspect of our consciousness – though under certain conditions, such as in chronic or terminal illness, it does in fact become *the* central and determining aspect of a life. It becomes life itself. However, in the genesis of anything pertinent to human beings, the genesis of language, consciousness, identity, or knowledge the pathic is at the intersection of the changes that occur. It is precisely because we are afflicted by the passions that we can truly know them only in and through experiencing them – we cannot know what passion talk pertains to without having felt the relevant passion ourselves. We cannot know what addiction really is unless we have experienced an addiction ourselves; we cannot know pain, unless we have felt and lived it; and we cannot know love unless we have felt it with our whole Being.

The pathic is unpredictable. We cannot anticipate aspects that are given to or come upon us, that is, which enter our lives by means of the pathic. Discoveries and (sudden) insights are but two of the phenomena where we make the pathic aspects of coming to know thematic. Christopher Columbus could not anticipate the Americas, which is why we say that he discovered them.[2] Because discovery and insight reveal something unknown, we often say that we have stumbled upon that which revealed itself in and through the discovery or insight.

There used to be a time in science education when "discovery learning" was the call of the day. Because it has become clear that (a) we cannot expect today's students to rediscover what is known to the sciences and (b) it is a waste of resources to wait until students discover what is already known by others in our culture, the approach has gone out of fashion. Yet there is some legitimacy to the use of the term. Every learning episode of something really new *is* an unanticipated discovery, for it could not have been anticipated on the basis of the current knowledge. Such knowledge is not a mere extension of knowledge implied what we already know but constitutes a new structure. This new structure emerges in unanticipated ways. Structure and its emergence are modeled by *catastrophe theory* (Thom 1981).[3] In catastrophe theory the points in which (a) radically new options for a system emerge or (b) radical transitions between two possible states of a system occur are called *catastrophes*. These two forms of catastrophes can be understood in terms of a diagram: by increasing a relevant parameter, a one-equilibrium system is driven through a bifurcation point, where a new, two-equilibrium system emerges (Fig. 11.1). The second form of catastrophe occurs when the system is driven from one equilibrium state into another by the proverbial wing beat of a butterfly. Catastrophes are crises or turning points, that is, they are moments characterized by *sudden* and unpredictable emergence and transition.

Crisis

Sudden transitions in life can be investigated as phenomena or experiences precisely in those instances that we tend to name *crisis*. The word derives from the

[2] The verb needs to be understood in terms of the tensions discussed in Chapter 2, as it implies the uncovering of something already existing but currently hidden from view. From the perspective of the knower, it does not make sense to theorize learning in this way, because the unknown cannot be aimed at.

[3] Catastrophe theory is a precursor and alternative to chaos theory. It is a mathematical theory with qualitative and quantitative dimensions that has been used to describe the evolution of new structures such as biological and developmental morphogenesis or the transitions between peace and war.

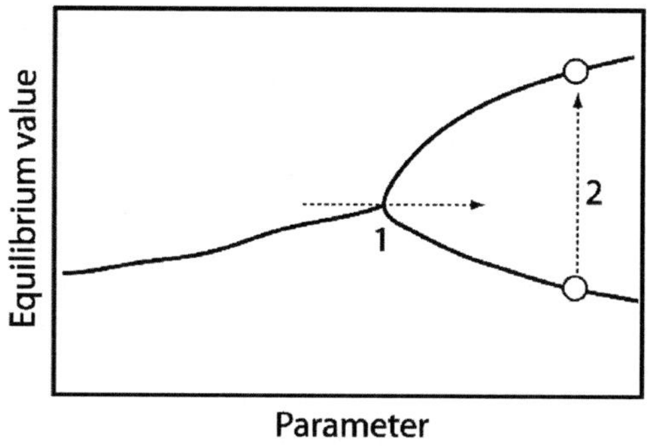

Fig. 11.1 Two types of crises exist in a system. **a**. When parameter increases to the point when the system changes from a one-equilibrium to a two-equilibrium system. **b**. When an infinitesimal variation causes the system to move from one to another equilibrium state.

Greek noun *krísis*, discrimination, decision, and the Greek verb *krín-ein*, to decide, to distinguish, which are based on the Proto-Indo-Germanic root *(s)krē-*, to cut. During crises, times of *unanticipated* transitions and the emergence of *unanticipated* structures, the pathic comes to be the dominant form of experience. During a crisis, therefore, we come to be confronted with the radically new, the foreign/strange. The crisis itself is syncopic, as it constitutes the changeover from the old to the new where it is impossible to say whether the instant belongs to the old or the new. Crisis therefore cannot be subsumed to either the old or the new form of knowing. Thus, in a crisis we do not only learn about some unanticipated aspect of life but also about the structure of crisis itself. It is not surprising, then, to hear philosophers suggest that an understanding of psychological phenomena – including cognition, learning – requires returning to the human condition. Such a return constitutes a task that "really only manifests itself fully in the context of pain, illness, and the human experience of lack which this involves" (Gadamer 1996, p. 87). This points our present investigation concerning the limits of the constructivist metaphor in the direction of the experience of passion, pathos, and suffering in human life. During crises, life reveals itself to itself in pathos – its mode of revelation *is* pathos. This revelation can occur only in the flesh, because of its ability to self-affect, that is, because it is endowed with agency (self-movement) and to sense this self-movement together with the affection that results from its senses. This revelation requires the flesh because there are no structures in the mind that could accommodate to meet the foreign/strange, that which cannot be anticipated because it is invisible and cannot be foreseen.

Suffering and passivity are central aspects of life; and patience in fact is relevant here in its multiple senses: (a) as uncomplaining endurance of pain, affliction, and inconvenience, (b) as forbearance under provocation, (c) as calm and self-possessed waiting, (d) as perseverance, and (e) as endurance and tolerance generally and generically. Patience has the same Latin etymological root *patī-* as passion and passivity, a passive form that may be traced to even older origin in the Proto-Indo-European *pē(i)-*, *pī-*, to harm, scold, put to shame. Another possible origin is the Proto-Indo-European root *k^w enth-*, to suffer, which has led to the Greek terms *pénthos* and *pathos*, affliction, misfortune, and passion. That is, since before time, patience not only means to undergo, to be subject to, but also obtains an agential dimension in that it is associated with an active taking up of the suffering, with perseverance, self-possessed waiting.[4] Precisely in patience – where we are *subject to* events that we cannot control by nature – will also shows itself at its best. That is, causality is suspended in crisis and decision. The concept of passion, rather than subjecting us to determinism, therefore allows us to understand the dialectic of freedom (willing) and necessity (having to). Passion allows us to understand the categories of freedom and necessity: to want to, to need to, to have to, to be able to, to be in position to, or to ought to do something. Grammatically, all of these are verbs are modes of the *subject*. Learning means becoming a subject, coming to understand the responsibility that comes with the categories of acting and suffering, patience, and decision. It is precisely the introduction of these categories that emphasizes Being as being-in-the-flesh not just a computing mechanism implemented in "wet-ware" (as distinct from hardware).

Experiencing (in) Crisis

Crises are periods and situations that we do not plan, construct, or intent in some other way. We are subject to them. We undergo transformations without being able to know beforehand what will be happening to us and what we will have learned once we are through such a crisis event. There is a form of discontinuity, and what follows cannot be anticipated or predicted based on what was and what we knew before the crisis. Using the analogy in Fig. 11.1, the equilibrium states to the right of the bifurcation point cannot be anticipated from the equilibrium state to the left of the bifurcation point; and the (vertical) transitions between the equilibrium points corresponding to a particular parameter are unpredictable by nature, as even unnoticeable influences can cause the changeover to the other equilibrium point.[5]

Crises lead to experiences and decisions that are without antecedent; experiences and decisions emerge from the crisis itself. My own understanding of the unity of living body and mind came during a crisis induced by many days without sleep and perhaps triggered by the consumption of marihuana. Moreover, this experience of a crisis has allowed me to understand the concept of crisis as a pathic aspect of life.

---

[4] "Before time" refers us to the time of pure Being, paradise before the snake and the apple from the tree of knowledge, when the representations required to conceive of time did not yet exist for our species. Time, as Husserl (1980) has shown, requires the making present of an absent presence, which occurs in the form of retentions, representations.

[5] This aspect of catastrophe theory has been used, for example, to model the sudden and unpredictable transition between peace and war in some historical cases.

During my late teens and early twenties, I experienced long periods of depression. At the time, I was mulling over, for example, the sense of life. But the more I attempted to solve the question in my mind, the more serious the depression became, arising to a great extent from the experience of a split between body and mind. But as much as my mind turned and twisted to deal with the questions of sense ("the meaning of [my] life"), the more it bit into its own tail, getting caught up in an infinite regress. Although meditation began to help me, the split between body and mind remained – perhaps not surprising given the going German ideology concerning the superiority of individuals who completed elite high schools and university.

Following the completion of my Masters of Science degree (physics), I hitchhiked all over Europe, then Canada and the US. In many places (and cars), I encountered opportunities to "toke" (to smoke and inhale! marihuana). While staying in a southern U.S. town with friends who had a substantial supply at their hands, resting up from months on the road, I began to read day and night for weeks on end with little or no sleep. The books tended to be on Zen Buddhism and quantum physics. Then, one afternoon, I had the strangest of experiences from which I emerged entirely transformed.

On this afternoon, in the same way as on other afternoons during that period, I begin by smoking a joint while reading one of Carlos Castañeda's ethnographic reports on the culture of the Yaqui Indians and their shamanist practices – including *The Teachings of Don Juan, A Separate Reality*, and *Journey to Ixtlan*.[6] As the drug takes effect, I *all of a sudden* have the sense that I am no longer breathing myself but something else outside of me is doing it for me. Or, rather, I am being breathed. Then a new sense emerges in an unanticipated fashion, the sense of being taut like the drumhead of a steel drum. It bulges outward toward one side, being a little larger than the shell that fixes and defines the outer edges of the drumhead. I am the drumhead, pulsating slowly. Each movement brings me closer to the normal resting state. I sense that if the drumhead-I moves through the equilibrium state, it will be my end. Death. A second image emerges, suddenly, existing side by side with the drumhead image. It is that of going through a singularity – the biblical camel that goes through the eye of a needle.[7] My whole body squeezed through a hole with zero extension. Death again. I can feel my whole living/lived bodily self resist. I do not want to die. But the vibrations toward the equilibrium state of the drumhead become stronger. I am moving/being moved closer to the

---

[6] There has been a tremendous debate about the veracity of these ethnographies. Although initially acclaimed critically, in part because of the academic credentials of the author, a graduate of the University of California at Los Angeles (UCLA), some scholars later suggested that the author made it all up; others say that the reports might be fiction but based on a substantial amount of veridical evidence. There also have been suggestions that these reports constitute a critique of the anthropological method.

[7] It is notoriously difficult to thread a needle, which has led to the biblical use of expressions such as making an elephant or camel go through the eye of a needle (e.g., "It is easier for a camel to go through the eye of a needle, than for a rich man to enter into the kingdom of God," Matthew 19:24). In popular culture, the aphorism is used to refer to difficult or impossible situations.

singularity where, as I anticipate, I will vanish. I resist. I do not want to die. But each movement occasioned by the unknown, imperceptible but *felt* outside force brings me closer to the state that I anticipate to be death. Then, all of a sudden emerges a question: "Why resist?" I sense that I am ready to die. I let go. I no longer remember what follows. I do not even remember loosing consciousness. I find myself again waking up. In finding myself waking up, I am finding my Self, my consciousness; but it is also a finding of something that exists against a ground that makes it possible in the first place. I am conscious against the unconscious state that preceded this instant.

Later, I notice that everything has changed for me at precisely that instant. For example, the poems I begin to write into my diary are marked by a radically different tone. Everything I have written before later comes to be collected under the title *Before the Great Divide.* The poems are characterized by an intellectualist approach to everything in life, all the while a realization is recognizable in the texts that this intellectual approach to life is at the origin of my (psychological) troubles. This is evident in one verse of a poem that I have written before this crisis that constituted the great divide: "Oh sober mind/ Barring the way/ To my SELF." The opening line of the stanza addresses the mind as something sober, to which, in the second and third line, are attributed the problem of finding the Self. The poetry and texts following the crisis, on the other hand, are marked by the absence of the opposition between body and mind. Thus, one verse of a poem written a few months after this experience reads like this: "Where there are no words/ space and time dissolved/ Being is/ The experience of unity." Here, decades before I consciously grasped the place of *re*presentation in the dehiscence of Being and beings, unity is attributed to the absence of words (i.e., *re*presentation).

This experience of crisis has changed me. It has happened to me. It is in and through (because of) the experience that I changed. I know and understand the world differently after the experience than I knew and understood the world preceding the crisis. Yet I have not engaged in any construction or reconstruction of knowledge. The images, contents of the experience, came to me: I have not intended them. I resisted them. I have not intended the crisis. I resisted it. Interestingly, the narrative also includes a tad of agency, when I decide to let go. But the general tone of the experience is one of passivity, being pushed through the eye of the needle, forced through the point of zero extension that is equivalent to death. But, as the episode shows, the instant of death is an opportunity of rebirth – though this was not known beforehand, as I had accepted to die. There is a different person emerging at the moment of death. Birth and death being two manifestations of the same thing: life is possible only with the experience of being born and dying (Bakhtin 1984). In this experience, my rebirth is the same as change in understanding.

It is important to note that this experience is not merely a marker of a change in self-understanding and an understanding of the human condition, but the content of the experience is an expression of the change that has occurred: crisis, death, and rebirth. Death as the movement through a singularity: we cannot think and intent the singularity and even less what – if anything – lies on the other side. The experience has the form of a crisis, and its content is the crisis itself. It is the end of a certain way of thinking and experiencing and the birth of a new way. The old and new "me" are different. But there is a continuity of the living body, which turns out to mark some sort of continuity across what we might term to be a "cognitive crisis." Again, this new understanding has not been the result of a construction or reconstruction. It is, if describable at all, more akin to a *sudden insight*, which, as its name suggest, happens to us unexpectedly.

A Phenomenology of Crisis

How should we understand learning in and through moments of crisis? Constructivism and objective methods fail to explain changes in knowing and learning precisely in those moments when a person is going through a crisis. This is so because the different dimensions of life, before the crisis, the crisis itself, and following the crisis no longer form a chain of cause and effect. The construction metaphor fails unless the constructing mechanism is outside that of which it constructs. But in the case of knowledge or consciousness, this is not the case. The mind that does the construction is also that which gets constructed. Such a system cannot explain how cognitive structures of higher complexity develop from less complex ones (Chapter 1) – in fact, the catastrophe theory explicated above has been used to model stagewise cognitive development (e.g., van der Maas and Molenaar 1992). How, then, should we go about understanding the learning in crisis and the learning about crisis? The German physician and physiologist Viktor von Weizsäcker (1973) has written about the problematic of something totally new arising from previous states. In this case, the unfolding of specific orders is more or less suddenly interrupted, when an entirely stormy event occurs. With and through it a new image of a different kind can emerge, the stable order of which obtains again a transparent and explainable structure that permits a causal analysis. It is impossible, however, to derive this new state in a simple manner from the former state. This would require explaining, based on temporal relations, the crisis as the joining element between the first and the third state; and this is impossible. Lacunae and shortcomings in causal explanations exist. But here the shortcomings are of different nature. The patient is the one who has the strongest impression of the crisis. S/he has, more than at any other time, a sense of being overwhelmed, of being torn on the inside, of discontinuity. Examples include instances of collapse, extreme dizziness, changes of consciousness as in schizophrenia, poisoning, depression, ecstasy, lust, and intoxication.

The author was led to his conclusions because individual patients gave consistent testimony to what was happening in/with them during the crisis. They had an increased inner awareness beyond the ordinary to live and perceive the critical process. They not only changed but also experienced the change itself. An example would be a patient, who, during the crisis, fantasized something apparently irrelevant but which incorporated the clearest sense perception of the structure of the crisis. Here, the crisis was re-translated into the language of common understanding. These are dreams or hypnotoid fantasies with the theme of "rebirth" so common in neuroses. In these, the patient experienced a particular spatial configuration as being reconfigured in a way impossible in spatial-graphical manner. For example, a patient thought that he had "to straighten" the kink in a curve "to produce a sphere." The figure had to reverse itself in infinity, invert itself before reappearing. The constraint to the impossible experienced by the patient is in this way the representation of a critical state: the crisis is a transition.

In this phenomenological analysis of crisis, we recognize the features of my own experience of the crisis that came to separate my former Self and self-understanding from the one that has characterized me after. Crisis, as can be seen from my example and the present analysis, is the crisis of the subject of experience. In and through the crisis, the subject experiences the death and rebirth of itself, the subject. The transitions denoted by the terms *death* and *rebirth* constitute discontinuities or singularities. This allows me to understand that the critical aspect of a crisis is not the transition between different orders but the experienced threats to the continuity and identity of the subject. This is quite evident in the fear that I experienced facing the image of having to move through the eye of the needle or through the null-point of the drumhead.

Death and rebirth are not singular experiences. We have in fact experiences that share similarities with the death and rebirth of the subject on a daily basis: when we fall asleep and re-awake. In these moments, my consciousness disappears only to reappear when I wake up in the morning. I fall asleep, consciously. I allow myself to fall asleep, from consciousness into a state of unconsciousness: "The truth of this immersion escapes and takes with it every kind of analysis" (Nancy 2007, p. 24). But this is a new day, a different day:

> The day in itself is the unequal, the singular, in the same way that the initial light was not and still is not anything other than the difference itself, the sharing of the primitive indistinction of a chaos, of a *khôra*, of a magma, of a springful profusion. The day always is another day, in general it is the other of the same. Tomorrow is another day, that is, another day and another different day. The passage to this other occurs via the equality of the night. All nights are equal. (p. 41)

During sleep, we are not conscious. And then there are instances just as we move from unconsciousness to consciousness, instances that are undecidable because we are both. The dream experiences are not of our construction – we are not the intentional agents of their construction. We receive them and they do not refer to something real. But they are real, because we remember them upon waking up and becoming conscious. There are many accounts of scientists who have woken up at night with the answer to some problem that they have pursued for quite a while but have not been able to solve while consciously thinking about them. I remember my own graduate studies where we had to solve problems. After working for days on some seemingly intractable difficulty, I repeatedly found myself waking up with the solution clearly in my mind. Because I was asleep, not conscious, I (my intentional I) could not have intentionally constructed the solution. New possibilities have emerged where the rational search has not yielded a solution.

Suffering

The passions take us beyond the limits of what constructivism can explain in terms of human knowing and understanding. This is so because we know and practically understand passion through the experience of passion. There is no other way of knowing what it feels to suffer than through suffering, there is no other way to know addiction than through the experience of addiction, there is no other way to know how joy feels than through the feeling joy, and there is no other way to know how it feels to be in love than through being in love. We are definitely not constructing this knowledge, because we have these experiences prior to any construction. It is precisely the living/lived bodily Self – which appears to the person without appearing *as something* – that points us to the originary experience that grounds knowing in a very different way. "From the perspective of intentionality, non-intentional experiences or real contents of experiences – whereby experiencing and experiences, sensation, and the sensed become one – are nothing else than formless and functionless materials that contribute nothing to the constitution of an object" (Waldenfels 1999, p. 40). From the intentionalist perspective, therefore, joy, suffering, and pain are nothing but qualities that cannot be ascribed as properties to some entity or process.

Suffering and pain, because we know them through self-affection, are important phenomena that can lead us to a better theory of knowing precisely because "I know it when I feel it" rather than because of a mental construction. It was precisely following an episode of suffering that I was given an understanding of the role radical passivity plays in life generally and in cognition more specifically. The experience was one of passivity more radical than any passivity we can think of, that is, construct. It is only through experience that we can (practically) understand and immanently know this form of radical passivity.

On a summer day in 2002, I have come to the university to teach my course on qualitative inquiry. In my office, a feeling of fatigue so strong as I have never experienced before surges within me. It is not only intense but also sudden – I am completely overwhelmed. Whereas I normally might have decided what to do next, at this instant I am no longer able to consider any thought. I experience a total absence of any form of intention. I no longer can do or want to do anything. I realize that I am aware of what is happening without being able to do anything for my intentionality to return. I cannot seek help or plan what to do next. Any capacity to intend has left me. I do not know how long I have been in this state. In the way I had done in the crisis episode, I abandon myself and let myself drop. Two hours later I find myself waking up on the floor in the middle of my office.

A few weeks later, I experience a similar episode. I am standing at home in my kitchen where I experience a surge of fatigue and a sudden absence of intentionality. I can see the couch in my family room only ten feet away from me but I am unable to walk over there to lie down. Lacking any intentionality, I drop to the floor and fall asleep, though I do not experience either falling or falling asleep. Again, I find myself later waking up in the middle of the kitchen floor.

The first time I was subjected to this experience I was terrified. I felt an absolute passivity, the inability to conceive of any next action, the inability to will, that is, to exercise intentionality. Up to that point in my life, I have always thought of myself in terms of agency, the capacity to act. I have always been a "doer," a person ready to act and to do so without delay. I am known to friends, acquaintances, and colleagues alike as a person with tremendous will power, which is so strong that repeatedly in my life it has taken me far beyond any threshold of pain that normally stops us from getting hurt.[8] But the year after two terrible bicycle accidents that involved concussions and episodes of unconsciousness, I began to experience chronic fatigue, to which I now attribute those episodes. Today I have learned to sense the arrival of such episodes. If I lie down, I fall asleep within seconds only to return to consciousness with a sensation not unlike that following a coma; if I do not lie down, I will err aimlessly around my home without regaining the intentionality required for acting.

In these episodes, passivity is not of the type where a person decides in his inner monologue, "I will do/say nothing." There is a dynamic at work much stronger than any will and exceeding any intentionality. This is an experience of being overcome by a force that is stronger than any imaginable force, stronger even than the intention to imagine something different. My account uses a common verb in its passive construction, *being overwhelmed.* The verb "to whelm" derives from a Proto-Indo-Germanic root *ku̯elp-*, to be curved, arched. To whelm over has the sense of turning over, invert. The verb thereby points us to an experience of being turned upside down or inside out. It is, in fact, a verb that has a long history in describing the (biblical) passions. The imagery that the verb evokes resembles my description in the section on crisis of the drumhead bulging to one side before being inverted to move to the other side.

Defenders of the constructivist metaphor might want to argue that the experience is constructed or that we can construct knowledge of such experiences. But this is not so, as we can see from the fact that physicians have no idea about and cannot make sense of my account. They cannot "construct" "meaning" of what I am saying, however hard I try to explain and however hard they try to make sense. The same has happened over and over again in another case of the passions, where I can sense that an episode of extreme fatigue is coming upon me. It tends to begin with the disappearing of focus while reading or editing. Then my eyes begin to jump over entire sentences, then several sentences at once and a disappearing of the ability to make sense of words.[9] I know that I have to get up from my desk and lie down only to be asleep within seconds. I wake up an hour or two later, with a

[8] Repeated instances that allowed me to exceed the threshold of pain occurred during rowing competitions or during long-distance hiking trips where injuries occurred far away from habitations, forcing long marches with pain prior to regaining civilization and medical assistance.

[9] Psychologists might be tempted to attribute this experience to "mindless reading," where the eyes move along sentences without "absorbing information" or "constructing meaning." But I am aware of what is happening, whereas mindless readers become aware of their decoupling only some time after the event.

slight headache and the same feeling I have had after returning from an episode of unconsciousness following a bicycle accident with concussion. I am thinking that the sense of the impending collapse into sleep must be something like the sense epileptics have preceding a seizure. As hard as I have tried, the various physicians and specialist doctors (e.g., neurologist, sleep specialist) cannot make sense of what I am saying – i.e., they fail to "construct" "meaning."

In this episode, we observe the description of an experience of discontinuity similar to that which appears in the section on crisis. Discontinuity includes moments of collapse and changes of consciousness. This includes instances of lapse of consciousness, such as in moments of falling unconscious and the process of falling asleep, that is, falling into a state without consciousness. Such discontinuity is happening to and upon us: it is not something that we could intend. The contents of my learning – i.e., how I know that an impending collapse into an episode of sleep is to occur – are not the result of a construction. It is in suffering such episodes repeatedly that I have been able to develop the immanent knowledge that another episode is making its way. Doctors cannot predict these episodes, they cannot predict when and why they will happen. But I know that an episode is impending because I can *feel* it from the way in which the event announces itself. Again, this has to be a passive grammatical form, as it is the event that announces itself rather than I who is constructing its arrival.

In intentionalist approaches to cognition and development (i.e., constructivism), learning occurs as the mind organizes itself to make the organism more adapted to its world. But seeking to control the world by better understanding it does not constitute the only means of learning. Therefore, a framework that theorizes learning and knowing only from an intentional stance is limited precisely because it does not and perhaps cannot explain the more general case. Illness and suffering, states that we tend not to desire and intend, are absolutely radical experiences that give rise to learning and understanding. Suffering and pain are not entities or properties I can *have* only to loose them again. But they are not entirely without me. Suffering and pain are neither objectively identifiable states of consciousness nor reflexively identifiable consciousness of states. Yet it is precisely during an extended illness – my physician tentatively denoted it as "chronic fatigue syndrome / fibromyalgia" – that I have come to understand a range of phenomena that our (Western) generally agential ideologies are incompatible with.[10] I have come to understand radical passivity, a form of passivity that is in excess of any decision to withhold actions and contributions to some ongoing event. If a person says, "I am shutting up and will not contribute any further to this conversation," then the non-contribution is itself the result of a decision. It is a form of

[10] While editing this chapter, I am between sessions in a sleep clinic designed to test different hypotheses about the origin of my fatigue. At this moment, there are indications that it might be obstructive sleep apnea, but only a second test in which the airways are forced open will reveal whether obstructive sleep apnea is the (only) cause of the sleep disturbances I experience. The effects of obstructive sleep apnea are hypo-oxygenation of the body and brain and (deep) sleep deprivation, interfering with the ability to recover from exercise, stress, and mental activity.

intended action: non-action as a form of action. Radical passivity, however, is not something one decides upon but something that one experiences as having come upon oneself.

The experience of absolute passivity has provided me with an opportunity to orient very differently to the questions of life generally and to knowing and learning more specifically. This also pertains to my thinking about science education for aboriginal students. In a culture of agency, where one can do and achieve anything one wants, the experience of passivity tends to go unnoticed and remains unthematized. For example, most members of a Western culture will find it hard to understand the difficulties members of Australian Aborigines or Canadian First Nations tribes experience to operate within a typically Western society. Living close by and traversing almost daily several Indian reserves and having taught aboriginal students in a local middle school, I know and have written about drop out rates and about the (non-) engagement of these students in the regular school curriculum. Prior to the experience described in this section, I was able to think about the observable passivity only in terms of agency. I was wondering why these students did not *want* to participate in their science, mathematics, and other lessons. The experience of absolute passivity, a passivity more passive than any willed passivity, has allowed me to think their passivity in new ways: These aboriginal students are subject to experiences of passivity that no consciousness has any control over.[11]

The experience of suffering, illness, and radical passivity provided me with a new ground for understanding important aspects of knowing and learning generally and of knowing and learning in science education more specifically. These are experiences of something that is both immanent and that we "know" in this manner, as immanent rather than *ek*static. This form of immanent knowing clearly is contrasted with the transcendent nature of explicit knowledge – socially constructed or otherwise. Even scholars concerned with knowing and learning tend to forget these important dimensions arising from our existence as living/lived bodily Selves that make our experiences possible in the first place. This is not only so in science education but also in other endeavors of human culture. For example, theories of speaking and language use in science classrooms tend to overlook that speaking is a possibility only because there is hearing. We can distinguish sounds because there is silence in the same way that we see a figure because of the diffuse nature of the ground (see Chapter 3).[12] Prior to my illness, I had not even thought about health. I learned to appreciate health against the devastating experience of illness and came to understand that "the mystery of illness bears witness to the

---

[11] Such an approach may also give us a better understanding of a family violence, sexual offenses, and other phenomena subject to criminal codes. In such situations, the offenders may find themselves absolutely passive with respect to their "drives" so that even if they are conscious of their experience, they cannot do anything about it. I understand them to be subject to their "drives" as I am subject to the absence of intentionality or to suffering, pain, and joy.

[12] This point was made thematic in the John Cage composition 4'33'', which involves a pianist playing 4 minutes and 33 seconds of silence.

great miracle of health, that it allows us to live in the happiness of forgetting, in a state of well-being, of lightness and ease" (Gadamer 1996, p. 87). In a similar manner, consciousness is salient against the moments of non-consciousness – the nightly experience of unconsciousness in sleep. All of these understandings are tied, for me, to the experience of radical passivity in illness.

Joy

In the philosophical literature, the notions of difference as such – i.e., *différance* – and its relation to singularity and language are important themes. However, the concepts are difficult: How can something be different from itself? What experiences led Arthur Rimbaud to say, "JE est un autre?" (*I* is an other)? How is this dehiscence of Self and Other (made) possible? What are its conditions? For me, the answers to these questions have come *suddenly*, unexpectedly, and unanticipatingly following an afternoon of lovemaking and the conversations that followed. In lovemaking, "being is caught up in the sensuous density of the now. It is the present moment that holds sway and is all-consuming" (Sheets-Johnstone 2009, p. 81). During such moments, time is suspended, future and past infinitely removed beyond the boundary of the clearing that defines consciousness. In such times, the "moment now is suddenly wild, overrunning its ground and pushing past and future to almost invisible horizons" (p. 81). Everything is Being: time, requiring thematizing consciousness, is suspended together with *re*presentation.

Experiencing Joy

We had gone to the mountain overlooking the vale. When arriving at a favorite spot, we sit down, taken by the view of the valley from that place. We begin to become absorbed in and with one another; we begin to make love. Every now and then a plane passes overhead, making thematizing consciousness appear only to allow it to lapse again into the ground. I remember merely sensing the possibility that someone looking down could see our naked bodies, our absorption in the embrace, and then consciousness disappears again. It is total absorption: no consciousness, no beings (things), no representation, no making present of the present or presence. Any notion of presence vanished. A phenomenological analysis of lovemaking shows that "in embracing the other, I embrace not a fragment but a sensuous density whose form is adumbrated in all the actual frontiers of the flesh" (p. 82). The other no longer exists for me in thematizing way, as the objectified object of my thoughts. Rather, "the other exists for me not as a series of parts touched but as an infrangible whole whose boundaries are beyond my flesh yet which my flesh nevertheless encompasses and extols" (p. 82).

But eventually we return to our senses. When we descend, there is nothing but an afterglow of an event that was not present. There is no memory, no presence of what had been present.

Experiences such as this one also have been characterized as *flow*. During flow experiences, linguistic (interpretive) consciousness is suspended. Colloquial expressions denoting the state of flow include "to be in the zone" or "to be in the groove." Generally, in flow there is a loss of the feeling of self-consciousness, of explicit interpretation, a merging of action and awareness, a lack of bodily needs, high levels of absorption in the activity, and a different sense of time, which can lead to a total suspension of time. During my time as a world-class rowing athlete, there were many instances where I "was in the zone." During races, this tended to lead to an absorption to such an extent that the even the competitors were no longer present in/to my consciousness. I was rowing in complete absorption until the bell announcing the finish line brought me back to consciousness.[13] Later, as a result of mediation and after having developed a Zen spirit to life generally, I have been able to experience flow over and over again, including everyday activities such as scholarly writing, gardening, cycling, teaching (e.g., Masciotra et al. 2007) – and love. Years ago, I wrote an entry in my personal notes that expresses being rapt by love, being love itself:

> A warm feeling is flooding me: I am thinking about you leaning into my arms, your slender body pressed against mine in the tight blue jeans, gray sweater, blue eyes. I feel like caressing you, shedding you with this warmth and feeling of tenderness, forever, in the drunken fever of night. Softness of skin, from which pours love like the smell of exotic worlds, mixed with the wild, herbaceous nature of my love. This drunkenness is not only experience of ourselves beyond the confines of cultures, words, & signs, but also incorporating all of them. The sandalwood incense is soaking everything with the forgetfulness in your arms and the scent of the orange tea is intensifying the horizon lit by the sun from below, while the sky above is matching my eyes.

Falling in love is a passion. Love is a passion, which we do not intend. It has the feel of flow, as we no longer engage in objectifying the other. If I were constructing the other, she would be the object of my constructive actions. Yet no such thematizing occurs. In fact, lovemaking teaches us the radical self-reflexive nature of all knowing. Thus, "the physical body can be lived as radical pleasure only insofar as a self-reflexive sensory modality exists: the physical body can otherwise not be thrust to the foreground of experience" (Sheets-Johnstone 2009, p. 83). This form of knowing in and of the body is more ancient and radical than any thematizing consciousness. It not only undergirds but also provides the resources for the emergence of thematizing consciousness. This is so because "carnal knowledge is knowledge of a pre-epistemological subject in the sense that the *hyle* carries its own meaning" (p. 87).[14] The experience of the physical body in lovemak-

---

[13] At the time, I denoted the beginning of this state as "closing the eyes." I literally do not know what happened between the closing of the eyes and the ringing of the bell at the end of the race.

[14] *Hyle* is Husserl's term for denoting matter, stuff, or substance that constitute the immanent data of our experience. Apprehension transforms these immanent data into something objectively transcendent: we perceive something *as* something (an object *as* being red, the redness of red).

ing constitutes "an existential-evolutionary elaboration of lived physicality" (p. 87). This physicality and the associated tactility thereby is the very foundation of human nature, including its individual and collective forms of knowing.

In lovemaking, I do not construct (thematize) the Other, though I may be aware of her presence. If I did construct the other, our relation would be instrumental: the Other as the intentional object of my actions. But I do not make her the object of reflection and reflective abstraction – there is no place for such thematization in Sheets-Johnstone's phenomenological analysis of lovemaking. In not objectifying the other, I also do no longer see her objectively. As soon as I get out of flow, aspects of life such as time become thematized again, being subject to a very different form of consciousness. This also was the case when the experience on the mountain became the topic of a conversation between my partner and me.

From Joy to Representation

> It is only from within the actually performed act, which is once-occurrent, integral, and unitary in its answerability, that we can find an approach to unitary and once-occurrent Being in its concrete actuality. (Bakhtin 1993, p. 28)

After returning from the mountain, she begins to talk about the afternoon, about its implication for planning the future. She attempts to put the experience in words and to place it in a sequence of other events that make our lives. This endeavor requires language, which is the foremost tool of retention, for bringing back a rapidly receding former present and making it present again in the current present. That is, such talk has the function to *re*present. Not all language is of that nature, and on evolutionary grounds it has not been thus. Children produce the sound [tʃɛə(r)] ("chair") but it is integrally related to the object we know as chair. There is no difference for the children, as Piaget and others have shown. This suggests that the sound-word does not *stand for* the object: it *is* part of the object's existence. As soon as the word and the object are separated, the identity disappears (with the simultaneous appearance of object permanence, see Chapter 2). In the same manner, as soon as we talk about past and future things, events, and phenomena, we bring them into the present. In making something present again, we also mark its non-presence. Representation is necessary because *Being* cannot be subject to thematizing consciousness unless it has receded into the past. As Husserl (1980) shows in his analysis of sound and the internal consciousness of time, it is only when available in retention that we can become conscious of the sound as sound and of time as time. Things (beings) come to stand transcendentally for phenomena that have become unavailable because they *have been* (as *Being*) in the distant or past. The use of these beings (*re*presentations) to achieve a renewed presence simultaneously means departure, separation, and distance. A dehiscence has occurred between the pure presence of our afternoon on the mountain and its renewed presence in her account of the event. We are now outside of the once-occurrent event where there is no longer any chance of finding the unitary and once-occurrent

Being in its answerability that Bakhtin (1993) exhorts us to find in the opening quote to this subsection. Even to understand the act, we have to take it from *within* the uniqueness of the situation rather than contemplating it as a fact from the outside and from a theoretical stance. We miss this uniqueness, its affective and deontological dimensions when we consider its cognized aspects, "its content (self-identical content) and not the moment of the actual and effective performance of the act" (p. 37). The more we move toward the theoretical understanding of an event expressed in language, the further away we move from the "once-occurrent, never repeatable, emotional-volitional and concretely individual" (p. 39). But it is precisely this that is important to me in understanding real cognition that matters to real living people in real lived situations. This "uniqueness of my participation in Being is the actual and effectual foundation of my life and my performed deed" (p. 41).

On that day, as soon as she begins to talk about the experience on the mountain, as soon as she thematizes the afternoon in language, the spell-like sense that has continued to vibrate in my body vanishes. Up until this moment, I had not even attempted to put into words and make sense of what has happened, the absorption in lovemaking. But before any sadness about the disappearance of the spell brought about by her talking could emerge and take hold of me, I realize – in the disappearance and with equal suddenty – that it is the putting into words, the very articulation of something that makes re-appear an event also makes it disappear as singular occurrence. In this way, the coming of the sign not only denotes the phenomenon but also becomes its tomb.[15] The singularity of an event, singular and therefore unnamed and unnamable, disappears in the recollection, the process of making it present again. I realize that as her account attempts to make present again the spell, the event, it actually pushes it forever away. As the beings become present, Being disappears. But Being was never present in consciousness, which, that afternoon, had withdrawn in and with our absorption. In making the event present again, placing it in front of both us to see and to reflect upon them, she has made the event disappear (as Being) at the very instant that she made it re-appear (in beings). This is so, as Parmenides has said millennia ago, because Being itself cannot appear as itself. That which appears is the reposing Being itself; the movement is only its appearance.

It is in this moment – in hearing her speak about the afternoon and what it means for our lives – that an understanding of the entire second part of *Sein und Zeit* dawns upon me as well as its exegesis in the work of Jacques Derrida. In that instant, I understand *all of a sudden* the term *khôra*, the displacement that comes with the process of writing, and the nature of the dehiscence of Being and beings. I understand the role of language in the associated displacement of Being and beings and I understand what Heidegger writes about when he articulates the temporal that temporalizes itself. The time that we use in the sciences – i.e., *ek*static and transcendent time – is born precisely in this displacement of Being and beings,

[15] The idea appears in Hegel's semiotics and subsequently is taken up in various places by Derrida, including *Glas* (Derrida 1974).

in the first making present of a present that is in the (immediate) past. But this form of time cannot assist us in understanding the moments of flow, whether we experience them in lovemaking, meditation, high-level sports, doing science education research, or science-related problem-solving. It was at that very instant that I came to understand – in an instance of sudden insight – how fourth- and fifth-grade students in a design curriculum I studied did not see, feel, or otherwise experience objective time to pass. Instead, when the teacher reminded them that the lesson – which had lasted the entire morning right up to the lunchtime break – has ended, they responded in utter surprise with the question, "The lesson is already over?"

It is precisely this shift between pure presence and the presence of the present that also provides us with an explanation for the experience of the appearance of the twin silos and the forgetting of the state that preceded it (Chapter 2). As I became aware of the twin silos, as they took on a presence in my consciousness, the previous state of Being was displaced. For me to take note of the twin silos, they had to be there, imaged on the retina. But I was not aware of them in the same manner that I may not be *thematically* aware of every oncoming car during that ominous seventh bicycle trip. As soon as the twin silos moved from the status of hylic (hyletic) data to becoming transcendent object – a process mediated by a new and itself transcendent being (representation) – they became permanent features in a permanent world, the permanence of which is brought about by representation. In becoming material entities in a geometric world, the twin silos also became independent Galilean objects, that is, objects independent of particular sensations to be discovered and seen by any other human being passing along this part of the country road.

We can learn from this section that there is a dehiscence between experience and what we can make thematic in language as experience. What language brings about is a displacement and replacement, a "replacement of the very place, in the place of the place: *khôra*. The tragedy and law of replacement is that it replaces the unique – the unique as the substitutable substitute. . . . And substitutable because unique" (Derrida 1996a, pp. 107–108). Bakhtin and Vygotsky tend to take the same position. Concerned with the function of language in communication and its relation to individual experience Vygotsky agrees with the linguist Sapir, citing a passage from the introductory chapter on language: "The elements of language, the symbols that ticket off experience, must therefore be associated with whole groups, delimited classes, of experience rather than with the single experiences themselves" (Sapir 1921/2004, pp. 8–9). Language comes to us from the other and, in our speaking, returns to the other. In using language to talk about experience, we not only bury and distance it, but also make it other than ourselves, other than what it had been in the utter subjectivity of our immanent Being in the pure presence of the event. Bakhtin (1993) comments: "The theoretical world is obtained through an essential and fundamental abstraction from," and we might say at the expense of, "the fact of my unique being and from the moral sense of that fact – 'as if I did not exist'" (p. 9). This abstract view of Being is "indifferent to the central fact – central for me – of my unique and actual communion with Being

(I, too, exist)" (p. 9). Precisely because of this indifference, the theoretical attitude to life is irrelevant to life: "It cannot determine my life as an answerable performing of deeds, it cannot provide any criteria for the life of practice, the life of the deed" (p. 9). This is so because the theoretical world "is *not* the Being *in which I live*, and if it were the only Being, *I* would not exist" (p. 9).

Coda

In this chapter, I show how a phenomenological study of different forms of passion allows us to understand how we know before we have language that separates us from our immanent knowing and Being as flux. I articulate very different forms of experience and the learning that emerged for me as a scholar of epistemology generally and of knowing and learning in science education more specifically. This chapter thereby exhibits the untenable nature of the constructivist claim that affectivity is unrelated to the structure of knowing. I initially articulated the role of passivity as a more basic category of knowing and learning than agency and intentionality for learning in science about five years ago. But the true import of the category of passivity and the relationship between Being and representation has not emerged for my until a close and careful analysis of the actual pathic experiences – lovemaking, suffering, being subject to a crisis – allowed me to understand knowing and learning in science in new and improved ways. It also has allowed me to understand the conditions under which the constructivist metaphor can hold up as well as those conditions – discussed throughout this book – under which it does not and cannot hold up. A re/constructed and re/situated metaphor of construction is offered in Chapter 13.

Chapter 12

From Incarnation to Responsibility

> In its Archi-intelligibility, Life comes to itself before all thought, accedes to itself before all thought. . . . Archi-intelligibility therefore suggests an Intelligibility that precedes everything that we have been understanding with this term since ancient Greece – which precedes all contemplation, all opening of a "space" to which a seeing could open itself. (Henry 2000, p. 125)
>
> Consciousness does not fall into a body – is not incarnated; it is disincarnation – or, to be more exact, a postponing of the corporeity of the body. (Levinas 1971, p. 140)

The examples and discussions throughout this book show that Kant, Piaget, the constructivists, and the embodiment/enactivist theorists all presuppose a subject, all presuppose that there is some organism working out problems in the way we know that human beings of today (sometimes) work out problems. But is this way of thinking useful on evolutionary grounds? On such grounds it is impossible for a subject to construct itself: "Oneself cannot make one*self*, it is already made in absolute passivity and, in this sense, already victim of a persecution that paralyzes any assumption that could wake up in it of posing it *for* itself" (Levinas 1990, p. 165, original emphasis). In this statement, Levinas rules out any possibility of an intentional Self to be the starting point of epistemology; he rules out the very possibility of constructivism as being a viable explanation of learning and development as a bottom up approach. Von Glasersfeld rules out questioning the nature and source of the Self by suggesting it to be a metaphysical issue. But we must ask the question of the origin of the subject of the self-constituting work that is presupposed in the constructivist metaphor.

What then does an evolutionary plausible epistemology have to look like? In this chapter, I further build on a recent argument for knowing and learning in mathematics (Roth 2010b) to suggest that it has to be an epistemology grounded in the living/lived body, the flesh, which can be affected and therefore can be the subject of as well as subject(ed) to actions. With the living/lived body, all knowing is incarnate first before it can be ekstatic, consciousness that stands out from experience (Being), and thereby stand for experience – even though this standing for always has to be from a (theoretical) remove. Consciousness is, in the words of Levinas, disincarnation, a postponing rather than celebration of the corporeity of

W.-M. Roth, *Passibility*, Classics in Science Education 3,
DOI 10.1007/978-94-007-1908-8_12, © Springer Science+Business Media B.V. 2011

the living/lived body. It therefore does not suffice to move to embodiment and enactivist theories, which focus on the ekstatic body, but we need to begin with the living/lived body that gives rise to consciousness that is at the heart of the constructivist metaphor.

Toward a First, Evolutionary Plausible Epistemology

How does the cognitive organism that appears in all of the intellectualist theories get off the ground when the very linguistic structures underlying the schema that are employed need to come about in the first place? The easy way out, which solves nothing, is a return to innatist theories, where everything that cannot be explained is attributed to the genes and to innate capacities. There are, however, three lines of research all of which propose theories that (a) are consistent with each other, (b) are plausible on evolutionary account, (c) build on the unity of emotion and movement as the building stone of the human psyche or Being, and (d) do not separate body and mind. The oldest of these approaches comes from the Russian (Marxist) social psychologist Alexei N. Leont'ev (1981), which was further developed by the German critical (Marxist) psychologist Klaus Holzkamp (1983). The dialectical materialist theory reconstructs the human psyche beginning with single-celled organism and develops an evolutionary account, consistent with the findings in biology and primatology, which could have been followed up to the emergence of hominids, where evolution eventually selected *Homo sapiens* and the rudimentary culture that it had available. In this theory, intellectualizing consciousness is a *consequence* of collective life and collective labor, reflecting material life at an ideal level. Without material life, no ideal life: the two are but different manifestations of life.

The two other approaches were developed independently by means of phenomenological analyses of life generally or of movement, especially dance movement. In both approaches, the capacity to move constitutes the form of immanent knowing and memory on which all other capacities are built. In the final analysis, we would not be able to speak or have developed language from initially primitive sound productions without the capacities to (a) move parts of the vocal tract and (b) push air through it to produce recognizably different sounds. Only a living/lived body can self-affect such that initially random movements turn into movement capacities; and only when there are movement capacities can something like movement *intention* emerge. Having the capacity to intend points us to the existence of a sense of "I can" that organizes the organism as a whole. All capacities are those of the "*I* can" rather than being disconnected capacities – as they are for Kant – so that they have to be pulled together by the mind. Emotions, rhythm, and tact all are intermodal phenomena that provide for an integrated organism rather than an organism that needs to construct the coherence among its parts. Corporeal-kinetic forms and archetypes are more fundamental organizing principles than the "(embodied) image schemas" that appear in the theories of

Kant, Piaget, the constructivists, and embodiment or enactivist scholars. These schemas already obtain (linguistic or language-like) mental structures that need to be explained rather than presupposed. The lines of research referred to all explain the emergence of such schema rather than presupposing them. All these lines of research are therefore consistent with the idea of carnal knowing as the condition for intellectual knowing; all these theories place primacy on human praxis as the origin and ground of knowing rather than the other way around. *Passibility* is the most fundamental capacity, for without it, there would be no sensory capacities that relate the subject to itself and to the world.

If we care to look, we can find passibility and radical passivity within and surrounding us at every instant of our lives. A leg fallen asleep teaches us about the relation of agency and passivity, about the passivity of the body, subject to the resistance and inertia that materiality provides to movement intention. When we intend to fall asleep, we need to let go of consciousness to be able to *fall* asleep (into a state of non-consciousness); waking up *happens* to us rather than being intended precisely because consciousness that enables intention is absent. We know pain, suffering, and joy because we *feel* it; and we feel these forms of passion because of passibility. From an evolutionary perspective, all of these features of life precede human forms of intellectual consciousness. There is, as Michel Henry suggests in the introductory quote to this chapter, an originary possibility for intelligibility – i.e., an *archi*-intelligibility – that precedes all intellectualizing thought and constitutes its foundation. A truly viable epistemology has to be plausible within an evolutionary perspective; it must not, as most epistemologies do today, provide teleological explanations that use our current forms of (culturally enabled) cognition to explain the evolution of cognition as if present-day human cognition were the *necessary* endpoint. This is the essence of Piaget's approach, which explains cognitive development in terms of the adult rationality as the necessary endpoint of natural development.

On evolutionary and cultural-historical grounds, we cannot theorize thought by presupposing the current structure of thought as the condition: Thought is a product rather than an antecedent of evolution. Thus, the question concerning the possibility of a first thought may be answered by a phenomenology concerned with understanding life. The one currently existing science aspiring to be a science (*-logy*) of life (Gr. *bíos*, life), biology, is not about life itself but about its manifestations. It is about molecules, genes, or physical and chemical processes. Thus, "that which never exposes itself in a treatment of biology is not only the coming of the book to objectivity but more essentially that which never comes to objectivity and remains irrepresentable" (Henry 2004, p. 193). Among these phenomena, we find "the effective power of prehension that turns the pages with one hand, the movements of the eyes of the reader and all the fundamental determinations of the absolute subjective body" (p. 193). Life itself is invisible originally, just as Being is invisible. It is in and through the archi-invisibility that Henry (2000) articulates in the introductory quote that Life becomes visible to itself in a process of "autogeneration as generation in itself of the First living Self in which it experiences itself and thus reveals itself to itself" (p. 125). Such self-experience requires a sensible-sensing (living/lived) body. This living/lived body has come to be denoted in the phenomenological literature as *the flesh*. This flesh allows knowing to be incarnate. The idea of the sensible-sensing (living/lived) body therefore implies the *interlacing* of the passive body, capable to be affected by others and itself, and the active, intentional body.

Interlacement

In our living/lived bodies, the material and ideal, the active and the passive, the different modes of sensing and feeling are interlaced and cross over. There are not different parts that must be coordinated, but there is one organism, one living/lived body aware of itself. Conscious awareness is but one of its modalities, a reflection of its life as a whole, but only one reflection among many. If we carefully look around ourselves, we can see evidence for the unitary nature of ourselves. I discovered this some 30 years ago. At the time, I was going regularly to watch classic movies from the 1940s. I began to notice that I had difficulties hearing what was said and, because the movies were old, first attributed my problems to the noisy soundtracks. But then one day I realized that I understood perfectly well while Humphrey Bogart was squarely looking at the camera, but I could no longer understand what he was saying when he was turned away. All of a sudden I realized that much of my "hearing" was in fact lip reading. Hearing and lip reading are one. Moreover, I realized that I had a hearing problem that went unnoticed before. Reading lips was a function that covered up for the diminishing capability of my ears. The same underlying shift across functions is responsible to a sensitivity that I discovered many years later: to the temporal shifting of audio and video tracks on television and magnetic storage media. As soon as there is the slightest shift, I do notice it – more so than others surrounding me, as I have learned from experience, because these others notice such shifts only when I point them out. Moreover, as soon as this is pointed out, most people will see that a movie is dubbed, for the voices do not fit the bodies and movements even if the dubbed voices are simultaneous with the lip movements of the actors.[1]

From a constructivist (the Kantian) position, the inputs from the different senses are coordinated by a powerful mind that brings coherence to an otherwise unordered world. Everything that we experience is said to be constructed – I heard this only days before writing these lines during a conference with several "high priests" of constructivism in mathematics education. It is precisely mind that constructs coherence and tests its construction for viability. But such experiences of

[1] Much later I found out that Merleau-Ponty (1996) provides an analysis of the same phenomenon, the dubbed movie, in support of communication as a unitary phenomenon. Thus, "we perceive it well during the projection of a dubbed movie where one makes skinny people speak with the voices of fat, the young with the voices of old, big people with the voices of the miniscule, which is absurd" (p. 70).

hearing that is really seeing, and seeing that really is hearing (understanding) shows that the two senses are not independent, they are two means of the same subject orienting and interacting with the sonorous world and its production. That is, the living body is at the origin of an interlacing, here the interlacing of the visible and the audible. Moreover, without movement, there is no sensing – as shown in Chapter 3 in the exploration with the drawing that can be seen as a hallway or as a truncated wedge (Fig. 3.4). Other forms of interlacing have been discussed in the philosophical and psychological literature; and this interlacing is precisely the condition for the emergence of any form of human knowing and learning rather than its result.

In Chapter 3, I use touch as a paradigmatic case to explore the role of the senses in the making of sense. Here I push this analysis further by drawing and building on analyses of touch that have become paradigmatic for the very possibility of knowing, because touch provides the possibility to study the relationship of the knower and the known, the sensing and the sensed in their simultaneity. Consider the exploration of unknown mouse pad surfaces with my left hand to find out why my optical mouse works on some but not on others. This hand, as we see in Chapter 3, explores an aspect of world that it comes to exteriorize once the sensed property of the pad is recognizably repeatable. Exteriorizing means making it stand out; the properties of the mouse pad surface are *ek*static in nature. The sensation no longer is a property of my sensing but a property of the world. In this case, my hand has learned to recognize a kind of surface. In fact, the surface has taught the sense so that it can *re*cognize the thing that has shaped the sensing perception. Because of the changed nature of perception, I now feel a mouse pad rather than having to interpret some raw input at my fingertips. The sensation has created an opening upon the world. But for the exploration to be able to teach me something, there has to be a fundamental relation between my movements, on the one hand, and what I touch, on the other hand. This then requires an interlacing of the internal sense of touching with the possibility to be a body that can be touched simultaneously (or in turn). We can experience such interlacing when we use the right hand to touch the touching left hand (Fig. 12.1). The touching left hand now becomes part of the same world that it is touching: it is both touching and of the order of the touched. The living/lived body (flesh) also is a material body among material bodies. But this is not all, for we have not yet addressed the self-sensing of the hand that senses.

Three types of experience are related to touch. There is the touch of the smoothness or roughness of the mouse pads, there is the experience of (the resistance of) my living/lived bodily Self moving part of itself to produce the sensation at my finger tips, and there is the touching of the touch as my right hand feels the left hand sliding over the mouse pad surfaces.[2] The sensation of the living/lived body moving itself intentionally also underlies perception, as I show in Chapter 3. The left hand felt by the right hand as something outside itself that can also be

---

[2] The origin of the intention in the power to act that is the result of self-affection has yet to be established.

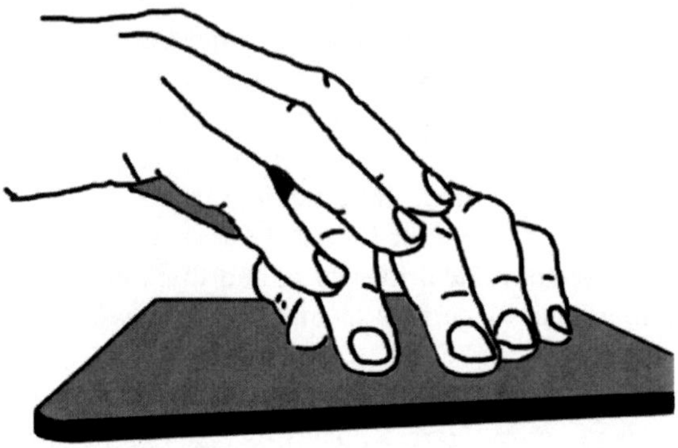

Fig. 12.1 The left hand is touching, thereby learning about the world, whereas the right hand is touching the hand that touches, the subject of knowing touching itself as an object of knowledge.

seen. That is, both seeing and touching are sensed from the inside having access to the same me as material body that explores the outside.

> We have to get used to think that everything visible is carved into the tangible, all tacit being is promised in a way to visibility, and that there is an encroachment, a crossing-over not only between the touching and the touched, but also between the tangible and the visible that is incrusted in it. . . . Because the same body sees and touches, the visible and tangible belong to the same world. . . . There is a double and crossed bearing of the visible in the tangible and of the tangible in the visible, the two maps are complete and yet they do not become confounded. (Merleau-Ponty 1964, p. 175)

It is precisely this crossing over that allows sighted people to have an image of the thing that they touch but cannot see. It is precisely the crossing-over that also lies at the apparent independence of the image of the world from the mode of its perception. But those who become sighted will have to acquire the crossing over between sight and other senses, because it has not been part of the "I can/will" that underlies all of the understanding of the world that we come to develop. The source of the incarnate "I can," in turn, is the result of an auto-affection that precedes all cognition, and therefore, all intent to construct anything that resembles knowledge of the world in the way that we know it.

When this auto-affection and crossing does not occur, the very behavioral and mental characteristics of human nature as a knowing being are absent, as seen in the work of a well known Russian psychologist concerned with educating deaf-blind children (Meshcheryakov 1979). Because these children are bereft of the capacity to interact with the world by means of the long-range senses sight and sound, they are condemned to a life of isolation unless special provisions are made that allow them to interact with others and to achieve the crossing-over present in sighted or hearing people. The Russian psychologist suggests that whereas these children have the capacity for mental development, they are bereft of a human mind prior to the special interventions in his clinic. These children did not take hold of any objects to be investigated by means of touch. They were not familiar with toys and had not developed an understanding what toys are. They manifested no (social) need to be in contact with other people. And they responded negatively to any attempts to touch them. It turns out that in the early stages of the development, these children *did not exhibit the intention* to knowing, they did not have the "What's this?" reflex. This reflex was an achievement that emerged after a tremendous amount of work in which the children came to compensate for the lack of vision and auditory capacities. That is, Meshcheryakov provides clear evidence contradicting the constructivist claim that intentional knowledge construction is innate. The deaf-blind children in Meshcheryakov's study did not engage in natural exploration of the world with their sense of touch but drop an unfamiliar object. It was only when they discovered an object as part of a purposeful action toward the satisfaction of a need that they actually came to develop an orientation to learning from the world by interacting with it. For example, a deaf-blind child learned about objects when adults placed a spoon in the child's hand and, taking it in their own hand, guided the child to eat from and with the spoon. The children *learned* to *re*/cognize objects as such that make up a world while adults guided their hands to perceive through touch with one hand what they and the adult guides were doing with the other hand. In these experiments, the crossover described by phenomenological philosophers was part of the training – though not made thematic as such by the psychologist – that led the deaf-blind children to develop a normal human mind.

Givenness of Intention

Intentions – such as intentions to act – should not be just taken as given, but their givenness should be explicated in an analysis of life. Thus, if I move my hand to find out about the surface texture of a mouse pad, the intention to find out more about the surface is grounded in forms of knowing that are immanent to my Being. The intention to seek information about the surface requires knowing that I can find something out by means of touch, and it requires knowing that I can move my arm and hand so that they reach the mouse pad. Further immanent knowing is required concerning the relationship of sensing and moving. Thus, I can have an intention only if my living/lived bodily Self already knows that it can move. This "I can move" is the result of originary non-intentional movements that establish the movement capacity, the immanent knowing of the "I can." Because these originary movements are not intended, they are not constructed: the intention is born immanently and the subject finds itself with the capacity to intend. The constructivist subject has its origin in this "I can."

In our exploration, the left and right hand may exchange roles of the touching and the touched. Merleau-Ponty (1964) uses this as the paradigm case for disqualifying the opposition of the touching/touched, sensing/sensible, seeing/visible, and so on. However, the powers inherent in the capacities of touching, seeing, smelling, tasting, and hearing themselves remain unexplained: the power of the *constitution* remains unthought. That is, what makes possible the intention to learn by means of the senses? The work with the deaf-blind children in the Soviet Union shows that this intention itself is the result of something else that is both the condition and precedes any intention to constitute (construct) a world (Meshcheryakov 1979). For Piaget and the constructivists, the orientational-investigatory reflex is innate. But the experience with the deaf-blind children shows that there is no innate unconditioned orientational reflex. Orientational analytic behavior comes into being as soon as the organism feels the influence of the environment – whether harmful or beneficial. And this environment as an entity independent of one's own material body but subject to its sensing capabilities is something that these children *have to learn* before they can develop a human mind. A totally unfamiliar entity does not stimulate the child into orientational investigatory behavior. Rather, slight variations of the stimuli known from prior practical engagement – eating with a spoon – provoke orientational behavior because new kinds of choices have to be made.

Intentional behavior presupposes a sense of "I can." If my living/lived bodily Self does not immanently know that it can move, it cannot have the intention to execute the movement. This sense, however, cannot found itself. There is no "I can" that brings itself into being in the form of an "I can 'I can'." Rather, the "I can" must emerge and result from an auto-affection that emerges with unintended and unintentional movements. This auto-affection has been made thematic as the auto-donation of life to itself. It is only because there is already an "I can move" that I can also intend to move my hands and arms to explore the mouse pads with respect to the texture of their surfaces. It is only because there is already an "I can feel" that I may intend to explore by touching. It is as if the organism endowed itself with an intention because it already is capable of moving its extremities in a coordinated way and because it already is capable of sensing. But it is not the organism that matters – it is life, realizing itself in the organism that gives itself to itself in such auto-affection (Henry 2000). The knowledge of these capacities is not explicit but immanent in the living/lived body. The emergence of these capacities requires an auto-affection, because the mind, not knowing that something like movement or sensing exists, cannot seek to move the body or sense a world. This immanent knowing comes from the living/lived body itself, in and through (passive) auto-affection during the originary and originating non-intentional movements that produce *kinetic melodies* that integrate individual motor acts into more complex movements (Luria 1976). These kinetic melodies, inscribed in our bodies by means of auto-affection, "constitute that basic, vast, and potentially ever-expandable repertoire of 'I cans'" (Sheets-Johnstone 2009, p. 255). But auto-affection does not mean (inherently conscious) self-presence. The living/lived body that knows of its capacities initially is not conscious of these capacities – e.g., a child speaks a language correctly even without knowing formal grammar and a golfer plays a perfect game even without knowing the laws of physics.[3] It is

[3] The night before I wrote this sentence, the Korean-American golfer Michelle Sung Wie shot the second "hole-in-one" of her career, that is, she "downed" a ball in one shot directly from the tee.

precisely because of the difference between self-affection and capacities, on the one hand, and intention and consciousness, on the other hand, that we get the radical uncertainty of actions in the way I describe it in Chapter 4.

Only a living/lived, sensible body – one that can sense itself moving because of auto-affection, one that has practical knowledge of the coordination that occurs with the sense of self-movement and sensation through external senses – can be a body of knowledge. Knowledge presupposes practical self-knowledge of the living, sensible body (flesh). The sensible body "*is the archi-tectonic of sense*" (Nancy 2006, p. 25, original emphasis). This conception contrasts the Kantian conception – also upheld by Piaget – that the mind and its schema are the origins of an architectonic unity. Here, tectonic derives from the Gr. *tektonikos,* pertaining to buildings, whereas archi- has a double sense, denoting both the primary (the main building of sense) and the original (the original building of sense). It is not that knowledge is merely embodied, some transcendental idea that has found a flesh to reside in.[4] Rather, sense as such arises from the structure of the living/lived body (flesh) and of all the senses it is endowed with (it has endowed itself with). The architectonic is thus due to the performative aspect of Being: "all the values of actual life and culture are arranged around the basic architectonic points of the actual world of the performed act or deed" (Bakhtin 1993, p. 54). Auto-affection, in turn, presupposes *affect*ability, *passibility*, the quality of being passible and the capacity for suffering, feeling joy, having sensations, and having emotions. Affectability and passibility only exist for the flesh; they constitute the condition for the kinds of incarnate knowing that characterizes human praxis. The unity and identity of the sensory phenomena are not realized by a synthesis of recognition in the concept, as an intellectualist approach would conceive of it, but this unity is built upon the unity and identity of the living/lived body (flesh) as a synergistic ensemble.

The experiment where one hand touches the hand that is touching something else (e.g., the mouse pad) also allows us to understand the emergence of an ekstatic world. Here, the left hand that is exploring the mouse pad is itself the subject of the touch of the right hand. The touching subject – which is not self-present to itself – also is subject to touch. It is by means of the right hand, the one that is touching, that the left hand comes to stand out, becomes ekstatic in the same way that the twin silos became ekstatic for me (Chapter 2): "In contact itself the touching and the touched separate, as if the touched moving off, always already other, did not have anything *in common* with me" (Levinas 1990, p. 137). In this movement, we get to know ourselves (left hand serving as pars pro toto) intellectually, as material body among material bodies. That is, we know (material) objects

---

It is unlikely that she could write and solve the differential equations that are required to describe the movement of her golf ball under the conditions she faced.

[4] This is the core of the critique Sheets-Johnstone (2009) articulates against the embodiment and enactivist approaches (i.e., theories articulated by Mark Johnson, George Lakoff, and Francisco Varela). As cited above, "conscious does not fall into a body – does not embody itself" (Levinas 1971, p. 140). Instead, consciousness is a form of transcendence, of disembodiment.

specifically and the world more generally only because of the touching that the right hand enacts upon the left hand, itself exploring the world. This is precisely what Meshcheryakov (1979) describes as central to the education of the deaf-blind children. They do learn to eat with a spoon and what a spoon is only if they also explore it and the hand that feeds them (that of others and later their own). It is precisely in the crossing-over movement that intellectual and intellectualizing knowledge becomes possible. In this movement, the world comes to stand out – while pure Being withdraws and is covered up by the emergence of beings.

The ekstatic world and the ekstatic left hand are of the same order, subject to the intentional exploration of the right hand. But this transcendence is given to us in the flesh. My own flesh (living body) comes to be established as a body: it is the outcome of the actions and sense of the flesh. This exemplifies that "my flesh cannot be constituted as body other than by a kinesthetic system, a system that already presupposes my flesh as an organ of perception" (Franck 1981, p. 47). No cognition (and learning) is possible if there is not already an auto-affection of the flesh, which allows the living flesh (i.e., Life) to know itself in immanent and ekstatic ways. This living/lived body is implicated in all perception and, permeated by the immanent rather than self-present "I can," which gives coherence to the multitude of experiences that we may have even with one and the same object. In fact, an object is the same because of the coherence achieved by the constancy of the living/lived body. Therefore, the "transcendent object gives itself in sketches, in a multitude of incarnated appearances"; as a result, anything that is offered "in this manner to my consciousness always is more than what appears in the strict sense, for it is the same thing in its flesh, that is, in its totality" (p. 58).

Passivity – The Ground of Agency

Agency is an evolutionary accomplishment rather than the starting point of life; agency cannot but emerge against passivity as its ground – much in the same way that any figure emerges from and requires the ground against which it appears. In the single-celled organisms where intentional movement first occurs, sensibility already has to exist, that is, the capacity to be affected and the capacity to "measure" needs. There is therefore a radical asymmetry that separates agency and passivity, the former enabled by and against the latter. Consciousness and intention, too, are not intended by a knowing consciousness but are received by the subject of action, subjected to the gifts of consciousness and intention. Throughout this book, I provide evidence and arguments to expose the blind spot of intellectualist approaches to cognition generally and knowing and learning in science education more specifically. Intellectualist approaches are fundamentally agential and have no place for the radical passivity that is integral part of our everyday lives. How this agency, the power to act, is given to the agent is never explained in and by intellectualist theories – some, as von Glasersfeld (1989b), relegate this issue to metaphysics. Others suggest that it is precisely the introduction of the subject –

whose agential and pathic capacities are grammatically expressed by verbs – that enriches biology with the pathic categories (von Weizsäcker 1973). Throughout this book I show that there is no agency without passivity, and in fact, agency is a capacity received by the subject who is but a welcoming host to this gift. Moreover, biological phenomena generally and psychological phenomena specifically cannot be explained as series of causally connected functions and facts that determine the engagement of the subject in the material and social worlds. Rather, throughout this book I highlight aspects of heretofore-untheorized dimensions of human knowing that require passion and passivity as an irreducible companion concept.

Communication as Paradigm

Communication is a good object for the study of the constitutive relation of agency and passivity because without communication generally and without language-use specifically, there simply would not be any ontology or epistemology: This *is* the concern of first philosophy. Communication presupposes alterity and intersubjectivity. It would be impossible if it had to begin with the individual, free subject (Levinas 1990). Moreover, as the following analysis shows, there is an ethical dimension underlying communication that is even more foundational than the "we" of intersubjectivity. Because it is a requirement for any philosophical talk about ontology and epistemology to have become possible: That is, communication constitutes a suitable paradigm for thinking about how humans have pulled off the challenge of creating any ontology and epistemology in the first place *without having to presuppose all the conditions that enable talk about such topics.* The existence of language as central feature of consciousness opens up other possibilities for thinking the relationship of agency and passivity. Thus, for example, speaking (producing sounds for communication) is possible because there is a capacity to hear (listen), that is, to be affected by sounds that arrive from outside the living/lived body. Even the shortest bit of transcript, such as the following excerpt from Fragment 3.1 recorded in the biology laboratory studying fish vision, allows us to understand some of the fundamental requirements for a theory of agency that is grounded in radical passivity.

Fragment 12.1 (excerpt from Fragment 3.1)

```
02 M:  a bit broad, ha?
       (0.75)
03 T:  its very broad, yea (0.45) its very hard to define a-
04     (1.32)
05 S:  strong=
06 T:  =straight line through it.
```

For me (M) to say anything at all intended for the other, this recipient is presupposed to be capable to be affected by the locutionary act (the fact of speaking).

This intended other, here Ted, has to open up and has to be willing to be affected (receive). In fact, the intent itself has to be available in the situation as a whole. Ted cannot know what is coming, and therefore cannot anticipate it and construct understanding. But in everyday hearing, we practically understand without having to engage in intentional interpretation as we might engage in the interpretation (exegesis) of a biblical text.[5] Conversely, the speaker does not know how the intended illocutionary act will affect the other. I might have had the intention to ask a question, but whether there will be a question–answer pair is still up in the air. The speaker is passive with respect to the ultimate effect that his own locution has on the other. This effect, the perlocutionary moment of the speech act is available to me, the speaker in turn 01 only in and through a subsequent response. I am radically passive with respect to the effect of my locutionary act irrespective of the intention (illocutionary dimension) I might have had. We know of the running apart of intentions and effect from everyday experiences, such as when one person intends to ask a question but the recipient feels offended.

Although the increasing pitch toward the end of the utterance might be heard as a question, in this fragment, turns 2 and 4 constitute a constative/confirmation pair. The locution "a bit broad, ha?" comes to be confirmed as a descriptive statement in the response, "It's very broad, yea, its very hard to define a . . . straight line through it" (turns 2 & 6). To receive the answer, an expression of the effect of the locution (i.e., the perlocutionary moment of the speech act), I have to open up to be affected as well. Such opening up means making myself vulnerable; and this vulnerability is addressed in the preference for confirmation, affirmation, and acceptance of the content of locutionary acts rather than disconfirmation, negation, and rejection. Agency and passivity are co-present even at the very microlevel of speaking in that any sounds produced require working against the resistances that the living/lived body exerts against change such that a produced sound is the result of agency and passivity "working" together. Here again, the capacity of being affected by sound precedes the possibility to use sound as a communicative medium. To speak, I have to develop the capacity to produce sounds (or other forms of signs). The very capacity to produce sounds, articulatory gestures; and aural perception is conditioned by the capacity for making sounds – a mutually constitutive relation beautifully revealed in the studies of Japanese to hear and pronounce the difference between /l/ and /r/.

Communication also is a good paradigm for thinking about how children come to know, for they participate in communicative situations prior to any consciousness and self-consciousness. Thus, communicative situations in which children participate prior to being conscious of themselves as independent Selves, constitute the

---

[5] Heidegger (1977), concerned with theorizing everyday understanding, therefore characterizes interpretation (as that of biblical texts) as a derivative, which has emerged from hearing, which constitutes primary understanding. Thus, birds, without knowing anything about language, consciousness, or texts *hear* another bird give a warning call. They do not have to stop and think what the call means – they do not even have the capacity to stop and reflect. Hearing-understanding is primary on evolutionary grounds and constituted the capacity on which interpretive hearing has evolved.

ground from which subjectivity emerges: subjectivity literally means being subjected (Lat. *sub-*, under + *jacēre*, to throw, to cast) to the other. In the perspective developed here, therefore, subjectivity "is the breakdown of the difference between Same and Other, the breakdown of intersubjectivity, and a proximity to the Other" (Wall 1999, p. 47). The sharing in and of the *with* has to exist from before and beyond Being, because it is the condition for the first sound-word not only to be spoken but also to be understood: language always already presupposes interlocutors. This epistemology is one acceptable on evolutionary and cultural-historical grounds. It allows conversations to emerge, and, with it, consciousness and knowing as we know them today, including communication about epistemology and ontology. Simultaneously with consciousness emerges ontology in everyday human praxis, our attention to things as things; and this praxis provides the ground for language to turn upon itself and therefore, for epistemologies (metaphysics) to emerge and evolve in the way we articulate them today.

Theorizing Agency and Passivity

Agency is irreducibly tied to passivity into a unitary but internally differentiated (therefore non-self-identical) phenomenon. The unit of agency and passivity can be understood and theorized taking passion generally and crisis specifically as a starting point (von Weizsäcker 1973). It is precisely in crisis that the pathic is opposed to the ontic, the factual as known from beforehand. In a moment of crisis, the dimension of the pathic rises to the tremendous heights of the ultimate power. The pathic covers such concepts as intention, expectation, surprise, danger, terror, safety, freedom, decision, and constraint. In a true crisis the new emerges unexpectedly and without any chance to anticipate it, because the encounter with the radically new is a pathic event. The pathic is the origin of want and have-to, agency and constraint. If we want (will) it, we can go further, exceed what we thought to have been able to do before, so that will extends what we can. But at the same time, there are situations, such as hysteria, where the patient no longer is able to intent.

> The problem of a hysterical plegia lies in part in the fact that the patient could if he only wanted. At least, this is the way in which an uninformed observer would judge. But here we are in contradiction with the patient, who says that he wants but cannot. Traditionally, the attempt has been made to remove this contradiction between the observer and the patient by saying that the patient is unable to want. Here, will was constrained by ability, and the state would have to be characterized by a limitation of the ability to will. (p. 271)

To think both phenomena – i.e., the ability to extend what the subject can do by means of the will and its inability to want ability – we require the pathic. In those moments of crisis that I describe and theorize in Chapter 11, I found myself precisely in this state where I no longer had the ability to intend, where I was unable to intend a decision about where to go next, or even simply to decide taking the few steps that separated me from the couch to lie down and rest or sleep. If action did not require decision, then what to do next would have been computable and would not even have required my thought: My mind would have computed what to do and I could have stepped to the couch and slept there. The decision is a tension between necessity and freedom, between having-to and wanting-to do something.

The pathic modalities are expressed as verbs. This inherently means that there is a subject. It is precisely the pathic that introduces the subject into our consideration of knowing and learning, doing, and participating. But introducing the subject also means breaking open the explanations of human actions in terms of causalities. It is precisely because human subjects make decisions – which are inherently non-computable – that we need to think thought and learning not in terms of the construction metaphor but in terms of the complementarity of agency and passivity, the former grounded in and emerging from the latter. But this pathic itself cannot be foundational: it is itself the result of alterity. Affection comes suddenly: "the blow of affection makes an impact traumatically, in a past more profound than all that I can reassemble by memory . . . dominate by the *a priori*: in a time before the beginning" (Levinas 1990, p. 140).

The analyses I provide throughout this book exhibit the centrality of passivity with respect to being in the world (e.g., having a conversation), learning, and the making of context. This passivity, however, has not been addressed in educational theorizing. Existing learning theories are aspect blind to this dimension of language, learning, and context and thereby fail to understand the fundamental constraints that learners are subject(ed) to. A similar absence makes it impossible for us to understand the difference between the planned and enacted curriculum: If agency were everything, then there ought not be a difference between the two forms of curriculum. However, students and teachers are both agential and passive with respect to the ways in which they bring the lived curriculum into being (see also Chapter 2): It is a collective process and product so that teachers also are *subject(ed) to* their conditions as much as they bring these about (and changes therein) qua agential subjects of curricular activity. The relation between teachers and students is built on language; but the "relationship of language presupposes transcendence, radical separation, the strangeness of the interlocutors, the revelation of the Other to me" (Levinas 1971, p. 45). This has implications for the way in which we have to rethink agency.

The most frequently cited cultural sociological theory links *agency* and *structure*, where structure itself consists of the dialectical relation of (material, social) *resources* and (personal) *schema* (Sewell 1992). That is, in acting (e.g., speaking) we employ resources (e.g., sound-words) and in the process form new schema; and our schema mediate what we can perceive as resources and how we perceive them. But without schema and resources, there cannot be anything like agency – there cannot be action without the (material) subject and the (material) object involved in the transaction. This theory, therefore, can be expressed as

$$\text{agency} \parallel \text{resources} \mid \text{schema} \qquad (12.1)$$

But seeing the world as populated by individual agency "and social structures leads inevitably to the assumption that structures at one end or the other are responsible for social order" (Rawls 1989, p. 165). Rather than understanding these terms as dialectically related, that is, as negations of each other (as in Hegelian dialectics), I understand the relation as dialogical on the sense that there is always cross talk between the two sides separated by the double line.

This relation expressed in 12.1 is asymmetrical: agency is by itself on one side of the relation ("||"), whereas there is another relation (resources | schema) on the opposing side. I show in this chapter that agency inherently stands in an irreducible and constitutive relationship with passivity. An improved theory therefore may be expressed as

$$\text{agency} \mid \text{passivity} \parallel \text{resources} \mid \text{schema} \qquad (12.2)$$

In this expression, agency and passivity presuppose each other just as resources and schema do in existing dialectical theories. It is not an opposition of two terms; rather, the two terms are but poles of an equivocal passage that Jacques Derrida, in various places refers to as *différance, khôra, trace,* or *margin*. If Bakhtin had been familiar with Derrida, he might have added "word" to this list. The concepts denote the delay between presence (*Sein, Être,* Being) and the presence of the present (*Seiendes*, *étant*, beings). The expression in 12.2 is consistent with other phenomenological analyses that deal with the first appearance of something that we could not anticipate because we did not know of its existence (e.g., the twin silos in Chapter 2): The existence of an essential passivity in the welcoming of the unknown, the radically new, the absolutely foreign. The order arising from interactions therefore is self-organizing because neither individuals nor existing structures can be its master.

In the social sciences generally and in science education specifically, theories that build on an agency | structure relation have come to have considerable currency. The present chapter shows that there is more to social situations than agency generally and to speaking (language use) specifically. There is also more to social situation than simple passivity, a form of action where someone *decides* to enact non-action. With speaking comes *radical passivity*: to listen, hear, and understand what the speaker is saying, the listener has to open up and expose herself to be affected by something that she cannot anticipate. It is precisely this capacity to be outside of myself in the world where I am open to be affected and modified that makes the world comprehensible and immediately endowed with sense (Bourdieu 1997). Conversely, to find out the answer to her intended question (which turns out to be realized in and by the listener's response), the original speaker has to abandon and expose herself in the same way. That is, to produce this social and societal situation as it unfolds (diachronic moments) and in its totality (synchronous moments), we require radical passivity as an integral part on the yonder side of and as the complement to structure, expanding agency to become an agency | passivity dialectic that complements the schema | resources dialectic that constitutes structure.

Thinking about language, learning, and context in education in terms of *agency | passivity || resources | schema* comes with tremendous potential for understanding what happens in our classrooms and what the particular constraints are in attempting to achieve specific learning outcomes. Future research has to work out how curriculum planning, teaching, and research on language, learning, and context have to be reconfigured to address and accommodate the essential passivity that is presupposed in and makes possible the already-theorized agency. Most importantly, passivity as articulated here does not refer to situations where someone decides not to speak – e.g., because someone else "silences" them, or because the person feels, as students from First Nations often feel in Western-style schools, that there is no space to get into the conversation. In such situations, there is an intent that orients a particular form of agency: not doing something that others in the situation already do, which is a derived form of passivity. Rather, as I show here, I am concerned with theorizing passivity that comes with speaking and hearing that is at the very heart of agency, the intentional moment realized in and by speaking. That is, I am concerned with what Levinas might call passivity more radical and more passive than any conceivable passivity. The first philosophy that Levinas (1990) articulates is based on phenomena that constitute the heart of the present book, including sensibility, sensuous lived experience, passivity, patience, vulnerability, joy (enjoyment), communication, and alterity.

Incarnate Acts and Responsibility

> [T]o *be* in life, to be *actually*, is to *act*, is to be unindifferent toward the once-occurrent whole. (Bakhtin 1993, p. 42)

To be is to act; to act is to affect something and someone. All acts are incarnate acts and, as such, are permeated by affective and deontological tonalities. Although "non-incarnated thought, non-incarnated action, non-incarnated fortuitous life" are also possible, each of these possibilities constitutes an "empty possibility" (p. 43). Non-incarnated consciousness is fortuitous. "And what is also fortuitous is the emotional-volitional tone of such an unincarnated thought – unincarnated in my answerability" (p. 44). As acts in a once-occurrent, continuously unfolding whole, incarnate acts are answerable. Being incarnate means being exposed – giving rise to the possibility of suffering, enjoyment, trauma, and crisis. Thus, "signification is only possible as an incarnation. The animation, the very pneuma of the psyche, alterity in identity, is the identity of a body exposed to the other, becoming 'for the other,' the possibility of *giving*" (Levinas 1990, p. 111). The incarnation here "is not a transcendental operation of a subject that is situated in the midst of the world it represents to itself; the sensible experience of the body is already and from the start incarnate" (p. 123). Accepting incarnation as a foundation of knowing and learning, therefore, constitutes a rejection of the constructivist subject concerned with representing the world to itself. Thus, "subjectivity of sensibility, taken as incarnation, is an abandon without return . . . the body as passivity and renouncement, a pure undergoing" (p. 127). Incarnation, therefore, also is a rejection of the embodiment and enactivist approaches to cognition, which do not and cannot conceive of the absolute passivity that comes with incarnation.

Thinking knowing and learning in terms of incarnation leads us to responsibility, as the incarnate act "presupposes my answerable participation" in the reproduction and transformation of society and life rather than "an abstracting from myself" (Bakhtin 1993, p. 18). Incarnation and responsibility are thereby thought together in a foundation of knowing action, the goal for learning oriented to life rather than school.[6] Thus, "it is only from within my participation that Being can be understood as an event" (p. 18). However, when we approach the question abstractly, in the contents of the said or done, we will not find it. That is, incarnation also allows us to think the responsible subject, students who understand themselves as being in charge of a world that – as ancient proverbs in many cultures suggest – they did not inherit from their parents but which they borrow from their children. This approach to ethics based on responsibility cannot be understood on the basis of a traditional, Kantian approach to ethics. Ethics considered in this way "is the breakup of the originary unity of transcendental apperception – that is to say, the beyond of experience" (Levinas 1990, p. 232).

Thinking knowing and learning in terms of incarnation, therefore, not only introduces to our considerations affect, emotions, motives, and motivation but also ethics. It no longer permits us to think of science education in terms of contents that are abstracted from anything that matters in this world. As a scientist, I am answerable for my production of the atomic bomb, Agent Orange, thalidomide, and genetically modified organisms. Incarnate knowing, which always is knowing-in-acting, is an ethical relation and a recognition of my Being as a "once-occurrent and never-repeatable, emotional-volitional and concrete individual" (Bakhtin 1993, p. 39). Thinking knowing by means of incarnation is a recognition of my "place in once-occurrent Being that is unique and never-repeatable, a place that cannot be taken by someone else and is impenetrable for someone else" (p. 40). Our participation in this world as a once-occurrent event involves for Bakhtin (as it does for Levinas) both the "moment of my passivity and the moment of my self-activity: (1) I find myself in Being (passivity) and I actively participate in it" (p. 41). Ultimately, therefore, this

> fact of *my non-alibi in Being*, which underlies the concrete and once-occurrent ought of the answerably performed ac, is not something I come to know of and to cognize but is something I acknowledge and affirm in a unique or once-occurrent manner. (p. 40)

My responsibility in and to Being, a responsibility for the other, therefore escapes representation and comes to be without exception and precedes any commitment. This responsibility is "preceding every free consent, every pact, every contract" (Levinas 1990, p. 141). That is, no epistemology that privileges mental constructions

---

[6] The Romans already knew that "non scholae sed vitae discimus" ("We learn not for school but for life").

standing for something else, *re*presentations, can account for this aspect of knowing (and learning) that comes with incarnation. These other epistemologies have to introduce ethics from the outside, make it an outside factor current conceptual change research does not even try to account for (e.g., Treagust and Duit 2008). But an epistemology grounded in the living/lived body, which is both subject of and subject(ed) to collective activity, inherently implies an ethical relation between Self and Other.

Part E

Epilogue

> It is this radical phenomenological precedence of life that thought forgets constantly when it takes itself for the principle of all knowledge, of everything that we can know, of everything that exists for us. This forgetting is all the more catastrophic when it concerns the body, or still that which is linked to it according to an invincible relation, the flesh – our flesh. (Henry 2000, p. 136)

In this concluding part, I begin by summarizing the shortcomings of intentionalist approaches to cognition generally and to the present forms of constructivism in science education more specifically. I then propose a way of situating it within a more encompassing theory. In this resulting theory, the intentional and constructive sides are complemented by the essentially passive and affective nature of being. I show how the new theory transcends the dichotomies between inside and outside, intra-psychological and inter-psychological, body and mind, or mind and world in conceiving of them of one-sided articulations of the world. When we think from a position of difference – rather than of sameness that underlies Kantian, Piagetian, and constructivist thought – then we come to understand that agency and passivity, body and mind, knowing and world, or theory and praxis are but one-sided manifestations of phenomena inherently different from themselves. Autonomy never comes without heteronomy, and heteronomy never without autonomy. *Intentional agency* that is the core of the constructivist self and mind never comes without *radical passivity*, *otherness*, and *undecidability* and it is only against agency that passivity, otherness, and undecidability are taking shape and become relevant.

Chapter 13

Sublating the Constructivist Metaphor

A sound theory of learning and development needs to be plausible on evolutionary (phylogenetic), cultural-historical, and ontogenetic grounds. That is, a theory aiming at explaining learning and development needs to be able to account for the transition from animal to human cognition, emergence of language when there is no language and linguistic consciousness, learning, the emergence of forms of Self, and so on. Intentionalist theories cannot explain these phenomena, because they take as explanatory premises (explanans) human-like intentions, language-like schemas, and an agential subject, each of which is a thing to be explained (explanandum). In this sense, the constructivist metaphor does not produce a viable theory precisely because it cannot explain how thought that continually constructs itself could have emerged in the first place.

The metaphor of construction is fallacious precisely when it deals with the way in which the absolutely new, the unseen, comes to be known, how it is brought into the realm of the visible – here an essentially passive aspect of life needs to be accounted for.[1] On the other hand, it is a useful metaphor when activity involves the known, which can be thought of as the visible. Learning as the coming of the visible from the invisible has to be thought in terms of passivity. As I show throughout this book, any learning that is associated with the passions, insight, or experiences of flow cannot be accounted for by means of a constructivist metaphor. This is so because this metaphor reduces every aspect of life to the visibility of intellectual consciousness. But, it is not so that life comes to be known by us because of thought, for thought is a product of collective activity. That is, as Karl Marx suggested long ago: "Consciousness cannot ever be anything other than conscious being, and the being of humans is their real life process" (Marx and Engels

[1] As far as scientists and science are concerned, the construction metaphor of conceptual change does not explain the phenomenon that the historian of science Thomas Kuhn (1970) described in *The Structure of Scientific Revolutions*. Thus, despite evidence to the contrary, scientists tend to hold on to old theories rather than abandoning them. Old paradigms die out when their defenders go rather than being reconstructed by former defenders.

W.-M. Roth, *Passibility*, Classics in Science Education 3,
DOI 10.1007/978-94-007-1908-8_13, © Springer Science+Business Media B.V. 2011

1969, p. 26).[2] Consciousness and thought are modes of life (*Sein*, *Being*), that is, life gives to itself thought allowing it to become aware of itself. In thought life reveals itself to itself – thought is the result of an auto-affection of life.

My intent for writing this book is not to throw out wholesale the constructivist metaphor and intentionalist theories. Rather, as the title of this chapter suggests, I am interested in understanding the constructivist metaphor, itself an accomplishment of human culture, within a larger framework of cognition. Within this larger framework, the constructivist metaphor will be part of a one-sided manifestation of thought, learning, and development that only works sometimes rather than all of the time. The philosopher Georg W. F. Hegel used the German verb *aufheben*, which has two mutually exclusive senses. On the one hand, it has the sense of "to keep," "to preserve." In this sense, the chapter title summarizes the contents of this chapter as keeping or preserving the constructivist metaphor. On the other hand, the verb has to sense of "to revoke," "to invalidate," or "to suspend." In this sense, the chapter title summarizes the chapter contents as an argument for ridding ourselves of the constructivist metaphor. In this chapter, I argue for the need to do both at the same time, that is, to preserve and to suspend the constructivist metaphor. I articulate under what conditions the metaphor of construction is useful and where it is not.

Limits to Intentionality

I already show in Chapter 11 that there are limits to intentionality in the sense that we do not know where it comes from. It is given to us, as we do not intend intentions. Every day we find ourselves with the intent to do this or that. Intention is given to us in total passivity. In some individuals (and sometimes), the problem of intentionality is different. It is not of the type of passivity that we make a decision about, for example, by saying (aloud or to ourselves) that we no longer participate in this or that. Non-participation or passivity is itself the result of an *active* decision. The kind of passivity that I am interested here, however, is of more radical nature. We are subject to it. We find ourselves in a situation and cannot go beyond; and no powers can get us out of the predicament because it is part of the nature of Being.

There are also limits to what intention allows us to do with respect to learning even in the case that we are willing students. In this book, I repeatedly refer to the fact that we cannot intend to learn that which we do not know. This might sound a little bit strange to some ears, which is reason enough for me to think about an

---

[2] Marx's statement is more ominous in his native German, where the sentence reads like this: "Das Bewußtsein kann nie etwas anderes sein als das bewußtes Sein, und das Sein der Menschen is ihr wirklicher Lebensprozeß." The two words that constitute the German word for "consciousness," that is, *bewußt* ("conscious") and *sein* ("to be," "being") are deployed differently in the same sentence thereby relating consciousness (Ger. *Bewußtsein*) and conscious Being (Ger. *bewußtes Sein*).

analogy to assist heretofore-unconvinced readers in thinking about the situation more deeply. Take the following experience. One day, I am taking a nap in a guestroom of our home. As I turn over, I all of a sudden notice a ticking sound that I have not been aware before. I look around and I see that it comes from an old alarm clock. But then I wonder about having noticed it only at that instant. I turn back to my original position – silence envelops me. I turn over, and I can hear the ticking. At this moment, a realization hits me[3]: I am sufficiently impaired in my left hear to no longer hear the ticking, whereas I can hear it with my right ear. The point with respect to learning and cognition is this: lying on my right ear, in an absolutely silent room, I could not have intended to hear or to learn about or to reflect on the ticking. The ticking simply does not exist at this instant. Even if a teacher were to ask me to focus on the ticking, it would not exist for me. (I know from my research in classrooms that in such cases students take anything as salient that they may take to bear on the current issue and as the thing talked about.)

We may now extend this situation to conceptual issues. How do I intend learning about a concept (in my analogy, the ticking) if I do not know it (hear it in my analogy)? How do I intentionally construct something where there is nothing yet for me? Pragmatist and phenomenological philosophers have already thought about this problem and have come to the same conclusion – I cannot intend the new language that I am in the process of evolving. I cannot even know what the language is going to do for me. This contrasts craftspeople, who, because of prior experiences, produce something that they have already an image of. A carpenter has an idea of what it means to end up with a house even if the particular house is different from other houses that s/he has previously built. My electrician has no problem taking my plan and use it to wire the lower level of my home exactly as I want it, because she has done so many times before. In contrast, someone who builds something new or brings something forth – and a student is engaged in poeisis as much as a poet – cannot know what s/he is in the process of making and what its purpose is. Thus, even

> someone like Galileo, Yeats, or Hegel (a "poet" in my wide sense of the term – the sense of "one who makes things new") is typically unable to make clear exactly what it is that he wants to do before developing the language in which he succeeds in doing it. His new vocabulary makes possible, for the first time, a formulation of its own purpose. It is a tool for doing something which could not have been envisaged prior to the development of a particular set of descriptions, those which it itself helps to provide. (Rorty 1989, pp. 12–13)

That is, only when the new vocabulary exists in its entirety can the creator of the vocabulary formulate its purpose. Prior to the development of this new vocabulary, this purpose could not be articulated. "Dabbling" and "messing about" may be much more appropriate metaphors for this process of development and learning than "constructing." Thus, the Copernican Revolution was not a rational decision based on observation and the working out of different working theses. "Rather, after a hundred years of inconclusive muddle, the Europeans found themselves

---

[3] Notice the passive construction of the sentence: As subject, I am also subject(ed) to this experience.

speaking in a way which took these interlocked theses for granted" (p. 6). The conceptual change involved is of tremendous magnitude, and a "change of this magnitude does not result from applying criteria (or from "arbitrary decision") any more than individuals become theists or atheists, or shift from one spouse or circle of friends to another, as a result of either of applying criteria or of *actes grauitis*" (p. 6).

But let us return to the analogy of hearing a sound. The situation is slightly different in another context, where we hear a sound but cannot make sense of it. For example, transcribers of videotapes from classroom research often hear that someone is talking but cannot decipher the word or words despite repeated listening to the tape recording. Now they may intentionally seek to understand that sound. For example, it is well known that given possibilities for what the word might have been, the transcriber may actually hear the sound as one or another of these possibilities. A transcriber may actually generate possible solutions to the problems, hypothesize possible words that have been uttered, and then test whether one or another is the one that has actually been uttered. This precisely is one of my strategies to recover difficult passages from a soundtrack.

We can learn from these examples that passibility, the capacity for sensation (and for suffering), precedes intentionality. We cannot intend learning (about) something if we do not already have the capacity to be affected. This also leads to the fact that learning *cannot be administratively planned* (Holzkamp 1992). The very fact that the thing to be learned is inaccessible to students prevents them from intentionally aiming at it. It is only because we hear that we can intend to find out something that we have noticed in the audible range. But hearing something is not a sufficient condition for the intention. In the same way that there is auto-affection when I move about, which leads me to learn about the power to act, there are phenomena of auto-affection that underlie the possibility to have sensation. For example, in the condition tinnitus – term tinnitus comes from the Latin *tinnīre*, to ring or tinkle – a person hears sounds (ticking, clicking, ringing beeping) that others do not. The sounds are generated in the hearing tract and are, in some instances, continuous whereas in others (in my own situation) are intermittent. Sometimes, as in objective tinnitus, the doctor can hear the ringing when using an instrument, but more often the condition exists in the subjective form. Tinnitus tends to increase with loss of regular hearing. The internally generated sounds and interfere with external sounds, which can no longer be heard properly.[4] Whether a sound can be associated with a "real" object or phenomenon, something that can be said to exist outside and independent of the particular ear, comes to be questioned in other types of situations as well. For example, we sometimes hear something only to discard the possibility of it having been a real sound when we cannot verify the sound or the presence of an entity (phenomenon) that could have produced the sound. (Children and fearful adults in a dark forest.)

---

[4] There are suggestions that one possible cause is a homeostatic compensation of a hearing loss that gives rise to hyperactivity in the auditory neurons.

Limits and Limitations of Construction

Intentionality is central to the constructivist metaphor in theories of learning and development. But there are limits and limitations to this metaphor that should make us question constructivism as the master theory that it is currently held to be in many educational contexts. A *general* theory is applicable to every *particular* case. But because it does not cover the general case, the constructivist metaphor cannot be used as a master theory. Constructivists acknowledge intention as a fundamental condition of development: "learners need to be active and to have a certain *intention* to learn" and they have to have an "active intentional role in the process of knowledge restructuring" (Treagust and Duit 2008, p. 301). But such orientations are possible only when they already know what a task aims at and when they already have all the competencies in place to actually achieve this task. If Rorty (1989) is right in saying that even a giant of culture is "typically unable to make clear exactly what it is that he wants before developing a language in which he succeeds doing it" (p. 12), how then should we expect just plain folks to be able to intent knowing prior to having the language as a whole? That is, a full description would be possible when a student, after having *successfully* completed a unit, undergoes a unit test with some word problems. The student recognizes what the problem is asking her and, using the information provided, goes about showing how to produce the solution. This student is in the same kind of situation that a carpenter might find himself, who has a building plan and, using his tools and the materials provided, puts up the wood framing of a house (in Canada, at least, where most houses use wood framing). But a student who does not know what the question is asking cannot engage in constructing such a solution. The student is in the same situation in which most Canadian construction workers would find themselves when asked to construct a *kraal* (e.g., at a construction competition).[5] Like the student asked to learn something that he has no idea about what it is, this construction worker would have to ask constantly whether s/he is right. Moreover, as Rorty points out, the tool metaphor is actually deceiving, because the learner is in a situation where the tools themselves do not yet exist. Learning a new language to learn about phenomena that the language enables means the tools, too, are in the making. The following two empirical cases – extracted from a database featuring experienced research scientists in think-aloud interview protocols about graphs culled from first- and second-year undergraduate courses – exhibit situations in which the constructivist metaphor is useful and when it is not.

The fragments derive from interviewee responses to a graph that features a standard population model (Fig. 13.1). It features a parabolic birthrate curve that intersects twice with a linear death rate curve. Fragment 13.1 is an excerpt immediately following the reading of the caption and visual inspection of the graph. It is quite evident that the biology professor produces an answer the contents of which

[5] A *kraal* – a South African English and Africans (Colonial Dutch) word apparently derived from the Portuguese term *curral*, corral – is a poor hut or hovel, an enclosed village, or an enclosure used in animal husbandry.

In the logistic model, birthrate and death rate are linear. Let's assume that the birthrate follows a quadratic function (e.g., $b = b_0 + (k_b)N - (k_c)N^2$), such that the birthrate and death rate look like in the figure. Such a function is biologically realistic if, for example, individuals have trouble finding mates when they are at very low density. Discuss the implication of the birth- and death rates in the figure, as regards conservation of such a species. Focus on the birth and death rates at the two intersection points of the lines, and on what happens to population sizes in the zones of population size below, between, and above the intersection points.

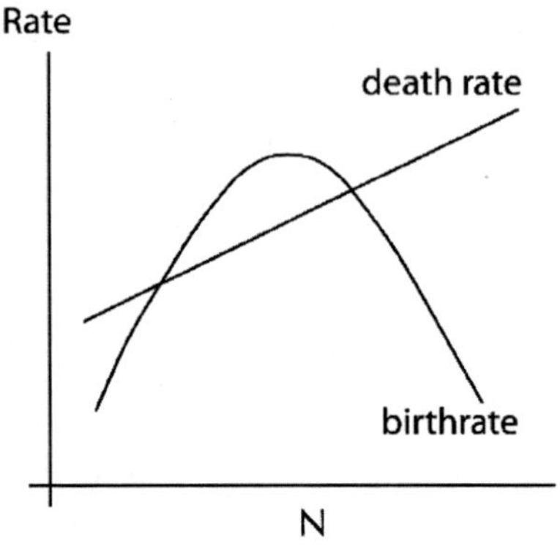

Fig. 13.1 This population graph is a standard feature of introductory ecology courses. It was used as a task in think-aloud protocols with practicing scientists.

are already anticipated in his response from the very beginning. That is, in the response as a whole, the ultimate aim of the response – unstable and stable population equilibriums – takes shape from the beginning of the response. The professor immediately picks out the first equilibrium point and articulates the relationship between birthrate and death rate to its left. The death rate is larger than the birthrate, which means that the population will go extinct. Then he focuses on "anything above that point" and, while pointing to the area between the two intersections, he articulates birthrates as exceeding death rates with the result of an increasing population.

Fragment 13.1

```
01 B:  well. okay. so, anything below this line ((draws line ver-
       tically through first intersection)), death rate is
       greater than the birthrate so that means the population
       size has to go this way ((draws arrow from intersection to
       the left)) which leads to extinction ((writes "extinc-
       tion")). anything above this point ((points to left inter-
       section)) means that you have birth greater than death,
       population growth to the point where birth and death equal
       each other again. so, in this range ((points to region be-
       tween the two intersections)) we=ve got population going
       this way ((draws an arrow from left to right)) and above
       here ((draws vertical line through right intersection
       point)) again death are greater than birthrates so the
       population goes this way ((draws an arrow from right to
       left with arrow head ending at intersection)).
```

In his answer, the professor does not refer to the unspecified "N" in his argument but immediately uses the term "population size" in his reading. That is, although this reading of the event appears to be straightforward, not all scientist participants in this study made the assumption that N is population size or population density. In fact, some biologists treated the abscissa as a temporal axis with developments from left to right (Roth and Bowen 2003). Moreover, when asked to provide a reading of the graph, second-year university students repeatedly asked if they could make the assumption that *N* decreases toward the left of the graph or, alternatively, if *N* increased as one goes from left to the right. That is, not only does the professor immediately focus on the intersection points and provide standard readings, he also reads all the parts of the task in a way that provide him with those resources that lead to the production of the standard answer. This standard answer is the one that a biology professor – who used this graph in his course that we videotaped in its entirety – would have accepted as correct.

In this instance, I deem it legitimate to use the construction metaphor. The professor seeks out and uses the resources provided in the think-aloud task and he *constructs* his answer. The resources are already recognized as resources, and the tools (language) he employs already are known tools. He can do so because he has anticipated the shape of this answer and then articulates all the resources and tools that are required. Because he knows what he needs to arrive at, he is in a position to monitor what he says and does. That is, he can enact the kind of metacognitive processes that intentional learning and conceptual change theories incorporate. These processes *require* the acting subject to know where he is going, that is, to have an intentional object. If the person does not know the intended object – new knowledge in the case of the learner, the question of the problem and the solution it anticipates – then the acting subject has nothing to go by for judging cognition. This is evident in the next fragment. That is, the construction metaphor works when we think of knowing, learning, development, and problem solving in science education in terms of the visible and therefore "visable," that is, suitable to be aimed at.[6] It does not work in cases where the thing to be learned is unknown, invisible, and therefore "invisable."

The following fragment derives from a database of think-aloud/interview protocols with physicists, who were asked to provide expert readings of structurally equivalent graphs from their native discipline and from biology. At the time, Annemarie has had over 30 years of experience and has received a university-wide award for excellence in teaching. Yet, as the fragment shows, she asks Daniel, an undergraduate student in her own department whether what she has said so far "is right then." After saying that she cannot get her head around what the task is asking her, Annemarie eventually begins. As she traces along the linear death rate (Fig. 13.2), she exhibits its increase from left to right. Then she traces the birthrate and describes it as "increasing faster" than the death rate. That is, unlike the biology professor, who compared the values of birthrates and death rates to the left and right of the intersection point, Annemarie makes salient the steepness or slope of each of the two curves and compares them.

6 The term "visable" and its opposite "invisable" are French neologisms that Marion (2004) creates on the basis of *viser*, to aim, in order to express that which cannot be aimed at because it is invisible. It is a challenge to the translator, because of the critique of the intentionality that it entails. The translators of different works of Marion (a) leave the term untranslated (i.e., *invisable*), (b) use "that which cannot be aimed at," or (c) translate it as "untargetable."

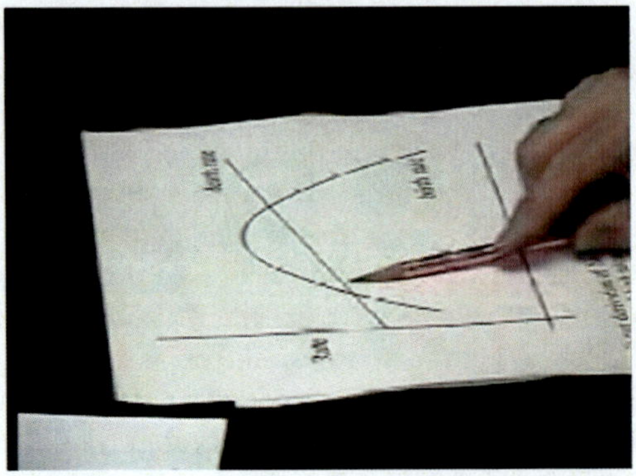

Fig. 13.2 Annemarie traces the death rate curve with the tip of her pen thereby making salient not only the line but also its slope.

Fragment 13.2

```
01 A:  SO (1.60) ((puts paper down on table)) here ((pencil to
       the graph, Fig. 13.2)) we have the (2.11) death rate
       increasing ((traces death rate)) (0.68) and the birth rate
       increasing ((traces birthrate)) and the birthrate is
       increasing (0.76) faster (0.87) than the death rate.
       (1.80) so they are both increasing but the birthrate is
       faster increasing than the death rate so presumably that
       means that the population is increasing.
02     (0.93)
03 A:  is that right then?
04     (0.96)
05 D:  u:m::
06     (0.41)
07 A:  round this region? ((circles the area left of the first
       intersection point))
```

Annemarie not only compares the slopes of the two curves but also articulates a temporal dimension from left to right that is to occur with the different *slopes*: "they are both increasing, but the birthrate is faster increasing than the death rate." She concludes: "So presumably that means the population is increasing." Here, the population is said to increase because the birthrate curve is steeper than the death rate curve; and, as further evidenced in other parts of the transcript, the evolution is temporal from left to right ("is increasing"). We can see that Annemarie, in contrast to the biology professor, makes salient the slopes of the two curves. Later during the protocol, it becomes clear that Annemarie confuses the birth*rates* and death *rates* with the slopes of the curves, which are the *rates* of the birthrate and death rate curves with respect to the population size (density).

After having produced this first description and drawn implications for the population, Annemarie stops. She then asks the undergraduate student whether "that" (which she had just produced) is right (turn 03). In this turn, Daniel is presupposed to know the correct response. In asking Daniel whether she is right Annemarie constitutes herself as lacking the knowledge and Daniel as the one who can assess whether what she has said is correct. Daniel hesitates, hedges (turn 05), thereby giving Annemarie the opportunity to make another attempt (turn 07). However, this time, she explicitly limits the extent of her question: the region below the first intersection point that she makes salient by means of a circling gesture. Repeatedly during the protocol, Annemarie asks Daniel whether she is right in saying what she has said immediately before. Why might she do so?

It is quite clear that Annemarie's work does not deserve the metaphor *construction*. The pieces she identifies are not those that are relevant for an answer. That is, the tools and resources required for the *construction* are unclear; what a correct result might look like is unclear, too. So Annemarie picks out what is salient to her, the slopes, articulates an implication, and, because she cannot know where this is taking her, she asks Daniel whether she is right. When he does not answer, she specifies the area that her question pertains to, as if she had not made her first question sufficiently clear. (Her specification in turn 07 tells us that she has heard the pauses and Daniel's hesitation as a problem in the question framing.) Because she does not know what an appropriate answer looks like – in the same way that my paradigmatic apprentice carpenters do not know what a kraal looks like – she cannot monitor her activity. She does not know whether she is right or whether what she has said brings her closer to the requested response. "Is that right then?" or "Am I right so far?" – which is another one of Annemarie's question during the session – also is a standard question of science students in situations where they are asked to learn something new. We might ask these questions: How can Annemarie expected to *construct* an answer if she does not already know what an answer looks like? How can a learner *intentionally* learn knowledge that she does not know what it looks like, feels like, or sounds like? How can a learner *intentionally* monitor her learning if she does not know where she is going?

The main problem with the intentional learning paradigm is that it thinks learning processes and outcomes from what is already known, the visible. That is, this paradigm takes what the teacher knows and what the curriculum articulates as that which is to be learned. Because the learning object and intention are visible to the teacher (researcher, psychologist), it is assumed that the student should see this object as well and therefore should be able to make it the object of intentional learning. This assumption would be equivalent, in my experience with the twin silos, to assuming that I should have seen the silos when in fact I had not during the first six trips. This orientation is typified in Piaget's research, which continuously tells us what a learner *cannot do* rather than what they do and why it is reasonable to do what they do given what they see and know.[7] But, how could I have intentionally learned about the twin silos during my first six trips along country road K342 when it did not even figure in my perception? How can I intend to learn, construct, or aim at something that does not exist as such in my mind? Saying that a teacher could have told me to watch out for the twin silos does not solve the

[7] It is typical for the psychologist to write this: "at 7 or 8 months a child has no idea of the permanence of objects, and does not dream of reversing a feeding bottle presented to him wrong way round" (Piaget and Inhelder 1967, p. 5).

problem. This presupposes that I already know "twins" and "silos." It does not help my paradigmatic construction workers to be told that *kraal* denotes an enclosure, living quarters, or village of the Zulu. To construct a kraal you have to know one, be able to recognize a kraal when you see one, or see it emerge as a result from your constructive activity. Once you know it you can also monitor your construction work, take something apart to redo it so that your work takes you closer to the anticipated and envisioned end result. This is why the biology professor could have easily caught an error in the construction of his answer. A very different approach, one that is plausible on evolutionary and ontogenetic grounds, thinks learning and development from the perspective of the invisible and the alien. It does not think learning and development in a teleological manner, oriented toward what is commonly accepted as practice in respective communities.

The Alien Invisible

Constructivist and other intellectualist theories conceive of learning in terms of what is already known. These theories explain the invisible in terms of the visible, the alien in terms of the familiar. Thus, many have noted the tendency towards finality in the work of Piaget, where the telos of development is adult rationality. Moreover, everything in Piaget's constructivism is explained in terms of cognitive development, and emotional and other forms of development are but forms of cognitive development. Affect is nothing but the fuel that makes intelligence work. When Piaget, von Glasersfeld, and others talk about me as a knowing self, they deprive me of everything that is particular (singular) in me, that makes me really me, as I experience myself. This is so because intellectualist theories never pose the question, "Who is doing the thinking/meditating?" How does one reintroduce the living, feeling person into cognition? It cannot be in the way current theories do it, for example, theories of conceptions and conceptual change – here, emotions, motivation, and other dimensions are added to cognition as externalities much like Ptolemy added epicycles to his theory of the universe whenever a new observation showed previous models to be inadequate. "The mistake of reflective philosophies is to believe that the meditating subject can absorb into its meditation, or appropriate without remainder, the object of its thought, that our being can be brought into our knowledge" (Merleau-Ponty 1945, p. 76). We cannot ever become entirely consciousness or reduce ourselves to consciousness only. This would be a trap, for consciousness could have never occurred. Consciousness presupposes the unconscious and non-conscious, against which it appears as a figure against ground. Geometry was founded against the background of the non-geometrical; science emerged on and against the ground of the non-scientific (Husserl 1939). Children's experiences in this world, as unscientific as they might be, need to be understood as the very ground on and against which some of them become scientists and Nobel Prize winners. That is, the non-scientific is not something to be eradicated but has to be something understood as the necessary prerequisite for the scientific to emerge.

On evolutionary and cultural-historical grounds, cognition could not have preceded thought and cogitation. Construction therefore cannot be the appropriate metaphor to explain how human forms of knowing have emerged. The problem that marks all currently dominant epistemologies is this: they confused two forms of seeing, understanding, and coming to understand:

> [T]wo forms of appearing so different as intentional seeing and that which permits this seeing to come to itself in the absence of all seeing: its pathetic auto-donation in absolute life. The Archi-intelligibility therefore does not only come before all conceivable intelligibility, but also it founds it and makes it possible. (Henry 2000, p. 130)

It is precisely this second form of coming to see (understand) that the constructivist metaphor does not explain – possibly because it falls on the blind spot of the constructivist metaphor – the process by means of which intentional seeing comes to be possible in the first place. This form of coming to see – as exemplified in the twin tower example and in the examples of Chapter 3 (the Dalmatian, the hallway/truncated wedge) – have to be understood in terms of the non-intentional givenness that permits seeing to come to itself in the absence of and prior to all seeing. The perceptual process itself is shaped thereby changing perception – rather than being a constant process in the way Piaget had thought (Gurwitsch 2010). There is no thought but the one given to itself in the auto-donation of Life. There is no visible that does not come from the invisible, in fact, that does not *require* the invisible, no thought that would not come from the unthought, the very absence of thought.

If we want to retain the idea of intention in learning in the face of the unknown, we have to embed it in a way of framing the learning problem from the perspective of the subject of learning. Because the subject does not know what it is supposed to learn, it cannot aim at it. The object of learning, from the perspective of the subject, is invisible; it constitutes something alien. Even if in the course of engaging in some learning task the subject had some sudden insight, it would still not be in the position to judge whether the discovered is the learning object that others intended for it. On the other hand, if the learning subject is asked to make sense of the task, what actually reveals itself to the subject cannot be anticipated and may be radically different from what the teacher intends. This is precisely what we observed in a study of learning in a German tenth-grade class studying chaos theory (e.g., Roth and Duit 2003).

On that day, the task required students to construct an explanation for the chaotic pendulum, the paradigmatic phenomenon of the curriculum where a pendulum with an iron bob is made to swing above three magnets (Fig. 13.3a). The students find out that they cannot predict above which of the three magnets the bob will come to rest wherever they start off the pendulum bob. The students are provided with an analogue, a plaster bowl with an inner surface constructed so that it obtains structural similarities with the magnetic field of the phenomenon to be explained (Fig. 13.3b). In one of the groups, the students discuss what happens to a

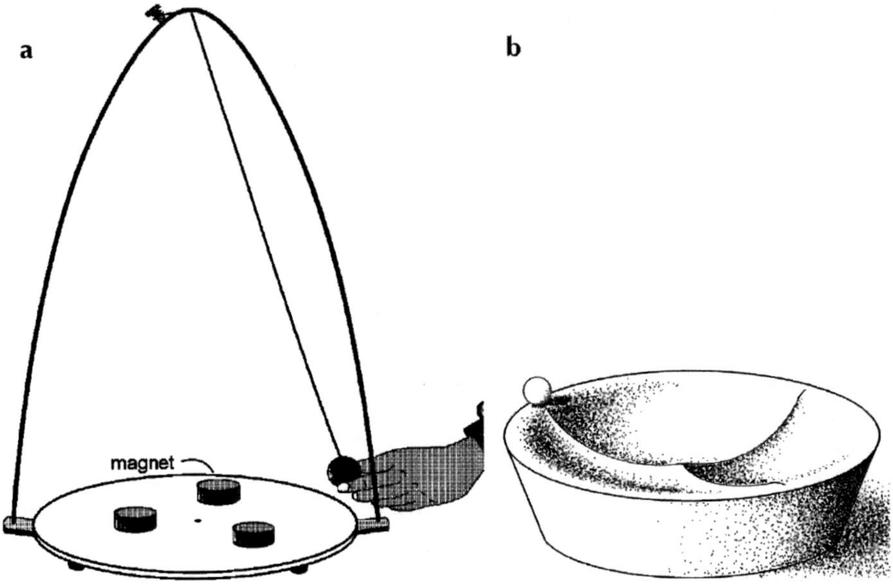

Fig. 13.3 Equipment used during a tenth-grade curriculum unit on chaos theory. **a.** The magnetic pendulum constitutes the paradigmatic phenomenon because students cannot predict the final resting point of the bob. **b.** The ridge structure in the plaster bowl models, for the person who knows, the ridge structure of the invisible (magnetic + gravitational) field within which the pendulum bob moves.

steel ball in the bowl when it comes to rest on one of the three ridges. Katrina then removes one of the three magnets to produce an analogue to the steel ball on one of the ridges in the plaster bowl. At that point, she brings up the Foucault pendulum, which, as she reports to have seen in a science museum, never swings in the same plane above a compass on the floor. Three other students in the group also have seen the Foucault pendulum in the same science museum. This leads the students to agree that the same effect that makes a Foucault pendulum change its orientation with respect to the compass dial below it also brings about the chaotic movement of the steel ball in the plaster bowl and the chaotic movement of the magnetic pendulum. They stop their investigation and tell the teacher that they have produced the requested explanation.

In this situation, the students do not know the standard explanation for the magnetic pendulum. The teacher has provided the plaster bowl because he thinks of and sees it as an analogy to the magnetic pendulum (i.e., same "deep" structure). But from the perspective of the learners, who do not know the standard explanation of the magnetic pendulum, the plaster bowl is not inherently an analogy. To be an analogy, one needs to know the structure that is common to the standard explanation of the magnetic pendulum and to the plaster bowl. But the structure relevant to an explanation of the magnetic pendulum is the (by the *teacher*) intended learning *outcome*. These students therefore are in the position of Rorty's poets, who do not know the purpose of the language until they have completely evolved it. It is not surprising, therefore, that any reasonable task outcome is as good as any other. The students also are in the same predicament as the fish biologists in Chapters 4 and 5. To know the deep structure of the magnetic pendulum phenomenon they need to see the relevant deep structure in the plaster bowl; and to articulate the relevant deep structure in the plaster bowl they need to know the relevant structure of the magnetic pendulum.

The artist Alberto Giacometti apparently once said, "'I do not know precisely what I see. It is too complex. So one has to attempt copying the visible to account for what one sees'" (Waldenfels 1999, p. 102). The artist not only exhibits the visible but also enables himself and others to see what there is to be seen. That is, in and through the artistic work, the artist comes to see rather than exhibits what is already seen. In a similar way, learning something new that is not already given in its entirety within existing knowledge means coming to see the invisible, that is, coming to see that which is invisible and therefore cannot be envisioned, aimed at, and worked towards. But how does something new reveal itself that has been invisible because it is alien? In dictionaries we can find definitions of the alien as the foreign and strange, far removed and inconsistent with our current understanding. An analysis of the process of seeing the invisible in the work of Wassily Kandinsky suggests that the new is related to the human capacity to make (draw) unforeseen lines before actually imagining and conceiving them (Henry 2005). This creation of a new idea from unanticipated and unforeseen gestural lines is apparent in the following fragment from the physics lecture in a third-year undergraduate course on thermodynamics that also features in Chapter 7.

The professor had prefaced this part of the lecture by saying that they ("we") could learn more about the phenomenon of cooling by means of adiabatic demagnetization as part of the magnetocaloric effect by looking at a graph. He slowly unfolds this part of the lecture, with a lot of pausing, gazing at the diagram as if he were pondering what comes next. Once all is said and done, the professor gazes at the diagram and says to the students, "I think I told you already that there is something wrong with this picture." In the fuller analysis presented in Chapter 7 I suggest that we can see in this lecture segment the evolution of communication and thinking (see also Roth 2010d). I suggest that the lecture content does not exist as a cognitive structure in the mind of the professor but rather, he develops the lecture content from some general unarticulated idea that concretizes itself as he communicates; and his communication evolves as his idea concretizes itself. In the following fragment, we see how a new line for the diagram emerges from hesitating hand movements.

After the announcement of a different way of understanding the magnetocaloric effect, he first produces a temperature (T)–entropy (S) coordinate system and draws two curves. One curve represents a zero magnetic field ($B = 0$) the other one a turned-on magnetic field ($B \neq 0$) (Fig. 13.4). The professor then begins an explanation, in which a substance is subjected to a magnetic field at constant temperature. He represents this in his diagram as a vertical line (i.e., constant temperature) from the $B = 0$ curve to the $B \neq 0$ curve. This is where Fragment 13.3 picks up. The professor gazes at what he has done so far, saying nothing (Fig. 13.4a). He

Fig. 13.4 A sequence of gestures prefigures the actual articulation of where a line needs to be placed in the explanation of a graphical representation of the magneto-caloric effect that a physics professor attempts to explain in turns 29–31 of Fragment 13.3.

moves his chalk-holding right hand closer as if he wanted to draw something (Fig. 13.4b), but then hesitates and utters "but when you" (turn 30). He pauses, produces a drawn-out interjection ("uh::") and then looks up and down in the diagram (Fig. 13.4c) while uttering "that when you when you then uh adiabatically" (Fig. 13.4d) (turn 31). As the transcript shows, he repeats himself, produces an interjection, and utters the term "adiabatically" slowly by drawing out individual syllables (turn 31).

Fragment 13.3 (Fragment 7.1b)

```
29  (1.79)
30  a:nd when you; (0.95) uh:::
31  that when you when you then uh a:dia:^batical[ly::] (0.53)
                                                 [((gesture hori-
    zontal))
32  demagnetize it, (0.44) it uh::::
33  (0.30) [y its 'that]
           [((draws line horizontal
34  [and so] its temperature is lowered.
    [((horizontal gesture
```

At the very end of uttering the adjective "adiabatically," the professor's hand begins to move horizontally continuing in the pause (Fig. 13.4e–g) ending the movement just before the next word "demagnetize" (turn 32). There is another 0.44-second pause, a drawn-out interjection ("uh:::") that lasts over 0.5 seconds, and another pause. Then, simultaneously with uttering "y it's that" he draws with his chalk a horizontal line that begins at the lower intersection of the vertical chalk line until it intersects with the $B = 0$ curve. In this, the chalk line comes to reproduce the line that we recognize as the one that the tentative movement had traced out earlier.

We can see here that the tentative movement arising from a situation marked by pauses and apparent hesitant and hesitating search for a next step in the demonstration constitutes a next opportunity. It is out of the hand movement that a new thought movement evolves. A new line emerges from the design of an unanticipated line traced out by the hand-arm movement across the existing diagram. To understand this production of the line, therefore, we must not seek an explanation in the visible. Rather, it is from the perspective of the invisible (unknown) that we have to think the visible (known). But as soon as the trace has made a new possibility visible, the invisible has withdrawn from that which is now visible usurps the situation, manifesting itself as if it had been there before. This therefore is precisely the same type of event that has occurred to me with the twin silos (Chapter 2), where something has emerged from the invisible and then manifests itself in a manner as if it had existed despite of me in the objective world.

From the perspective of the knowing subject, the new line becomes visible in the hand movement so that we have to understand the line not from the a posteriori perspective of what is visible but from the perspective of the invisible, the alien that cannot be anticipated precisely because it is unknown. The invisible is *invisable*. The hand movement provides the gaze with a possible trace. It transmits to the "gaze its own movement as the imprescriptible condition to be able, precisely, to follow with one's eyes the ascent of the unseen in it to the visible" (Marion 2004, p. 43). The new possibility shows itself from itself in the hand movement, having arisen from the strictly unseen to become a "purely visible ectype" (p. 38), an archetype that has come to stand out (*ek-*), has become *ek*static. If the professor had already imagined or conceived of the line that he draws in turn 33, all the gesturing between turns 29 and 31 would have been unnecessary and gratuitous. There is in fact research attributing to gestural hand and arm movements no other role than producing activation potentials for linguistic consciousness to be mobilized (e.g., Roth 2002). But who then produces the gestures if the conscious mind does not do it, if the conscious mind requires the gestures to become conscious of something? How can the absolutely new emerge other than from the unknown, which recedes and therefore remains inaccessible with the coming of the known?

Thinking learning in terms of the invisible allows us to understand future knowledge as the unforeseeable – that which after the fact will be the newly known. The unseen and unforeseeable cannot be anticipated or aimed at, so that any learning, from this perspective, is understood as a donation. The newly known has given itself to the knowing subject, who could not have aimed at what it now knows because the latter has arisen from the unknown and (in advance) unknowable.

When Is Construction?

We now have all the elements for understanding the range of application of the construction metaphor. It operates and works in the purely visible, among objects and things that are already seen. What is visible becomes resource in the constructive effort because even the end result can be anticipated, that is, seen; and because it can be seen it can be anticipated, it is foreseeable. On the other hand, the constructivist metaphor is a non-viable approach for understanding learning from the perspective of the learner when it comes to the learning of the (radically) new (concepts). With this articulation, we are then precisely in the situation that Rorty describes in the quote on page 247, which says that even creative people cannot anticipate what they want to do until they have a language for doing it. For the painter, this language consists of the perceptual types – color and form – that are used as resources in making a painting. For the professor in the preceding episode, it is a gesture that brings into the visible the possibility of a line that heretofore was invisible. Different lines are possible, as shown in other parts of the lecture, where the professor gestures different possibilities before fixing one of them as chalk line on the chalkboard.

This analysis shows how the visible emerges from the previously invisible, and, therefore, the knowable from the unknown. To see what makes itself be seen, to see what the newly seen gives requires standing back so that the new can stand out, be *ek*static. Such standing back occurs over and again in the physics lecture, including in the instant featured in the preceding section (Fig. 13.4a) when the professor moves toward the chalkboard only after having starred at it for a while from afar. He does so again after the newly envisioned and drawn line has been fixed on the chalkboard. Such "stepping back indicates above all that the one who has just physically put the color or lines on the surface of the canvas did not know, at the moment of effecting it, what he did" (Marion 2004, p. 44). We do not have to stand back – "to take a better look" – if we already know what there is to be seen and known. The producer of a line does step back "since, in order to see its effect, he mast detach himself from his work, in order to learn, afterward, what visible appears there" (p. 44). That is, learning occurs by reflecting on what has done, and this doing has occurred without prior knowledge of what it was. At the moment of reflection, the elements that constitute the seen and known are at the verge of being. This is where the construction metaphor might begin to have its place. Given the possibility of the new line prefigured in the gesture, the professor can begin to reflect upon it. But prior to the gesture, there was not yet anything to reflect about or with. Prior to the gesture, the construction metaphor is inoperative. Emergence is a much better metaphor at this stage.

This perspective takes us further in the understanding of the subject of learning. It is evident that the radically new cannot be intended. It gives itself to the learner, who receives the new knowledge as a gift, who is a welcoming host, much in the same way that I received the twin silos as a gift, being a welcoming host to this fact of the world. Even the most advanced painters have to admit that "despite all his work, it is not he who put in the work on the painting but the painting itself, which . . . opens itself to the visible on its own initiative" (p. 44). Paraphrasing Marion we therefore can say that *to learn means to await donation.* Whatever labor the learner engages in provides opportunities for the visible to emerge from the invisible and figure (known) from the ground (unknown) that constitutes the transition between the visible and invisible.

Returning to another example provided earlier of the Dalmatian (Fig. 3.1) allows us to think up an image of the relationship between the learner and the thing learned. Constructivism tells us that the learner constructs any new knowledge based on what it already knows; this knowledge is tested for fit in the world. When we consider the experience of the Dalmatian that all of a sudden rises from the invisible – unseen and therefore unforeseen – the constructivist metaphor no longer holds up. When learners, for whatever reason and however instigated, engage with the ink splotches, it is not merely their own activity from which the unforeseen arises into the visible. Rather, they eyes move guided by the lines and colored splotches. The outside world guides the eyes, so that their movement is not entirely due to themselves. Once visible, the Dalmatian has given itself in the sense that it has been the drawing that guided the eyes allowing the newly seen to become visible. In fact, it is movement that is at the origin of the newly seen just as it was movement that allowed a new line to emerge in the case of the professor. In the case of the Dalmatian, it is the movement of the eyes that traces out the possibility of that which comes to be seen and shows itself to have been there all along. This movement has two aspects, one that follows the outlines and the color splotches and the other that moves away from them. Without the continuous movement to and away from a potential figure, no figure can emerge; but the figure itself emerges from the tracing. Whatever appears as figure, whatever becomes visible, does so by standing out – i.e., is an *ectype* – over and against everything else that remains ground. This ground stands between the invisible and the visible, it entails the visible (figure), for, as I show in Chapter 3, without the eye moving between figure and ground no figure can emerge. That is, coming to see the Dalmatian involves the eye movements that trace the figure and the eye movements that make the trace stand against the ground. From these two movements, the "ectypes triumph over the unseen by escaping from the background" (Marion 2004, p. 39). As a result, perception itself is reconfigured and finds the Dalmatian every time the eye is directed toward the figure. Learning, therefore, is the triumph of the known, the concept, against the unknown: *the known can be known only against the unknown,* just as the figure can be seen only against the ground. The ground itself shows nothing but nevertheless is a requirement for the figure, which therefore always already entails the ground (which therefore always *trails* the figure). But this ground is not added to the figure, background is not added to ectypes: "The ectypes originate as from their most intimate unseen and, henceforth, the most foreign" (p. 38). This ground constitutes the intermediate area, the space, *khora*, between the strictly unseen and that which is seen in the light of the clearing.

Orders of the Visible

The phenomenology of the visible may provide us with a much better metaphor than constructivism because it asks us to consider the invisible. Moreover, a genetic (historical) perspective requires us to take the perspective of the invisible as a metaphor for how the cultural and individual known has emerged unforeseeably from the unknown (invisible) via the potentiality that comes with the ground (practically known). Perception has two dimensions: *recognizing perception* and *seeing perception* (Waldenfels 1999). "Recognizing perception" describes those instances where we see something *as* something. The something is already given and we come to see it as something – for example, students see a book on a table *as* the result of the equilibrium of two opposing forces. This type of seeing is heteronomous, for the students are entirely familiar with the phenomenon and they are familiar with the idea of forces. The students come to see what they already know; they can easily relate it to another situation such as arm wrestling where their hand does not move when the opponent's and their own forces are equal. This is the case that the constructivist metaphor is able to describe, the seeing of something as something. In this instance, everything is given, the thing to be known and the knowledge; the learning consists in using the known and applying it to a new already given (visible) situation. The order is that of the known. For example, after students have seen an explication of Newton's third law for the case of a book resting on a table (Fig. 13.5a), they are likely able to construct an answer to the question of applying the law to the case of arm wrestling for the situation where the two hands and arms remain do not move (Fig. 13.5b). The laws of perception do not derive from the newly seen but from that which is already known. This is the Platonic case, where *ideas* – Gr. *idéa*, look, semblance, form or *eidós*, form – refer to that which is seen *as* seen. Kant writes that we see what the mental concepts prefigure as that which can be known; and his approach characterizes learning theories to this day.

A very different issue is the case of "seeing perception." In this case, perception is autonomous because the laws of seeing derive from the ground from which the seen (figure) springs forth. A seeing perception is possible only when the what and how of perception does not already exist prior to the arrival of the newly seen (content, mode), that is, when the new is really new from the perspective of the learner. In seeing perception, the what and how of seeing are shaped by the ground from which the invisible shows itself only to retreat. It is in this interaction that the newly seen and the new ways of seeing emerge. Thus, for example, once we see the new we recognize it because our ways of seeing – our perceptual modes – have emerged simultaneously. Once we have learned to see the Dalmatian in Fig. 3.1, we always see it; we no longer have to look for a more or less extended amount of time until we see it. We do not have to interpret the perceptual field. In the sciences, for example, students may see for a first time the fact that a phase transition takes time and come to understand it as an instance of a change in entropy; and they understand the change of entropy from this observed fact. Here,

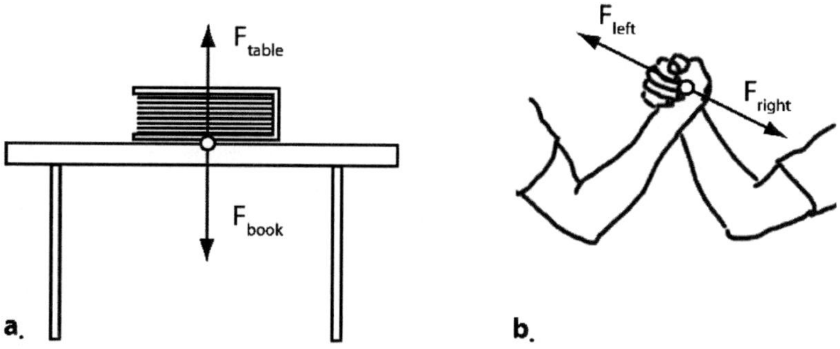

Fig. 13.5 **a**. A standard example to illustrate Newton's third law. **b**. Example of arm wrestling to which Newton's third law can be applied.

too, the what and how of seeing co-emerge. This new way (mode) of seeing and the new what of seeing can then be redeployed in recognizing seeing.

In my two paradigmatic examples, the twin silos and the Dalmatian, the two modes of perception rapidly follow each other. First, a newly visible comes to be seen when the setting (landscape, ink on white paper) shapes a new way of seeing *for this* context. As soon as the newly visible has arisen from the ground, recognizing perception assigns the seen to something already known: we see the figure *as* twin silos or *as* Dalmatian (dog). The invisible is part and partial of the visible world; and between them lies the ground as the spacing, *khôra*, where the invisible crosses into the visible. Learning, as I suggest in and with this book, has an essentially pathic ground, which we need to theorize as such in order to understand how the known can emerge from the unknown.

By understanding *seeing perception* in terms of the new what and how that are enabled in the relation between the subject and its world, with the emergence of a new perceptual and behavioral order, we also return to the role of movement in the constitution of the world as it appears to us. We have seen how initially random movements come to form corporeal kinetic archetypes, kinetic melodies, that can be replayed given some intentional impulse, energy, and intensity. This repeatability exhibits itself in stable forms, as we see in Chapter 3, where the perception of two-dimensional lines as cubes is shown to arise from such kinetic archetypes. Movement is also at the origin of the newly seen in the case of the Dalmatian dog; and it is at the origin of the new lines drawn by the physics professor in interaction with the chalkboard drawing. Here we see how the actual drawing of the line repeats a movement already enacted in the gesture. This points us to the role of repeatability in consciousness and learning.

Repeatability, however, throws perception outside itself, makes it *ek*static. In fact, it does so in two ways. First, repeatability allows us to find lines or colors again and again, which leads to the presence of an object that is stable in and through this repeatability. This repeatability arises both from the structure of the feeling/felt body and the world, the former's movements being shaped by the structures of the latter. Second, precisely because seeing perception is repeatable, because it repeats a structure in the world, it also is repeatable between (human) subjects, who, given their structurally equivalent living/lived bodies, also may repeat the lines of the world in their own movements. Repeatability therefore can also be understood as a doubling of the gaze: in my gaze, being able to repeat the seen, the gaze of the other is already implied. Thus, "without the repeatability *as something* and the figuration/withdrawal *of something* we would literally see *no thing*, that is, see nothing at all" (Waldenfels 1999, p. 155).

Repeatability, *ek*stasis, and sharedness therefore all are integral part of the objectivity of the world and the inherent shareability of its structure. Such structure requires *ek*stasis, which is born and guaranteed by repeatability. That is, my living/lived bodily Self inherently shares its world with the Other and does not constitute an isolated, solipsist, disembodied mind that the constructivist metaphor paints. My living/lived body concretizes life in the same way that your living/lived body does; my subjectivity is a concrete realization of subjectivity as such. This sharedness returns us to the anonymity of perception, which lies outside of ourselves – the phenomena we come to see and "see" are given to us.[8] Perception therefore never is completely in control of and over itself – or we would not be able to escape the learning paradox. We are able to move beyond the currently known and exceed its structural level precisely because of the absolutely new perceptual contents and processes that *seeing perception* can produce.

The anonymity of perception is not so difficult to understand from the perspective of cultural-historical activity theory. Thus, it has been suggested and empirically shown that higher cognitive functions emerge from, are the results of, and have been first social relations (Vygotsky 1978). Seeing the world in the way of an expert – as reading or speaking – has been a social relation first. This, therefore, makes seeing something that can be taught by producing the kinds of social relations in which seeing first emerges. This, again, is not a strange idea at all, as every parent will remember the first instances of reading with their child or children. An integral part of these first readings are images, allowing parents and children to engage in a pedagogy of seeing, where the children learn to see the world in particular ways. Their processes and contents of seeing co-emerge in and from social relations. Perceiving, as reading or speaking, are public events that are learned in and through participation in such public events. They are external first and therefore constitute something anonymous and alien, in the same way that the letters are alien to the small child, who might wonder how the ink splotches are related to what adults say and how to yield anything repeatable at all.

Perceptual processes are invisible, making the work of seeing unseen. This is here in the case of the truncated wedge/hallway, where the eyes engage in movements that make the phenomena seen in the way they are seen. Yet this movement is invisible in normal perception. That is, there is another form of doubling of the

---

[8] An anthropology of reading in science, for example, is not concerned with this or that individual's reading – e.g., the "meaning" a person "constructs" of an online text – but with the witnessable features of an anonymous praxis of reading observable in and realized by any individual's reading of a (science) text (e.g. Roth 2010a).

gaze, which both makes visible what is seen and invisible its own seeing. Because the processes of seeing tend to be inherently invisible, only their effects can be seen in the actual participation with practitioners. Thus, that which distinguishes an expert from a novice social (sociological) analyst cannot be shown explicitly but shows itself obliquely when the novice participates with the expert in the analysis of actual cases (Bourdieu 1992). This invisibility and anonymity of perception also lie at the origin of the dehiscence between living/lived work and its accounts, which may exist in the things produced (e.g., what is perceived) or in the instruction (e.g., what to do to). Precisely because this work is invisible, it withdraws itself from repeatability and therefore also from representation. It is for this reason that no process can be instructed directly and without remainder (see Chapter 4). The instructed have to find in their own actions that which the instruction teaches them to see or do. Together with the seen/done also emerge the corporeal kinetic forms that lead to repeatability and historicity, and, with these, the objectivity of the seen/done.

Seeing perception is concerned with the alien, which, because of its nature, relates it to being affected and therefore to affect. The alien is the source of desires, fears, and threats. It is only "in disregard of such affective relations that the image reduces itself to neutral, cognitive image schemas, which are devoid of any image aura" (Waldenfels 1999, p. 135). Learning, inherently dealing in and with the alien, is affective, with the desires to know, fears to fail, or threats to identity.

The visible does not go without the invisible: It is in the perceptual ground that the invisible crosses over into the visible to become figure. There is therefore never figure without ground. But we cannot understand the visible/figure from the perspective of the visible/figure, because in this case we would reason teleological ways, the visible/figure being prefigured in the world (realism) or in the genes (innatism). The visible is not constructed, because the disembodied mind (perception) would never know whether it is awake or in a dream. But we do *feel* the difference between images of the world and images in the dream. It is therefore from the invisible and ground that we have to think the visible figure. In the same way, we have to think the known from the perspective of the unknown, the familiar from the perspective of the alien. It is precisely in the spacing between the two extremes – *khôra*, the dehiscence of Being and beings – that learning and development occur. This learning and development cannot be understood in terms of intentionality and the construction metaphor, that is, in terms of agency, but it requires a beginning with the pathic and the passions. This is why *passibility* constitutes the (off-) limits of the constructivist metaphor.

References

Abraham, N. (1995). *Rhythms: On the work, translation, and psychoanalysis*. Stanford, CA: Stanford University Press.

Anderson, J. R., Greeno, J. G., Reder, L. M., & Simon, H. A. (2000). Perspectives on learning, thinking, and activity. *Educational Researcher, 29* (4), 11–13.

Aristotle. (1986). *De anima*. Harmondsworth: Penguin.

Bakhtin, M. M. (1981). *The dialogic imagination*. Austin, TX: University of Texas Press.

Bakhtin, M. M. (1984). *Rabelais and his world*. Bloomington, IN: Indiana University Press.

Bakhtin, M. M. (1993). *Toward a philosophy of the act*. Austin, TX: University of Texas Press.

Bakhtine, M. [Volochinov, V. N.] (1977). *Le marxisme et la philosophie du langage: essai d'application de la méthode sociologique en linguistique*. Paris: Les Éditions de Minuit.

Bereiter, C. (1986). Toward a solution of the learning paradox. *Review of Educational Research, 55,* 201–226.

Blanchot, M. (1959). *Le livre à venir*. Paris: Gallimard.

Bourdieu, P. (1980). *Le sens pratique*. Paris: Les Éditions de Minuit.

Bourdieu, P. (1992). The practice of reflexive sociology (The Paris workshop). In P. Bourdieu, & L. J. D. Wacquant, *An invitation to reflexive sociology* (pp. 216–260). Chicago: University of Chicago Press.

Bourdieu, P. (1997). *Méditations pascaliennes*. Paris: Seuil.

Brüstle, C., Ghattas, N., Risi, C., & Schouten, S. (2005). Zur Einleitung: Rhythmus im Prozess. In C. Brüstle, N. Ghattas, C. Risi, & S. Schouten (Eds.), *Aus dem Takt: Rhythmus in Kunst, Kultur und Natur* (pp. 9–27). Bielefeld: Transcript.

Chi, M. T. H. (1992). Conceptual change within and across ontological categories: Examples from learning and discovery in science. In R. Giere (Ed.), *Cognitive models of science: Minnesota studies in the philosophy of science* (pp. 129–186). Minneapolis, MN: University of Minnesota Press.

Collins, R. (2004). *Interaction ritual chains*. Princeton, NJ: Princeton University Press.

Csikszentmihalyi, M. (1988). The flow experience and its significance for human psychology. In M. Csikszentmihalyi & I. S. Csikszentmihalyi (Eds.), *Flow in consciousness* (pp. 15–35). Cambridge: Cambridge University Press.

Davidson, D. (1986). A nice derangement of epitaphs. In E. Lepore (Ed.), *Truth and interpretation* (pp. 433–446). Oxford: Blackwell.

de Man, P. (1983). *Blindness and insight*. Minneapolis, MN: University of Minnesota Press.

Derrida, J. (1967). *De la grammatologie*. Paris: Les Éditions de Minuit.

W.-M. Roth, *Passibility*, Classics in Science Education 3,
DOI 10.1007/978-94-007-1908-8, © Springer Science+Business Media B.V. 2011

Derrida, J. (1968). La différance. In *Tel Quel: Théorie d'ensemble* (pp. 41–66). Paris: Éditions du Seuil.
Derrida, J. (1972). *Marges de la philosophie*. Paris: Les Éditions de Minuit.
Derrida, J. (1974). *Glas*. Paris: Galilée.
Derrida, J. (1984). *Otobiographies: L'enseignement de Nietzsche et la politique du nom proper*. Paris: Galilée.
Derrida, J. (1991). *Donner le temps: 1. La fausse monnaie*. Paris: Galilée.
Derrida, J. (1992). *Points de suspension: entretiens*. Paris: Galilée.
Derrida, J. (1995). *The gift of death*. Chicago: University of Chicago Press.
Derrida, J. (1996a). *Le monolinguisme de l'autre ou la prothèse d'origine*. Paris: Galilée.
Derrida, J. (1996b). *Résistances de la psychanalyse*. Paris: Galilée.
Derrida, J. (2007). *Psyche: Inventions of the other vol. 1*. Stanford, CA: Stanford University Press.
Edwards, D., & Potter, J. (1992). *Discursive psychology*. London: Sage.
Franck, D. (1981). *Chair et corps: Sur la phénoménologie de Husserl*. Paris: Les Éditions de Minuit.
Freud, S. (1999). *Gesammelte Werke Band XVII*. Frankfurt: Fischer.
Gadamer, H-G. (1996). *The enigma of health: The art of healing in a scientific age*. Stanford, CA: Stanford University Press.
Garfinkel, H. (1967). *Studies in ethnomethodology*. Englewood Cliffs, NJ: Prentice-Hall.
Garfinkel, H. (1996). Ethnomethodology's program. *Social Psychology Quarterly, 59*, 5–21.
Garfinkel, H., Lynch, M., & Livingston, E. (1981). The work of a discovering science construed with materials from the optically discovered pulsar. *Philosophy of the Social Sciences, 11*, 131–158.
Garfinkel, H., & Sacks, H. (1986). On formal structures of practical action. In H. Garfinkel (Ed.), *Ethnomethodological studies of work* (pp. 160–193). London: Routledge & Kegan Paul.
Greeno, J. G., Collins, A. M., & Resnick, L. B. (1996). Cognition and learning. In D. Berliner & R. Calfee (Eds.), *Handbook of educational psychology* (pp. 15–41). New York: MacMillian.
Gregory, R. L. (1994). *Even odder perceptions*. London: Routledge.
Gurwitsch, A. (2010). *The collected works of Aron Gurwitsch (1901–1973) vol. III – The field of consciousness: Phenomenology of theme, thematic field, and marginal consciousness*. Dordrecht, The Netherlands: Springer.
Hegel, G. W. F. (1979a). *Werke Band 3: Phänomenologie des Geistes*, Frankfurt: Suhrkamp.
Hegel, G. W. F. (1979b). *Werke Band 8: Enzyklopädie der philosophischen Wissenschaften im Grundrisse*. Frankfurt: Suhrkamp.
Heidegger, M. (1954). *Was heißt Denken?* Tübingen: Max Niemeyer.
Heidegger, M. (1977). *Sein und Zeit*. Tübingen: Max Niemeyer.
Heidegger, M. (2000). *Gesamtausgabe Band 7: Vorträge und Aufsätze*. Frankfurt: Vittorio Klostermann.
Henry, M. (1965). *Philosophie et phénoménologie du corps: Essai sur l'ontologie biranienne*. Paris: Presses Universitaires de France.
Henry, M. (1985). *Généalogie de la psychanalyse*. Paris: Presses Universitaires de France.
Henry, M. (1999). Material phenomenology and language (or, pathos and language). *Continental Philosophy Review, 32*, 343–365.
Henry, M. (2000). *Incarnation: Une philosophie de la chair*. Paris: Seuil.
Henry, M. (2003). *Phénoménologie de la vie I: De la phénoménologie*. Paris: Presses Universitaires de France.

Henry, M. (2004). *Phénoménologie de la vie III: De l'art et du politique*. Paris: Presses Universitaires de France.
Henry, M. (2005). *Voire l'invisible: Sur Kandinsky*. Paris: Presses Universitaires de France.
Holzkamp, K. (1979). Zur kritisch-psychologischen Theorie der Subjektivität II: Das Verhältnis individueller Subjekte zu gesellschaftlichen Subjekten und die frühkindliche Genese der Subjektivität. *Forum Kritische Psychologie, 5*, 7–46.
Holzkamp, K. (1983). *Grundlegung der Psychologie*. Frankfurt: Campus.
Holzkamp, K. (1992). Die Fiktion administrativer Planbarkeit schulischer Lernprozesse. In K.-H. Braun & K. Wetzel (Eds.), *Lernwidersprüche und pädagogisches Handeln* (pp. 91–113). Marburg: Verlag Arbeit und Gesellschaft.
Holzkamp, K. (1993). *Lernen: Subjektwissenschaftliche Grundlegung*. Frankfurt: Campus.
Hsu, P.-L. (2010). Thinking dialogically about thought and language. In W.-M. Roth (Ed.), *ReUniting sociological and psychological perspectives* (pp. 155–165). Dordrecht, The Netherlands: Springer.
Husserl, E. (1913). Ideen zu einer reinen Phänomenologie und phänomenologischen Philosophie. In *Jahrbuch für Philosophie und phänomenologische Forschung, I*, 1–323. Halle a.d. Saale: Max Niemeyer.
Husserl, E. (1939). Die Frage nach dem Ursprung der Geometrie als intentionalhistorisches Problem. *Revue internationale de philosophie, 1*, 203–225.
Husserl, E. (1991). *Cartesianische Meditationen und Pariser Vorträge*. Dordrecht, The Netherlands: Kluwer Academic Publishers.
Husserl, E. (1980). *Vorlesungen zur Phänomenologie des inneren Zeitbewußtseins*. Tübingen: Max Niemeyer.
Husserl, E. (1991). *On the phenomenology of the consciousness of internal time (1893–1917)*. Dordrecht, The Netherlands: Kluwer Academic Publishers.
Husserl, E. (2006). *Späte Texte über Zeitkonstitution (1929–1934): Die C-Manuskripte*. Dordrecht, The Netherlands: Springer.
Jordan, K., & Lynch, M. (1993). The mainstreaming of a molecular biological tool: A case study of a new technique. In G. Button (Ed.), *Technology in working order: Studies of work, interaction, and technology* (pp. 162–178). London: Routledge.
Jordan, K., & Lynch, M. (1998). The dissemination, standardization and routinization of a molecular biological technique. *Social Studies of Science, 28*, 773–800.
Kant, I. (1956a). *Werke Band II: Kritik der reinen Vernunft*. Wiesbaden: Insel.
Kant, I. (1956b). *Werke Band IV: Schriften zur Ethik und Religionsphilosophie*. Wiesbaden: Insel.
Kant, I. (1960). *Werke Band I: Vorkritische Schriften bis 1768*. Wiesbaden: Insel.
Kant, I. (1964). *Werke Band VI: Schriften zur Anthropologie, Geschichtsphilosophie, Politik und Pädagogik*. Frankfurt: Insel.
Kuhn, T. S. (1970). *The structure of scientific revolutions* (2nd ed.). Chicago: University of Chicago Press.
Lazarus, R. S. (1989). Constructs of the mind in mental health and psychotherapy. In A. Freeman, K. M. Simon, L. E. Beutler, & H. Arkowitz (Eds.), *Comprehensive handbook of cognitive therapy* (pp. 99–122). New York: Plenum.
Leonard, W. H. (1989). Using inquiry laboratory strategies in college science courses. *Research Matters 8904*. Accessed October 27, 2009 at, http://www.narst.org/publications/research/inquiry.cfm
Leont'ev, A. N. (1981). *Problems of the development of the mind*. Moscow: Progress.
Leontjew, A. N. (1982). *Tätigkeit, Bewusstsein, Persönlichkeit*. Köln: Pahl-Rugenstein.
Levinas, E. (1971). *Totalité et infini: essaie sur l'extériorité*. La Haye: Martinus Nijhoff.
Levinas, E. (1990). *Autrement qu'être ou au-delà de l'essence*. Paris: Librairie Générale Française.
Luria, A. R. (1976). *The working brain*. New York: Basic Books.

Lynch, M., Livingston, E., & Garfinkel, H. (1983). Temporal order in laboratory work. In K. D. Knorr-Cetina & M. Mulkay (Eds.), *Science observed: Perspectives on the social study of science* (pp. 205–238). London: Sage.

Maine de Biran, P. (2001). *Œuvres tome vii/2: Essai sur les fondements de la psycholgie*. Paris: Librairie Philosophique J. Vrin.

Marion, J.-L. (2004). *The crossing of the visible*. Stanford, CA: Stanford University Press.

Marx, K., & Engels, F. (1969). *Werke Band 3: Die deutsche Ideologie*. Berlin: Dietz.

Masciotra, D., Roth, W.-M., & Morel, D. (2007). *Enaction: Toward a Zen mind in learning and teaching*. Rotterdam: Sense.

Merleau-Ponty, M. (1945). *Phénoménologie de la perception*. Paris: Gallimard.

Merleau-Ponty, M. (1964). *Le visible et l'invisible*. Paris: Gallimard.

Merleau-Ponty, M. (1996). *Sens et non-sens*. Paris: Gallimard.

Meshcheryakov, A. (1979). *Awakening to life: On the education of deaf-blind children in the Soviet Union*. Moscow: Progress.

Ministry of Education British Columbia (MoE BC). (2008). Science grade 10: Integrated resource package 2008. Victoria, BC: Author. Accessed August 6, 2010 at, www.bced.gov.bc.ca/irp/sci102008.pdf

Nancy, J-L. (1993a). Éloge de la mêlée. *Transeuropéenne, 1*, 8–18.

Nancy, J-L. (1993b). *The birth to presence*. Stanford, CA: Stanford University Press.

Nancy, J-L. (2000). *Being singular plural*. Stanford, CA: Stanford University Press.

Nancy, J-L. (2006). *Corpus*. Paris: Éditions Métailié.

Nancy, J-L. (2007). *Tombe the sommeil*. Paris: Galilée.

Nancy, J-L. (2008). *The discourse of the syncope: Logodaedalus*. Stanford, CA: Stanford University Press.

Naville, E. (1874). *Maine de Biran: sa vie et ses pensées*. Paris: Librairie Académique.

Nöth, W. (1990). *Handbook of semiotics*. Bloomington, IN: Indiana University Press.

Pascual-Leone, J. (1996). Vygotsky, Piaget, and the problems of Plato. *Swiss Journal of Psychology, 55*, 84–92.

Piaget, J. (1967). *Six psychological studies*. New York: Random House.

Piaget, J. (1970). *Genetic epistemology*. New York: W. W. Norton.

Piaget, J. (1981). *Intelligence and affectivity: Their relationship during child development*. Palo Alto, CA: Annual Review Press.

Piaget, J., & Inhelder, B. (1967). *The child's conception of space*. London: Routledge & Kegan Paul.

Pines, A. L. (1985). Toward a taxonomy of conceptual relations. In L. H. T. West & A. L. Pines (Eds.), *Cognitive structure and conceptual change* (pp. 101–115). New York: Academic Press.

Pintrich, P. R., Marx, R. W., & Boyle, R. A. (1993). Beyond cold conceptual change: The role of motivational beliefs and classroom contextual factors in the process of conceptual change. *Review of Educational Research, 63*, 167–199.

Posner, G. J., Strike, K. A., Hewson, P. W., & Gertzog, W. A. (1982). Accommodation of a scientific conception: Toward a theory of conceptual change. *Science Education, 66*, 211–227.

Rawls, A. W. (1989). Language, self, and social order: A reformulation of Goffman and Sacks. *Human Studies, 12*, 147–172.

Ricœur, P. (1990). *Soi-même comme un autre*. Paris: Seuil.

Rimbaud, A. (1951). *Œuvres complètes*. Paris: Gallimard.

Rorty, R. (1979). *Philosophy and the mirror of nature*. Princeton, NJ: Princeton University Press.

Rorty, R. (1989). *Contingency, irony, and solidarity*. Cambridge: Cambridge University Press.

Roth, W.-M. (1998). Starting small and with uncertainty: Toward a neurocomputational account of knowing and learning in science. *International Journal of Science Education, 20*, 1089–1105.

Roth, W.-M. (2002). Gestures: Their role in teaching and learning. *Review of Educational Research, 71*, 365–392.

Roth, W.-M. (2004). A phenomenological anthropology of texts and literacy. In A. Peacock & A. Cleghorn (Eds.), *Missing the meaning: The development and use of print and non-print learning materials* (pp. 253–266). Houndmills, Basingstoke, Hampshire: Palgrave.

Roth, W.-M. (2008). The nature of scientific conceptions: A discursive psychological perspective. *Educational Research Review, 3*, 30–50.

Roth, W.-M. (2009). Radical uncertainty in scientific discovery work. *Science, Technology & Human Values, 34*, 313–336.

Roth, W.-M. (2010a). An anthropology of reading science texts in online media. *Semiotica, 182*, 409–442.

Roth, W.-M. (2010b). Incarnation: Radicalizing the embodiment of mathematics. *For the Learning of Mathematics, 30*(2), 2–9.

Roth, W.-M. (2010c). *Language, learning, context: Talking the talk.* London: Routledge.

Roth, W.-M. (2010d). Thinking and speaking: A dynamic approach. In W.-M. Roth (Ed.), *ReUniting sociological and psychological perspectives* (pp. 113–143). Dordrecht, The Netherlands: Springer.

Roth, W.-M. (2011). *Geometry as objective science in elementary classrooms: Mathematics in the flesh.* New York: Routledge.

Roth, W.-M., & Bowen, G. M. (2003). When are graphs ten thousand words worth? An expert/expert study. *Cognition and Instruction, 21*, 429–473.

Roth, W.-M., & Duit, R. (2003). Emergence, flexibility, and stabilization of language in a physics classroom. *Journal for Research in Science Teaching, 40*, 869–897.

Roth, W.-M., Gabel, D., Brown, L., & Rice, D. (1992). A combined method of INDSCAL and cognitive mapping for describing changes in students' cognitive structure. *Journal of Educational and Psychological Measurement, 52*, 769–779.

Roth, W.-M., & Lawless, D. (2002). When up is down and down is up: Body orientation, proximity and gestures as resources for listeners. *Language in Society, 31*, 1–28.

Roth, W.-M., McRobbie, C., Lucas, K. B., & Boutonné, S. (1997). The local production of order in traditional science laboratories: A phenomenological analysis. *Learning and Instruction, 7*, 107–136.

Roth, W.-M., & van Eijck, M. (2010). Fullness of life as minimal unit: Science, technology, engineering and mathematics (STEM) learning across the life span. *Science Education, 94*, 1027–1048.

Sapir, E. (1921/2004). *Language: An introduction to the study of speech.* Minneola, NY: Dover.

Sartre, J.-P. (1956). *Being and nothingness*. New York: Gramercy Books.

Sewell, W. H. (1992). A theory of structure: Duality, agency and transformation. *American Journal of Sociology, 98*, 1–29.

Sfard, A. (2008). *Thinking as communicating: Human development, the growth of discourses, and mathematizing*. Cambridge: Cambridge University Press.

Sheets-Johnstone, M. (2006). Essential clarifications of "self-affection" and Husserl's "sphere of ownness": First steps toward a pure phenomenology of (human) nature. *Continental Philosophy Review, 39*, 361–391.

Sheets-Johnstone, M. (2009). *The corporeal turn: An interdisciplinary reader.* Exeter: Imprint Academic.

Silverman, H. J. (1980/1997). *Piaget, philosophy, and the human sciences*. Evanston, IL: Northwestern University Press.

Sørensen, K. H., & Levold, N. (1992). Tacit networks, heterogeneous engineers, and embodied technology. *Science, Technology, & Human Values, 17*, 13–35.

Steffe, L. P. (1991). The learning paradox: A plausible counterexample. In L. P. Steffe (Ed.), *Epistemological foundations of mathematical experience* (pp. 26–44). New York: Springer.

Stoppard, T. (1975). *Travesties: A play.* New York: Grove Press.

Suchman, L. A. (1987). *Plans and situated actions: The problem of human-machine communication.* Cambridge: Cambridge University Press.

Tabachneck-Schijf, H. J. M., Leonardo, A. M., & Simon, H. A. (1997). CaMeRa: A computational model for multiple representations. *Cognitive Science, 21*, 305–350.

Thom, R. (1981). Worüber soll man sich wundern. In K. Maurin, K. Michalski, & E. Rudolph (Eds.), *Offene Systeme II: Logik und Zeit* (pp. 41–107). Stuttgart: Klett-Cotta.

Todorov, T. (1984). *Mikhail Bakhtin: The dialogical principle.* Minneapolis, MN: University of Minnesota Press.

Treagust, D., & Duit, R. (2008). Conceptual change: A discussion of theoretical, methodological and practical challenges for science education. *Cultural Studies of Science Education, 3*, 297–328.

van der Maas, H. L., & Molenaar, P. C. (1992). Stagewise cognitive development: An application of catastrophe theory. *Psychological Review, 99*, 395–417.

Varelas, M., Pappas, C. C., Kane, J. M., Arsenault, A., Hankes, J., & Cowan, B. M. (2008). Urban primary-grade children think and talk science: Curricula and instructional practices that nurture participation and argumentation. *Science Education, 92,* 65–95.

Vološinov, V. N. (1973). *Marxism and the philosophy of language*. Cambridge, MA: Harvard University Press.

von Glasersfeld, E. (1989a). Cognition, construction of knowledge, and teaching. *Synthese, 80*, 121–140.

von Glasersfeld, E. (1989b). Facts and the self from a constructivist point of view. *Poetics, 18*, 435–448.

von Glasersfeld, E. (2001). Scheme theory as a key to the learning paradox. In A. Tryphon & J. J. Vonèche (Eds.), *Working with Piaget: Essays in honour of Bärbel Inhelder* (pp. 141–148). East Sussex: Psychology Press.

von Uexküll, J. (1973). *Theoretische Biology*. Frankfurt: Suhrkamp.

von Weizsäcker, V. (1973). *Der Gestaltkreis: Theorie der Einheit von Wahrnehmen und Bewegen*. Frankfurt: Suhrkamp.

Vygotsky, L. S. (1978). *Mind in society: The development of higher psychological processes*. Cambridge, MA: Harvard University Press.

Vygotsky, L. S. (1986). *Thought and language*. Cambridge, MA: MIT Press.

Vygotsky, L. S. (2002). *Denken und Sprechen.* Weinheim, Germany: Beltz.

Waldenfels, B. (1999). *Sinnesschwellen: Studien zur Phänomenologie des Fremden 3.* Frankfurt: Suhrkamp.

Wall, T. C. (1999). *Radical passivity: Levinas, Blanchot, and Agamben*. Albany: State University of New York Press.

Wittgenstein, L. (1921). Logisch-philosophische Abhandlung. *Annalen der Naturphilosophie, 14*, 185–262.

Wittgenstein, L. (1958). *Philosophische Untersuchungen/Philosophical investigations* (3rd ed.). New York: Macmillan.

Wittgenstein, L. (1968). Notes for lectures on "private experience" and "sense data." *Philosophical Review, 77*, 275–320.

Wittgenstein, L. (2000). *Wittgenstein source: Bergen text edition.* Accessed September 22, 2010 at, URL htpp://www.wittgensteinsource.org.

Woelert, P. (2007). Kant's hands, spatial orientation, and the Copernican turn. *Continental Philosophy Review, 40*, 139–150.

Index

A

Accommodation, 32, 44, 93, 114, 200, 209, 240
Accountability, 26, 28, 39, 73, 84, 87, 95
Action
  situated, 27, 71–73, 75–76, 78–79, 83–85, 138
Actualization, 75, 137, 179
Adhocing, 76
Agency, 12, 15, 17–19, 21, 26–27, 34, 48, 51, 54–58, 68, 105, 139, 153–156, 176–178, 181, 187, 189, 206, 209–210, 212, 216–218, 224, 227, 234–240, 243, 245, 265
Alignment, 27, 197–198, 200–202
Alterity, 20, 108, 112, 152, 235, 238, 240
Anaximander, 24
Anthropology, 180, 264
Apophansis, 36, 48
Archetype, 67, 86, 199, 203, 226, 259, 263
Aristotle, 39, 51, 125, 129
Assimilation, 7, 114
Attunement, 131, 160, 163, 169, 176, 187, 195–197, 199, 201
Auto-affection, 17–18, 21, 46, 52, 54, 56, 67–68, 85–86, 88, 90, 104–105, 230, 232–234, 246, 248
Auto/biography, 105, 108, 138, 145–146, 152, 154
Auto-donation, 48, 68, 86, 232, 255
Auto-movement, 68
Autonomy, 6, 106, 108, 128, 131, 153, 243, 262
Awareness, 17, 30–31, 38, 88, 103, 128, 138, 142, 213, 220, 228

B

Bakhtin, M. M., 3, 9, 23, 27, 36, 52, 71, 75, 106, 108, 111, 119, 123, 131, 141, 143, 145, 149, 157, 161–162, 185, 196, 205–206, 212, 221–223, 233, 239–241
Beat, 117, 134, 198, 202, 208
Bedeutung, 159, 163–164
Befindlichkeit, 160, 197
Bewußtsein, 246
Birth, 27, 34, 42, 132, 212, 214, 250, 252
Blake, Q., 26
Blanchot, M., 156
Blind spot, 1
Body
  living/lived, 4–5, 8–9, 17–18, 20, 23, 27, 39, 54, 67–68, 74, 80,

88, 90, 92, 96, 98, 122, 128, 134, 138, 177, 179, 182–183, 186–187, 192, 195, 197, 201, 225–226, 228–229, 232–236, 242, 264
movement, 136, 194, 203
orientation, 88, 136, 198–199
position, 88, 136, 190, 197–198, 200
Bourdieu, P., 5, 9, 75, 177–178, 239, 265
Breakdown, 76, 180, 237

C
Causality, 4, 6, 83, 210
Clearing, 14, 69, 219, 261
Cognitive science, 88, 101
Columbus, C., 26, 208
Communication, 11, 43, 88, 107, 112–113, 121, 127–128, 132, 134, 136, 138–139, 144, 155, 163, 167, 169–170, 187, 196–197, 200, 223, 228, 235–236, 240, 257
Competency, 11, 76, 94, 124–125, 149, 180, 196, 249
Consciousness, 3, 6, 8, 21, 23, 27, 30–32, 34–41, 52, 67, 86–89, 91, 101, 103–104, 112, 128, 130, 134, 138, 143–144, 154, 160, 169, 177, 179, 182, 186, 194–196, 198, 205–207, 212–214, 216–223, 225–227, 233–236, 240, 245–246, 254, 259, 263
Contradiction, 80, 107, 125, 237
Control, 15–16, 57–58, 73, 80, 84, 146–148, 150–151, 155, 157, 177, 185–187, 189, 195, 210, 217–218, 264
Copernicus, N., 45, 123, 125, 247
Corporeal-kinetic form, 199, 203
Crisis, 237, 240
Culture
cultural-historical, 8, 10, 14, 16, 38, 90, 109, 111, 119, 122, 131, 144, 157, 160, 169, 171, 194, 196, 199, 207, 227, 237, 245, 255, 264
reproduction, 26, 119, 125, 157, 195–196, 199–201, 241
transformation, 7, 97, 119, 125, 157, 200–201, 241
Western, 119, 218

D
Dalmatian, 26, 49, 255, 261–263
Darstellung, 109
Dasein, 153, 177
Death, 4, 15, 27, 34, 42, 84, 157, 211–212, 214, 249–252
Decision, 17–18, 76, 183, 206, 209–210, 217, 237, 246–247
Deconstruction, 9–10, 161
de/construction, 3
Dehiscence, 23, 27, 35–36, 40, 75, 88, 103–104, 128, 138, 212, 219, 221–223, 265
Deixis, 167
Deontology, 3, 206, 222, 240
Derrida, J., 9, 25, 27, 37–38, 52, 102–103, 105, 107–109, 129, 152–156, 162–163, 167, 171, 183, 206–207, 222–223, 239
Descartes, R., 67, 89
Différance, 27, 35–36, 52, 102–105, 108, 183, 219, 239
Discovery, 23, 95, 102, 142, 208
Disembodiment, 44, 233
Disposition, 16, 26, 178
Donation, 37, 259, 261
Dostoyevsky, F., 131
Duality, 20, 28, 89–90, 92
Dwelling, 40–41, 109, 146, 169

E
Écriture, 27, 37
Ectype, 259, 261
Effacement, 38
Eksistence, 32, 94, 100, 102
Ekstasis, 35
Ekstatic, 20
Embodiment, 3, 5–6, 20, 85, 179, 225–227, 233, 241

Emergence, 3, 8, 13, 18–21, 26, 28, 31, 34, 41, 47–49, 54, 69, 77, 89, 94, 99, 104, 115, 126, 134, 157, 160, 166, 176, 180, 187, 202, 205, 208–210, 213, 220, 222, 226, 229, 232–234, 237, 245, 254–255, 259, 261–265
Emotion, 4–5, 8, 15–16, 20–21, 24, 108, 124, 143, 155–156, 179–182, 185–190, 193–202, 226, 233, 241, 254
Emotional-volitional, 75, 185, 196, 222, 240–241
Empathy, 155, 197
Empiricism, 130
Enactivism, 3, 5, 20, 179, 202, 225–227, 233, 241
Entfernung, 177
Epicycle, 6, 14–15, 179, 181, 254
Epistemology, 5, 7, 12, 17, 20, 23, 29, 39, 43, 85, 179, 195, 206, 224–225, 227, 235, 237, 241
Ereignis, 38–40
Étant, 239
Ethics, 122, 235, 241–242
Ethnomethodology, 72, 87, 180
Être, 239
Existence, 4, 6, 12, 26, 30, 32, 34–36, 49–50, 66, 90, 92, 94–95, 98, 100, 102, 136–137, 143, 152, 156, 189, 192, 218, 221, 226, 235, 239
Explanandum, 85, 245
Explanans, 85, 245
Exteriority, 4, 31, 40, 100, 105
Externality, 4, 86, 130, 157, 179, 196, 254

F

Familiarity, 9, 13, 47, 68, 101, 109, 112, 117, 130, 149, 161, 165, 171–172, 175–177, 180, 230, 239, 254, 262, 265
  unfamiliar, 29, 31, 41, 52, 161, 167, 171–172, 176, 231–232
Flesh, 3, 5–6, 14, 17–20, 27, 49, 54, 56, 67, 85, 108, 179, 183, 209, 219, 225–229, 233–234, 243
Flow, 34–35, 40, 94, 101–104, 131, 133, 220–221, 223, 245
Foreign/strange, 8–9, 25, 69, 161, 167, 171, 176–177, 209
Freedom, 6, 210, 237–238
Freud, S., 4, 103

G

Galileo, G., 28, 33, 35, 91, 94, 96–101, 103, 223
Garfinkel, H., 9, 73, 75, 84, 87, 95
Gaze, 15, 46, 48, 55, 67, 91, 133, 198–199, 259, 264–265
Geflecht, 171
Gesture, 88, 90, 117, 128, 134–136, 138, 174, 188–189, 192, 194, 196–197, 199–200, 236, 253, 258–260, 263
Giacometti, A., 26, 257
Gift, 37–38, 207, 235, 260
Givenness, 36, 68, 91, 123, 155, 207, 231, 255
Gloss, 50, 62–63, 74, 80, 82, 84, 97, 116, 118, 120, 129, 132, 134, 197
Gurwitsch, A., 7, 21, 28, 32, 40–41, 46, 255

H

Habitus, 178
Health, 160, 218
Hegel, G. W. F., 102, 137, 152, 180, 182, 196, 206, 222, 246–247
Heidegger, M., 5, 9, 19, 24–25, 27, 38–40, 103–104, 107, 109, 153, 155–156, 160, 163, 166, 169, 171, 176–177, 180, 197, 222, 236
Henry, M., 4, 9, 13–14, 28, 33, 35, 47, 49, 68, 72, 85–86, 89, 100, 103, 109, 179, 181, 203, 205, 225, 227, 232, 243, 255, 257
Heraclitus, 24

Heteronomy, 106, 128, 153, 243
Historicity, 265
Holstein, 48, 50
Homo sapiens, 94, 187, 189, 226
Hospitality, 207
Husserl, E., 9, 12, 34, 39, 42, 71, 81, 91, 103, 108, 121, 126, 207, 210, 220–221, 254
Hyle, 42, 220, 223

I
Idealism, 4, 177
Identity, 6, 144–145, 152–155, 207, 214, 221, 233, 240, 264
  dis-identification, 155
  identification, 66, 154–155, 183
  non-self-identity, 3, 10, 36, 105, 108, 138, 152, 158
Ideology, 105–106, 144, 178, 206, 211
Idiosyncracy, 168–169
Illness, 11, 44, 98–99, 182, 207, 209, 217–218
Illocution, 144, 236
Illusion
  Müller-Lyer, 86–87
  optical, 86
Immanence, 4, 18–19, 40, 47, 49, 56–57, 67–68, 72, 74, 84–88, 103, 128, 138, 217–218, 220, 223–224, 226, 231–232, 234
Incarnation, 4, 21, 43, 183, 225, 228, 230, 232, 240–242
Incommensurability, 87
Insight, 102–103, 107, 175, 208, 213, 223, 245, 255
Intellectualism, 3–7, 9, 40, 130, 142, 144, 177, 179–182, 185, 194, 198, 202, 205, 212, 226, 233–234, 254
Intellectualist, 3–4, 9, 130, 142, 144, 179–182, 185, 194, 198, 202, 205, 212, 226, 233–234, 254
Intelligibility, 118, 121–122, 125, 145, 148, 154, 160, 169, 227, 255
Intentionality, 3, 14, 17, 23, 27, 34–35, 67–68, 91, 93, 100, 144, 157, 183, 187, 189, 207, 215–216, 218, 224, 246, 248, 251, 265
International Phonetics Alphabet, 168
International Phonetics Association, 39, 44, 168
Intersubjectivity, 50, 235, 237
Intonation, 116, 118, 120, 133, 153, 166, 174, 185, 197
Ipseity, 155–156, 181

K
Kandinsky, W., 257
Kant, I., 4–6, 8, 13, 21, 23–24, 39, 44–46, 52, 67, 85, 91, 108–109, 121, 142, 152, 164, 167, 180, 183, 185, 187, 225–226, 228, 233, 241, 243, 262
Kepler, J., 125
Khôra, 27, 38–39, 103–104, 214, 222–223, 239, 263, 265
Kinesthesia, 54, 56, 67, 86–87, 188, 202, 234
Kinetic melody, 67–68, 72–73, 86, 202, 232, 263
Kraal, 249, 253–254

L
Language, 5, 9, 12–13, 19–20, 27, 34, 36, 39, 41, 48, 75, 100, 105–109, 111–114, 118–119, 121–126, 128–131, 137, 139, 142–144, 146, 152–157, 159–163, 166–172, 175, 177–178, 199, 207, 214, 218–219, 221–224, 226, 232, 235–237, 239–240, 245, 247, 249, 251, 256, 260
Lao Tzu, 28
Learning paradox, 5, 7, 14, 26, 85, 264
Lebenswelt, 91–92

Leont'ev, A. N., 14–15, 160
Levinas, E., 3, 6, 101, 142, 144–145, 156–157, 207, 225, 233, 235, 238, 240–241
Life, 3–4, 6–7, 11–13, 15–16, 19, 21, 23, 34–36, 40, 48, 52, 73–74, 78, 84–86, 90, 93, 105–106, 108, 116, 119, 122, 131, 143, 146, 152–155, 157, 159–161, 165, 168, 174, 178–181, 185, 189, 196, 198, 202, 205–213, 215–216, 218, 220–223, 225–228, 230–234, 240–241, 243, 245, 255, 264
Locution, 144, 235–236
Logodaedalus, 24
Logos, 24, 39, 41, 131, 180
Love, 14, 19, 186, 205, 208, 215, 219–220

M
Maine de Biran, P., 9, 27–28, 67–68
Manifestation, 14, 26, 34, 38, 78, 84, 86, 103, 108, 130, 152, 158, 168, 179–180, 182, 186, 189, 195, 200, 209, 212, 226–227, 243, 246, 259
Margin, 239
Marion, J.-L., 9, 13, 50, 251, 259–261
Marx, K., 6, 8, 51, 180, 226, 245–246
Mêlée, 157
Memory, 33–34, 37, 49, 56, 67–68, 102, 104, 114, 119, 129, 134, 203, 220, 226, 238
  core dump, 119, 121, 129, 132–134, 139, 148
  immemorial, 49, 104, 203
Merleau-Ponty, M., 9, 32, 38, 46, 51–52, 54–55, 88–89, 128–130, 136, 141, 186, 189, 194, 200, 228, 230–231, 254
Metaphor, 3–5, 9–10, 14, 17, 19, 23, 25, 27, 29, 32, 36, 43–44, 51, 74, 85, 90–91, 99, 103, 107, 109, 122–123, 129, 136, 143, 173, 178–179, 185, 193, 202, 209, 213, 216, 224–226, 238, 245–246, 249, 251, 253, 255, 260–262, 264–265
Metaphysics, 4–5, 7–8, 17, 20–21, 24, 36, 103, 142, 160, 164–165, 178–180, 186, 225, 234, 237
Mind
  conscious, 40, 68, 73, 113, 133, 259
  mental framework, 93, 127, 171
Monad, 108, 121–122
Mood, 163, 196–197
Motivation, 14–16, 20, 179–180, 182, 186, 188–189, 195, 241, 254
Multiplicity, 45
Mundaneity, 4, 12, 75, 139, 160, 176

N
Nancy, J.-L., 3–4, 9, 24, 43, 109, 142, 152, 157, 214, 233
Neuron, 56, 248
Newcomer, 27, 100, 174–175
Newton, I., 125, 262
Nietzsche, F., 154
Non-intellectualist, 3
Non-intentional, 3, 35, 91, 215, 231–232, 255

O
Object/motive, 16
Once-occurrent, 23, 71, 75, 111, 119, 152, 156, 185, 196, 221, 240–241
Ontogeny, 12, 90, 245, 254
Ontology, 63, 235, 237
Other, 8, 15, 20, 68, 111–112, 141–144, 146, 152–156, 160–162, 219, 221, 229, 237–238, 242, 264

P
Pain, 9, 19–20, 44, 57, 98–99, 144, 153, 180, 182, 208–209, 215–218, 227

Parmenides, 24, 38, 222
Participative thinking, 75, 119, 131, 196, 198, 201
Passivity, 5–6, 9, 17–20, 25–26, 36, 43, 54–55, 57, 68–69, 85, 89, 101, 105, 108, 125, 128, 134, 144, 152–157, 180, 183, 202, 206–207, 209, 212, 215–219, 224–225, 227, 234– 241, 243, 245–246
  radical, 17–20, 28, 69, 106–108, 112, 122, 144, 155, 183, 215, 217–218, 227, 234–235, 239, 243
Pathos, 51–54, 56, 58, 68, 91, 179–181, 199, 206–210, 224, 235, 237–238, 263, 265
Patience, 144–145, 209, 240
Perception, 21
Performative, 3, 19, 167, 198, 233
Perlocution, 145, 236
Permanence, 80, 90, 93–94, 99, 221, 223, 253
Phylogenesis, 245
Piaget, J., 4, 7–8, 10, 12–13, 21, 25, 29, 32, 41, 44, 46, 54, 85, 88, 92, 98, 121, 179–181, 205, 207, 221, 225, 227, 232–233, 243, 253–255
PISA, 11
Pitch, 64, 116, 118, 153, 190, 192, 197–201, 236
Plato, 27, 38, 103, 262
Poet, 108, 141, 153, 247
Poetry, 212
Possibility, 5, 7, 9, 12–15, 17, 21, 34, 39, 44, 49, 52, 68, 76–77, 80, 83, 89, 94–95, 105–106, 112, 121, 143, 146, 152, 156–157, 162–163, 179, 181, 186–187, 207, 214, 218–219, 225, 227, 229, 235–236, 240, 248, 259–261
Practice, 3, 5, 15, 27–28, 75, 77, 80, 95, 160, 164, 166, 175, 224, 254
Pragmatism, 8, 20, 72, 109, 160–161, 164, 200
Proposition, 23, 113, 159
Prosody, 136, 200–202
Psyche, 4, 89, 131, 187, 226, 240
Psychoanalysis, 37, 155
Psychology, 5, 9, 14, 16, 24, 26, 54, 85, 98, 101, 176, 180, 185, 187–188, 196, 207

R

Radical passivity, 17–20, 28, 69, 106–108, 112, 122, 144, 155, 183, 215, 217–218, 227, 234–235, 239, 243
Reason, 5–6, 25, 45, 48, 51, 58, 65, 67–68, 74, 93, 116, 118–120, 157, 183, 185, 187, 196, 246, 261, 265
  Vernunft, 6
Reflex, 231–232
Regress, 94, 141, 211
Repeatability, 99, 263–265
Repetition, 44, 105, 131, 162
Reproducibility, 92
Resource, 5, 12, 16, 107, 120–121, 198, 200–201, 260
Responsibility, 12, 122, 196, 210, 241
Revelation, 162, 209, 238
Rhythm, 188, 197–198, 200–202, 226
Ricœur, P., 138
Rimbaud, A., 20, 108, 138, 141–142, 145, 219
Ritual, 199, 201

S

Saccade, 57
Sapir, E., 131, 139, 223
Sartre, J.-P., 3, 5, 54, 56, 143
Schema, 4, 18, 32, 44, 54, 85, 88, 90, 93, 179–180, 202, 226, 233, 238–240, 245, 265
  sensorimotor, 18, 44, 54, 85, 88, 93

Seiendes, 27, 38, 239
Sein, 27, 39, 222, 239, 246
Self, 5, 20, 102, 108, 119, 134, 138, 141–145, 152–154, 156, 212, 214–215, 219, 225, 227, 229, 231–232, 242, 245, 264
Selfhood, 20, 106
Self-movement, 52, 54, 209, 233
Self-understanding, 157, 212, 214
Sensation
  dull, 57
  tactile, 54
Sense
  impression, 34–36, 51, 91, 94, 99, 183
  nonsense, 163, 178
  practical, 27, 85
Signification, 20, 33, 72, 109, 131, 144, 154, 160–161, 163, 171, 176, 189, 192, 197, 199–200, 207, 240
Singularity, 4, 77, 111, 119, 121, 131, 139, 145, 148, 153–156, 211–214, 219, 222, 254
Sinn, 163
Society, 13, 119, 122, 145, 157, 171, 180, 195–196, 199, 218, 241
  societal, 9, 13, 107, 119, 125, 160–161, 166, 169–170, 176, 178, 181, 199, 239
Spatiality, 4
Speech, 27, 39, 111, 118, 128–131, 133–136, 144, 153, 185, 189–190, 192, 197–199, 201–202, 236
STEM, 10–11, 14, 17, 25, 29, 41, 88
Stimulus, 232
Structure
  mental, 18, 27, 40, 43, 45–46, 48, 88, 98, 106, 111, 119, 124, 127, 132, 160, 227
Subjectivity, 17, 20, 39, 50, 119, 121, 168, 207, 223, 237, 240, 264
Sublation, 206, 246
Suffering, 20, 153, 180, 182, 205, 207, 209, 215, 217–218, 224, 227, 233, 240, 248
Synchrony, 198–201

T
Teleology, 25, 227, 254, 265
Temporality, 4, 35, 39, 104, 134, 197
Temporalization, 104
Temporization, 35, 39
Theory-praxis gap, 150
Thinking in movement, 86, 135, 138
Thrownness, 19
Trace, 36, 138, 169, 239, 259, 261
Transcendence, 13, 23, 40, 71–72, 178–179, 214, 218, 220, 222–223, 233–234, 240–241
Transference, 155
Truth, 126, 167, 214

U
Uncertainty, 18, 20, 23–24, 27, 59, 63, 71, 83, 95, 97, 134, 233
Unconscious, 4, 7, 103, 212, 217, 254
Undecidability, 8–9, 20, 36, 52, 109, 112, 171, 180, 182, 188, 194, 197, 214, 243
Understanding
  Verstand, 6
Unit of analysis, 145
Utterance, 64, 84, 101, 115–116, 118, 131, 144, 149, 167, 189–190, 202, 236

V
Valence, 188–190, 194–195, 197–200
Verb
  transitive, 13, 19, 37, 206
Viability, 7–8, 17–20, 23, 27, 44, 84, 88, 90, 98, 105, 141, 169, 183, 225, 227–228, 245

Von Glasersfeld, E., 7–8, 13, 17–18, 28, 44–45, 88, 93, 98, 108–109, 111, 121, 141–143, 152, 162–163, 165, 167–169, 180, 225, 234, 254
Vorstellung, 109, 164
Vygotsky, L. S., 6, 8–9, 14, 38, 88, 111, 129–131, 134–139, 165, 169, 176, 178, 180–182, 187, 189, 196–197, 223, 264

W
Wittgenstein, L., 20, 47, 58–59, 109, 119, 124, 159–160, 162–165, 167, 169–171, 175, 178

Z
Zen, 211, 220